LASER EXPERIMENTS FOR CHEMISTRY AND PHYSICS

Laser Experiments for Chemistry and Physics

Robert N. Compton

and

Michael A. Duncan

OXFORD
UNIVERSITY PRESS

OXFORD

UNIVERSITY PRESS

Great Clarendon Street, Oxford, OX2 6DP,
United Kingdom

Oxford University Press is a department of the University of Oxford.
It furthers the University's objective of excellence in research, scholarship,
and education by publishing worldwide. Oxford is a registered trade mark of
Oxford University Press in the UK and in certain other countries

First Edition published in 2016

Impression: 1

Published in the United States of America by Oxford University Press
198 Madison Avenue, New York, NY 10016, United States of America

British Library Cataloguing in Publication Data
Data available

Library of Congress Control Number: 2015939606

ISBN 978-0-19-874297-5 (hbk.)
ISBN 978-0-19-874298-2 (pbk.)

Printed and bound by
CPI Group (UK) Ltd, Croydon, CR0 4YY

Preface

Like most scientists, the authors of this textbook have been greatly influenced by their academic backgrounds and mentors. Duncan and Compton were both undergraduate students in small liberal arts colleges (Furman University and Berea College, respectively) in which student research was a vital part of their educational experience. Performing undergraduate research was of equal importance to course work. Both authors began their research careers at the time of the development of the laser and were witness to many incredible advances in the development and applications of this new technology.

In graduate school, Compton was particularly influenced by the first year experimental research course required of every incoming graduate student (theory or experiment) in the Department of Physics at the University of Florida. In addition to experiments, Professor Alex Green included glass blowing and machine shop fundamentals in this course. Although it is not a required course, a similar experience is offered to the physics and chemistry graduate students at the University of Tennessee (UT) by Compton. Some of these experiments are included in this book. In a few instances, the laboratory experiments resulted in an area of research. For example, in the undergraduate laboratories students were required to record and analyze a Raman spectrum of CCl_4 at room temperature. Due to the thermal rovibrational population, the Stokes lines are very broad at room temperature. To resolve the isotope lines of CCl_4, one of the students in the advanced physics laboratory (Darrin Ellis) recorded a Raman spectrum of CCl_4 *submerged* under liquid nitrogen in a Styrofoam cup. Due to rovibrational cooling, the spectrum of CCl_4 under liquid nitrogen shows fully resolved isotopic contributions due to the ^{35}Cl and ^{37}Cl isotopes (see Chapter 18, Figure 18.3). A number of papers have appeared illustrating the advantages and utility of Raman Under Nitrogen (RUN). As a graduate student in 1962 working under G. S. Hurst at the Oak Ridge National Laboratory, Compton was privileged to purchase and use one of the first commercial (Bendix Corporation) time-of-flight mass spectrometers (TOF-MS). His thesis employed electron impact ionization, but after the development of the pulsed laser Compton integrated many home-built TOF-MS instruments into laser ionization experiments at ORNL and UT. In 1979, colleagues J. A. D. Stockdale and Compton formed a scientific instruments company in Oak Ridge known as Comstock, which manufactured and sold TOFs as well as electrostatic energy analyzers (ESAs). Many of these instruments have been used by researchers around the world to record multiphoton ionization mass and photoelectron spectra. This company has employed a number scientists, engineers, and machinists, which has enriched their lives as well as science in general.

As an undergraduate at Furman, Duncan learned to solder from Professor Lon Knight and together they built a vacuum system for matrix isolation experiments and EPR studies of unusual radicals. At Rice University in the group of Richard Smalley, he was given the task of designing and building their first time-of-flight mass spectrometer. With virtually no electronics background, Duncan was guided by the electronics shop to the classic paper by Wiley and McLaren (*Rev. Sci. Instrum.* **26**, 1150 (1955)). After many design pitfalls and burned-out circuits, he eventually made a working instrument and incorporated it into their supersonic molecular beam machine. Other experiments and a misaligned Nd:YAG laser led inadvertently to the discovery of the laser vaporization method used to produce gas-phase clusters of metal atoms and other refractory elements. It was this laser vaporization method and the same mass spectrometer that were later employed by Jim Heath and Sean O'Brien in the Smalley group to discover C_{60}. At Georgia, Duncan has continued to design and implement new versions of time-of-flight mass spectrometers into his research, and has also developed new experiments for the undergraduate labs using these instruments, many of which are included in this book.

The authors have employed many kinds of modern lasers and laser techniques in their research and teaching laboratories. Compton was one of the first to study and employ nonlinear optical processes in atomic and molecular research and some of this early work is found in the chapters describing multiphoton ionization, stimulated electronic Raman scattering (SERS), and third harmonic generation (THG). Duncan has employed photoionization, laser desorption analysis of materials, and mass-selected ion photodissociation measurements throughout his research. Both authors have developed advanced physical chemistry laboratories at their respective universities and are deeply indebted to the numerous students who have passed through these laboratories, many of whom are now directing their own physical chemistry or physics laboratories at other universities.

This book is designed to introduce researchers to the breadth of available experiments in chemical physics that can be integrated into physics and chemistry laboratories. The authors have tried to include chapters that are fundamental to lasers and spectroscopy and provide theoretical background to the experiments described. For example, the chapters on properties of light, diffraction, etc., and rovibrational spectroscopy should be read before considering the experiments describing IR and Raman spectroscopy, optical activity, and other laser-based experiments. We hope that the enjoyment of laser experiments and their role in advancing chemistry and physics can be appreciated by a new generation of scientists through the experiments presented here.

Robert N. Compton, Knoxville, TN
Michael A. Duncan, Athens, GA

Acknowledgments

The authors are grateful to the many graduate and undergraduate students who have worked in their labs over the years and have contributed to the experiments described here. Special recognition is due to the many Teaching Assistants who implemented these experiments in the undergraduate and graduate laboratories at the University of Tennessee and the University of Georgia. MAD would like to thank Professors Lionnel Carreira and Allen King, who shared ideas and provided feedback for lab experiments early in his career at Georgia. RNC is indebted to Professor Charles Feigerle for collaborations on the study of MPI of molecular iodine, to Dr. Donald Armstrong, Dr. Stewart Hager, and Dr. Jeffery Steill (Sandia Laboratory) for assisting with recording the numerous FTIR spectra, and to Dr. James Parks for sharing an experiment from the advanced physics laboratories at UT. Special thanks are also due to former masters degree student Darren Ellis for performing the first RUN spectrum in our laboratories at UT.

We would also like to acknowledge the special help in preparing this manuscript from Mr. Jonathan Maner (proofreading), Ms. Alex Orlowsky (figure preparation), Mr. Joshua Marks (photography), Mr. Timothy Ward and Mr. David McDonald (computational work), and Mr. Ivan Geigerman (figure preparation). We are also grateful to Drs. Sharani Roy and Jay Agarwal for a critical assessment of Chapter 15 on quantum chemistry calculations. Dr. Paul Siders is gratefully acknowledged for contributions to Chapter 22 on Fermi resonances.

Table of Contents

Part I

Introduction to Light, Lasers, and Optics

Elementary Properties of Light

Introduction

The experiments described in this book involve the use of laser light in a wide range of applications. The present chapter is an elementary introduction to the properties of light that will serve to make these experiments more understandable. This chapter, along with Chapter 2, is certainly not intended to replace a rigorous course in physical or geometrical optics; rather it is a summary of some of the important concepts and formulae in optics applied to the use of lasers in chemical physics research.

All electromagnetic (E&M) waves propagate with the same velocity in free space (which is defined as c) and differ only in frequency, ν, and wavelength, λ, through the relationship $c = \lambda\nu$. There is one report of E&M waves having a wavelength of 1.9×10^7 miles.[1] At the other extreme, the Compton Gamma Ray Observatory (GRO), a NASA satellite, has been used to detect gamma rays with wavelengths as small as 40 femtometers (40×10^{-15} meters), approximately the size of a proton.

The description of electromagnetic radiation is carried out using a classical/quantum mechanical wave/particle treatment. Certain phenomena are best described using waves and others can only be explained by treating light as particles called photons. All light phenomena can be explained by treating light as photons; however, in most cases the wave picture is easier to employ. The energy of a photon is Planck's constant h times the frequency, ν ($E = h\nu$). Photons have momentum ($p = h/\lambda$) but no mass. In 1890 Heinrich Hertz first produced radiowaves ("Maxwellian waves") in the laboratory and it was many more years before Marconi put them to good use in communications. We now know that any material object (gas, liquid, or solid) having a temperature above absolute zero (0 K) will emit photons. For so-called "black-body" radiators the temperature is related to the photon energy $h\nu$ and emitted light intensity $I(\nu,T)$ by the Planck Black-body radiation formula:

$$I(\nu, T) = \frac{2h\nu^3}{c^2} \frac{1}{\left(e^{\frac{h\nu}{kT}} - 1\right)} \tag{1.1}$$

The Planck radiation law represents the power per unit area per unit solid angle per unit frequency emitted from a black body in thermal equilibrium.

Laser Experiments for Chemistry and Physics. First Edition. Robert N. Compton and Michael A. Duncan.
© Robert N. Compton and Michael A. Duncan 2016. Published in 2016 by Oxford University Press.

This function has a maximum (i.e., setting the derivative $dI(\nu,T)/d\nu = 0$ and solving for ν_{max} from equation (1.1)) corresponding to $h\nu_{max} = 2.82 \ kT$. Alternately one can write this in the form of the Wien's Displacement Law as $\lambda_{max} T = 2.8977685(51) \times 10^{-3} \ m \cdot K$.

The Planck radiation law is known to accurately describe the emission from stars, the heater eye on an electric stove, and even the cosmic background radiation left over from the Big Bang creation of the universe. The Cosmic Microwave Background (CMB) consists of photons left over from the Big Bang, which through multiple collisions have established a thermal "cosmic black-body" equilibrium. In Figure 1.1 the microwave cosmic background radiation data is fitted to a Planck radiation law prediction (equation 1.1) giving a temperature of 2.725 K. The data were recorded using the Far-Infrared Absolute Spectrophotometer (FIRAS) onboard NASA's Cosmic Background Explorer (COBE) satellite. [2] It is remarkable that the experimental data points lie within the line (theory) in Figure 1.1.

The black-body temperature for the data in Figure 1.1 was determined from a fit to equation (1.1), but this can also be estimated using Wien's displacement law $h\nu_{max} = 2.82 \ kT$ or alternately $\nu_{max} = 58.79 \ (GHz/K)T$. Using $T = 2.725 \ K$ gives $\nu_{max} = 160.2 \ GHz$.

It is impossible to make a representation of the full wavelength range of E&M waves. There are regions of the spectrum yet to be explored at both the long and short wavelength regions. Figure 1.2 is an attempt to show on one scale the range of the spectrum that science has presently explored.

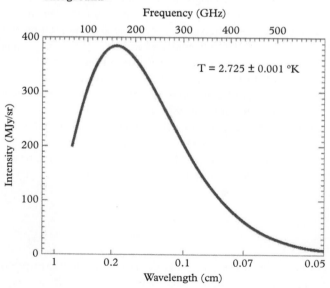

Figure 1.1 *Cosmic microwave background spectrum fit to a Planck radiation law function (equation 1). The data points are smaller than the line drawn using the Planck law. The deviations are less than 0.30% of the peak brightness, with an rms value of 0.01%. (This COBE/FIRAS image was kindly provided by COBE Science Team/NASA.)*

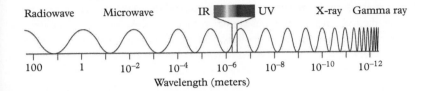

Radiowave   Microwave   IR   UV   X-ray   Gamma ray

100   1 10^{-2} 10^{-4} 10^{-6} 10^{-8} 10^{-10} 10^{-12}

Wavelength (meters)

Figure 1.2 *Illustration of the known full electromagnetic spectrum of light from radiowaves to gamma rays.*

Table 1.1 *Approximate wavelength, frequency, energy, and temperature corresponding to Figure 1.2.*

E&M	Wavelength cm, except as noted	Frequency Hz	Energy eV	Energy cm^{-1}	Energy kcal/mol	Temp. K
Radiowave	> 10	~ 10^9	< 10^{-5}	< 0.1 cm^{-1}	< 2.9×10^{-4}	< 0.03
Microwave	10 – 0.01	~ 10^{12}	10^{-5} – 0.01	0.1 – 100	3×10^{-4} – 3×10^{-1}	0.03 – 30
Infrared	0.01 – 7×10^{-5}	~ 10^{14}	0.01 – 1	10^2 – 10^4	3×10^{-1} – 29	30 – 4100
Visible	7×10^{-5} – 4×10^{-5} 700 – 400 nm	~ 10^{15}	~ 1 – 3	$(1.0 – 2.5) \times 10^4$	29 – 71	4100 – 7300
Ultraviolet	4×10^{-5} – 10^{-7} 400 – 1 nm	~ 10^{16}	~ 3 – 10^3	2.5×10^4 – 10^7	71 – 2.9×10^4	7300 – 3×10^6
X-ray	10^{-7} – 10^{-9}	~ 10^{18}	10^3 – 10^5	10^7 – 10^9	10^4 – 10^6	3×10^6 – 3×10^8
Gamma ray	< 10^{-9}	> 3×10^{19}	> 10^5	> 10^9		> 3×10^8

Maxwell was the first to show that visible light represents but a narrow component of the entire electromagnetic spectrum. From this figure we see that visible light represents only one octave out of almost 60 octaves of the known electromagnetic spectrum. This wavelength range is further described in Table 1.1 in terms of wavelength, frequency, energy, and temperature.

The visible spectrum ranges in wavelength from approximately 400 to 700 nm. The sensitivity of the human eye to visible light depends upon the observer and the time of day. Figure 1.3 shows the relative sensitivity curves for the *standard observer* as defined by the International Commission on Illumination. The International Commission on Illumination, which was established in 1913 and based in Vienna, Austria, is the international authority on light, illumination, and color.

A complete description of the sensitivity of the eye can be found in the book by Williamson and Cummins.[3] As mentioned before, visible light is but a narrow and somewhat ill-defined region of the spectrum. The curves in Figure 1.3 are ill-defined in the sense that the relative sensitivity of the eye for humans can only be considered for a *standard observer*. The standard curve also varies under different light levels because of the sensitivity difference of the rods and cones covering the retina. The rod system is about 1000 times more sensitive than the

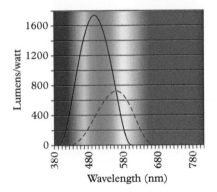

Figure 1.3 *Relative sensitivity of the human eye of a standard observer as a function of wavelength under conditions of daylight vision (photopic, dashed line) and night vision (scotopic, solid line).*

cones and as a result the relative color sensitivity differs for night vision (scotopic) and daylight vision (photopic) as shown in Figure 1.3. An interesting experiment to demonstrate the varied sensitivities of the human eye is to have a number of people with their eyes dark-adapted observe the heaters on an electric stove as the stove is turned on in the dark. Some individuals will notice that the stove begins to appear red before others. The onset of first observing red light depends on the sensitivity of each individual to long wavelength light.

Maxwell's Equations

In 1873, Maxwell considered the accumulated body of experimental observations concerning electricity and magnetism uncovered by Gauss, Faraday, Biot-Savart, Hertz, Ampere, and others and combined all of these empirical laws into one unifying theory of electricity and magnetism. Maxwell's equations can be written for a medium of dielectric constant ε and charge density ρ in which a current J flows as

$$\nabla \cdot \vec{D} = \rho \qquad \text{Gauss' Law}$$

$$\nabla \cdot \vec{B} = 0 \qquad \text{Gauss' Law of Magnetism}$$

$$\nabla \times \vec{H} = \vec{J} + \frac{\partial \vec{D}}{\partial t} \qquad \text{Ampere's Circuit Law} \tag{1.2}$$

$$\nabla \times \vec{E} = -\frac{\partial \vec{B}}{\partial t} \qquad \text{Faraday's Law of Induction}$$

in the MKS (meter–kilogram–second) system of units, where $\vec{D} = \varepsilon \vec{E}$, $\vec{B} = \mu \vec{H}$, ε and μ are the electric permittivity and magnetic permeability of the medium, and ∇ is the gradient operator. It is easy to show from Maxwell's four equations above that electromagnetic waves consist of perpendicular $\vec{E}$ and $\vec{B}$ fields traveling in a direction mutually orthogonal to $\vec{B}$ and $\vec{E}$ with a velocity equal to c in a vacuum. Since $\varepsilon = \varepsilon_0$ and $\mu = \mu_0$ in a vacuum and the divergence of the electric field is zero we can write

$$\nabla \cdot \vec{E} = 0$$

$$\nabla \times \vec{B} = \mu_0 \varepsilon_0 \frac{\partial \vec{E}}{\partial t}$$

$$\nabla \times \vec{E} = -\frac{\partial \vec{B}}{\partial t} \tag{1.3}$$

Upon taking curl (∇_x) of $\vec{E}$ twice and collecting terms it is easy to show that

$$\nabla \times \nabla \times \vec{E} = \nabla \left(\nabla \cdot \vec{E} \right) - \nabla^2 \vec{E} \tag{1.4}$$

Substituting $\nabla \times \vec{E} = -\frac{\partial \vec{B}}{\partial t}$ into the left side of equation 1.4 and replacing $\nabla \times \vec{B}$ by $\mu_0 \varepsilon_0 \frac{\partial \vec{E}}{\partial t}$ results in the wave equation

$$\nabla^2 \vec{E} - \mu_0 \varepsilon_0 \frac{\partial^2 \vec{E}}{\partial t^2} = 0$$

in which the velocity of the wave is $v = \frac{1}{\sqrt{\mu_0 \varepsilon_0}}$.

Also, since $\nabla \cdot \vec{E} = \nabla \cdot \vec{B} = 0$, both $\vec{E}$ and $\vec{B}$ are orthogonal to the direction of the propagation, **k**. Likewise since the curl of $\vec{B}$ and $\vec{E}$ are proportional to the two derivatives of $\vec{B}$ and $\vec{E}$, respectively, $\vec{E}$ and $\vec{B}$ are easily shown to be orthogonal to each other. Thus electromagnetic waves consist of $\vec{E}$ and $\vec{B}$ waves traveling in a direction **k** in which all three vectors are orthogonal. The velocity in a vacuum is denoted c from *celeritas* the Latin word for speed. Chapter 7 is devoted to determining the speed of light from a measurement of ε_0 as well as timing the speed of light pulses over a known length. First we examine a brief early history of the estimates of the speed of light.

Speed of light

Galileo first considered measuring the speed of light in much the same way that the speed of sound was first determined, by positioning two observers with synchronized clocks on separate mountaintops each with a lantern that could be shuttered. Measuring the time delay would allow a measurement of the speed of light c (or more appropriately c/n_{air}). We now know that human hands and brains are far too slow to record such a short time because of the small distances involved. However, in 1676 a Danish astronomer named Olaf Röemer used astronomical distances to estimate the speed of light in a similar manner. Röemer noted that the time elapse between lunar eclipses of Jupiter became shorter as the Earth moved closer to Jupiter and became longer as the Earth and Jupiter drew farther apart. He attributed this anomalous behavior to a finite speed of light and arrived at a value of approximately 2.14×10^8 m/sec. Given the uncertainty of the distance between the Earth and Jupiter at that time, the derived speed is quite close to the presently accepted value of 2.99792458×10^8 m/sec. This historic experiment showed for the first time that light has a finite speed. Later scientists replaced the shuttered lanterns of Galileo with mechanical rotating shutters. For example, in 1849 Fizeau used the tooth-wheel chopper method of determining the speed of light and arrived at the value of 3.15×10^8 m/sec. Foucault and later Michelson replaced the chopper with a rotating mirror to obtain much more accurate values (e.g. $c = 299{,}792{,}458$ cm/sec) using ever increasing light paths. (In Chapter 7 we describe pulsed laser methods to determine the speed of light.)

Fifty years after Röemer's estimates of the speed of light, the British astronomer James Bradley used the aberration of starlight to determine this quantity. Since the Earth is moving ($v \approx 30$ km/sec) relative to the fixed stars a telescope

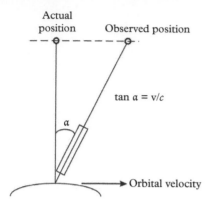

Actual position   Observed position

$\tan \alpha = v/c$

α

Orbital velocity

Figure 1.4 *Illustration of the velocity of light determination by Bradley in 1726 using the aberration of starlight. The orbital velocity refers to the velocity of the Earth around the Sun and relative to the stars. Thus the velocity changes sign every year.*

needs to be tilted slightly to observe a star much like we tilt an umbrella while walking in the rain to prevent getting wet. When a star is directly overhead we can easily see that the velocity aberration of the starlight produces a tilt angle of $\tan^{-1} v/c$ (see Figure 1.4):

For all such stars the angle $\alpha = 20.5$ arc seconds (0.0568 degrees). This is known as the constant of aberration $k = 20.49552''$. From the aberration of starlight Bradley determined

$$c = \frac{30 \times 10^3 \text{ km/sec}}{\tan(20.5 \text{ arcsec})} = 3.02 \times 10^8 \text{ m/sec}$$

This value is very close to the accepted value of c.

If an E&M wave is moving through a medium with an index of refraction n the velocity is c/n. The velocity of this wave is also $v = \lambda \nu$ where λ is the wavelength and ν is the frequency of the wave. Since energy is always conserved, the frequency in one medium is the same in another medium, however the wavelength of an E&M wave traveling in a vacuum is reduced in a medium by $1/n$, i.e., λ/n. The ratio of the speed of light in a vacuum to the phase velocity of light in a medium is defined as the index of refraction $n = c/v$. The index of refraction depends upon the medium and the wavelength of the E&M wave. From Maxwell's equations, the speed of light in a vacuum is $c = \frac{1}{\sqrt{\mu_0 \varepsilon_0}}$ and is a universal constant. Since c, μ_0, and ε_0 are constants, μ_0 is arbitrarily taken to be $4\pi \times 10^{-7} \frac{\text{mkg}}{\text{coul}^2}$ and $\varepsilon_0 = 8.854878164 \times 10^{-12} \frac{\text{s}^2\text{C}^2}{\text{m}^2\text{kg}}$; when c is defined as 299,792,458 m/sec. In Chapter 7 a simple experiment designed to measure the capacitance between two plates is used to determine the speed of light.

Since the velocity of light in a vacuum is equal to the product of its wavelength and frequency ($c = \lambda \nu$), one of the most accurate methods to determine the speed of light is to measure the wavelength and frequency of a light beam. A very simple but pedagogical method to demonstrate this is to employ a candy bar and a commercial microwave oven. Most microwave ovens operate at a frequency of 2.45 GHz (2.45×10^9 Hz) and a wavelength of 12.23 cm. This produces a standing wave inside the microwave and hot spots occur separated by $\lambda/2$ or ~ 6.1 cm. Placing a Hershey bar in the microwave for a few seconds will produce melted indentations at a distance of one-half wavelength or ~ 6.1 cm. Students have performed this exercise and obtained values close to the known velocity of light using this crude method. Data from one student gave a separation between the melted spots of 6 cm for a microwave operating at 2.45×10^9 Hz giving the speed of light to be 3.0×10^8 m/sec. A clever student could surely devise a method that would produce higher resolution of the hot spots resulting in a more accurate value for the speed of light using this "sweet" experiment.

Light as a particle

Up to this point light has been described as a wave, whereas we know from Einstein's Nobel Prize (1921) winning explanation of the photoelectric effect that light also has properties characteristic of a particle. The particles of light are called photons, which are related to their momentum and energy through Plank's constant, h, i.e., $p = h/\lambda$ and energy $E = h\nu$. The velocities of these particles are still equal to *c in vacuo*.

Another important property of the photon is its intrinsic angular momentum. If one determines the intrinsic spin angular momentum of a photon along the direction of propagation the value will always be either $+1\hbar$ or $-1\hbar$, where $\hbar = h/2\pi$. Since the photon possesses unit intrinsic angular momentum, it is classified as a Bose particle (aka "boson") and obeys Bose–Einstein statistics. If the photons are $+1\hbar$, we call them right-circularly polarized (RCP) and if $-1\hbar$, left-circularly polarized (LCP). Thus, for a chiral (optically active) medium the index of refraction (and therefore the phase velocity) is different for RCP and LCP light. From the wave perspective, RCPL and LCPL consist of two $\vec{E}$ and $\vec{B}$ waves, each of equal magnitude and at right angles to each other, which are retarded (RCPL) or advanced (LCPL) by $\lambda/4$ relative to each other.

In the discussion of Maxwell's equations above we treated monochromatic waves as those with definite frequency and wavelength. In practice, even the most monochromatic laser or most sharply tuned radio transmitter consists of a finite spread of frequencies or wavelengths. A sinusoidal wave traveling in one dimension (x) can be represented by

$$u(x, t) = e^{i(k \cdot x - \omega t)} \tag{1.6}$$

where the wavevector k is related to the angular frequency $\omega (2\pi\nu)$ and phase velocity v. As a wave travels through a medium, it can experience dispersion and dissipation. A medium is dispersive if the phase velocity is different for each frequency component of the wave and dissipative if the wave is attenuated. Each frequency present may also be attenuated by a different amount.

A continuous monochromatic wave which is "chopped," i.e. turned on and off for a time Δt, will experience a spread in frequencies given approximately by the uncertainty principle

$$\Delta E \Delta t \approx h, \text{ or } \Delta\nu \approx \frac{1}{\Delta t} \tag{1.7}$$

Thus a femtosecond laser pulse inherently contains a wide spread in frequencies. The effect of turning a light on and off is illustrated in Figure 1.5. Since the A(k) function resulting from the A(t) function shown in Figure 1.5 is sharply peaked about k_0 (ω_0) we can expand the frequency $\omega(k)$ about ω_0, i.e.,

$$\omega(k) = \omega_0 + \frac{d\omega}{dk}\Big|_0 (k - k_0) + \cdots \tag{1.8}$$

and within this approximation we can define a group velocity as the velocity of the peak of the undistorted shape in A(k)

$$v_g = \frac{d\omega}{dk}\Big|_0 \tag{1.9}$$

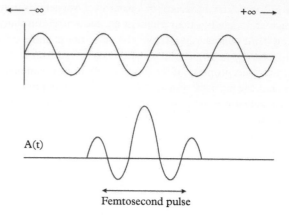

Figure 1.5 *Illustration of turning on and off (chopping) a pure sine wave that extends to + and −∞. Due to the uncertainty principal (ΔEΔt ~ ℏ) the resulting pulse contains a superposition of many frequencies.*

The energy transport occurs with the group velocity and thus information is carried at v_g as well. We can also write

$$\omega(k) = \frac{ck}{n(k)} \qquad (1.10)$$

and the phase velocity is given by

$$v_p = \frac{\omega(k)}{n(k)} \qquad (1.11)$$

The group velocity can be written

$$v_g = \frac{c}{n} - v_g \frac{\omega}{n} \frac{dn}{d\omega} \qquad (1.12)$$

$$\text{or } v_g = \frac{c}{n} - v_g \lambda \frac{dv}{d\lambda} \qquad (1.13)$$

For most media $\frac{dn}{d\omega} > 0$ and $n > 1$ (normal dispersion) and therefore the velocity of energy flow is less than c. The regions of anomalous dispersion $\frac{dn}{d\omega} < 0$ (i.e., negative) and $\frac{dn}{d\omega}$ are usually large as well and the group velocity defined by equation (1.9) is greater than c. In this case the solution is physically unrealistic and the concept of a group velocity is no longer valid. Happily, we do not have energy traveling faster than c and special relativity is safe, for now.

The properties of light are fundamental to Einstein's special theory of relativity. The two postulates of special relativity are: (1) the laws of physics are the same in all inertial reference frames, and (2) the speed of light in free space is the same for all observers, regardless of their motion relative to the light source. The second postulate was beautifully demonstrated in an experiment by Sadeh in 1963.[4] In this experiment, two oppositely directed 0.51 MeV photons were produced from "in flight" annihilation of fast moving singlet positronium (an electron/positron pair with opposite half-integer spins). A positron moving at a speed close to that of light picks up an electron from a material at x = 0 forming singlet positronium. The fast moving positron picks up an electron at rest and the velocity of the e⁺e⁻ pair in the center-of-mass frame is approximately equal to $c/2$. The positron quickly decays into two 0.5 MeV gamma ray photons with opposite direction and helicity. The two photons are detected as shown in the simplified experimental setup in Figure 1.6.

It was found that the two gamma rays reach the photon counters at the left or right at the same time when the annihilation occurs at equal distances between the two photon counters, i.e. the velocity of light is independent of the velocity of the source of light relative to an observer. This experiment represents a vivid demonstration of the validity of the second postulate of special relativity. Dryzek and Singleton [5] proposed a similar experiment which can be used in advanced undergraduate laboratories. This apparatus could be used to place limits on alternative theories of special relativity that have transformations other than the Lorentz transformation.

Let us now consider the properties of circularly polarized light. In doing so, we discuss the so-called wave–particle duality. To explain the diffraction and interference of light it was necessary to consider light as a wave. Conversely, the experimental discoveries of the photoelectric effect and Compton scattering, along with observations of light interactions at very low light levels, required a particle description of light. The currently accepted theory of light combines these two wave–particle concepts into a single "quantized electromagnetic field theory," i.e., Quantum Electrodynamics or QED. This theory is so complete that the QED is often referred to as Q. E. D. (*Quod erat demonstrandum*). A very readable and enlightening description of this theory is found in the Alix G. Mautner Memorial Lectures by Richard Feynman.[6] Thus, we describe light as particles (photons) having no rest mass and always traveling at the speed of c *in vacuo*, and whose probabilities of observation are predicted by the quantum mechanical wave equation. However, to discuss polarization we often revert to the wave description as a convenient mental crutch.

The massless photon has unit intrinsic spin and its spatial exchange properties obey Bose–Einstein statistics. Photons moving in a straight line at the speed c carry a spin angular momentum of $\pm\hbar$ depending on the projection of the spin on the direction of propagation. Thus a single photon can be thought of as having right- or left-handed helicity and can alternately be called right- or left-circularly

Figure 1.6 *An electron–positron pair (singlet positronium) traveling from left to right with speed c/2 decays exactly half way between the two detectors into two 0.51 MeV gamma rays (photons). Since the positrons depicted here have their spins in opposite direction, the net spin angular momentum is zero and the two outgoing photons are necessarily of opposite helicity. The two oppositely directed photons arrive at the two detectors at precisely the same time (equidistant from x = 0) verifying the second postulate of special relativity.*

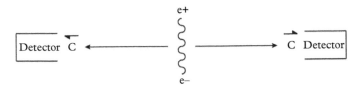

polarized light (CPL). Unpolarized light, which is often called natural light, can be viewed as an equal (racemic) admixture of these circularly polarized photons. Linearly polarized light (LPL) consists of an electromagnetic wave in which the E vector is confined to a spatially fixed plane, the plane of oscillation. The wavevector changes direction randomly and rapidly for unpolarized light. LPL results when there is a definite phase relationship between left- and right-circularly polarized beams. There are many experimental tools for the preparation of circularly polarized light (see for example [7] and [8]). Figure 1.7 illustrates the basic ideas of making RCPL and LCPL. More details about the polarization of light are given in Chapter 2.

In summary, photons possess energy $E = h\nu$ and have linear momentum given by the Heisenberg relation $p = h/\lambda$. Each photon also carries $\pm 1\hbar$ unit of spin angular momentum. The physical existence of linear momentum is seen in many different experiments: photoelectron spectroscopy, Compton scattering, etc. The angular momentum carried by CPL is more difficult to detect. However, a demonstration of the transfer of mechanical torque from circularly polarized photons was reported over 75 years ago.[9]

CPL is used in many areas of physics and chemistry. Particularly important in chemistry is the measurement of circular dichroism. Circular dichroism is the difference in absorption for LCPL and RCPL. The differential absorption of left- and right-handed light as a function of wavelength also leads to the rotation of linearly polarized light or optical rotary dispersion. We will briefly summarize the standard methods used in many studies involving circular dichroism. In doing so,

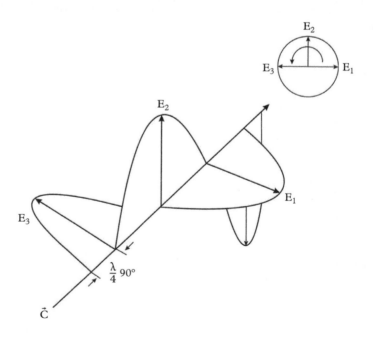

Figure 1.7 *Two linearly polarized and mutually perpendicular* E *vectors which are out of phase by* $\Delta\varphi = \pi/4$ *are combined to produce a circularly polarized light (CPL) wave. In the case shown the light wave* E *vector appears to be rotating counter-clockwise to the observer and is called RCPL. If* $\Delta\varphi = -\pi/4$, *the light is LCPL.*

we rely upon the wave theory of light to describe these methods. A quarter-wave plate (see Chapter 2) consists of a thin sheet of anisotropic material in which the speed of light of one component of LPL possesses a different speed than its perpendicular component. If the plane of polarization of LPL is aligned at 45° with respect to the optical axis, one component can be retarded or advanced relative to the other. If the thickness of the sheet is chosen to advance (retard) one wave by a relative distance of one-quarter wavelength of the light RCPL (LCPL) can be produced. Using this method a specific quarter-wave plate (i.e., thickness) is required for each wavelength. The Soliel–Babinet compensator alleviates this problem by providing two sliding wedges whose thickness can be adjusted to give one-quarter wavelength retardation for a desired wavelength. Finally, the most useful device for producing circularly polarized light consists of passing linearly polarized light through a single Fresnel rhomb in which the plane of polarization is adjusted to +45° with respect to the optical axis to produce RCPL. Rotation of the plane of polarization of the incident LPL to −45° results in LCPL. This method produces CPL for a wide range of wavelengths. One disadvantage is that the propagation direction of the CPL is displaced from that of the incident light after passing through the Fresnel rhomb.

We illustrate the two most often employed methods of producing and analyzing CPL with reference to insects that reflect only circularly polarized light in nature. Many beetles are known to re-emit or reflect only left-circularly polarized light. The American scientist A. A. Michelson studied this phenomenon in some detail in the early 1900s. For example, the Scarab Beetle (*Scarabaeus sacer*) and the jeweled beetle (*C. gloriosa*) reflect only LCPL, as reported in many studies (see reference [10]). The common June Bug (Scientific name: *phjllophaga* sp. of order Coleopteran) also gives off only left-circularly polarized light. In Figure 1.8

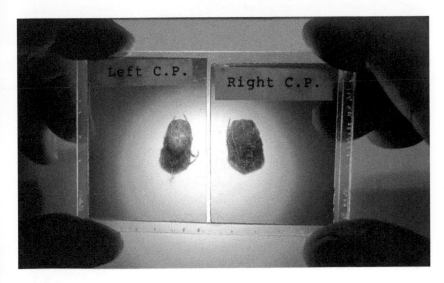

Figure 1.8 *The light from two June Bugs observed using two circularly polarizing sheets constructed from a quarter-wave plate with polarizing sheet whose plane of polarization is oriented at + 45° (LCPL) or − 45° (RCPL) with respect to the optical axis of the quarter-wave sheet.*

a crystal plastic sheet (Edmund Scientific) which is a quarter-wave retarder that is approximately a quarter-wave plate over the visible light spectrum coupled with a linear polarizer is employed to observe the green LCPL reflected from a June Bug. If CPL passes through this quarter-wave retarder, the emerging light will be linearly polarized at an angle which can be used to determine the sense of the CPL using a linear polarizer (Polaroid sheet). The two pictures show a sheet of linear polarizer material with its plane of polarizations at ±45° with respect to the optical axis of the retarder sheet. Clearly the June Bug is observed to only reflect LCPL. The same June Bug shown in Figure 1.8 has been analyzed using a single Fresnel Rhomb and a Glan Taylor prism which also shows only LCPL. The rhomb/prism combination is a better detector of CPL than the quarter-wave plate but it is considerably more expensive. A June Bug is a most useful insect to keep around the laboratory to check for circularly polarized light and to determine the sense of the polarization optics!

In recent years another exotic beetle has invaded North America and is killing millions of ash trees. The emerald ash borer (EAB), *Agrilus planipennis* (Coleoptera: Buprestidae), was first discovered in North America in Detroit, MI, in 2002 and has made its way into the Smoky Mountains. The exact reason why these insects evolved to employ LCPL is not yet known. Some researchers have speculated that CPL allows them to detect mates while hiding from prey. The Emerald Ash Borer is difficult to detect at low densities especially in areas where it has just migrated. The effectiveness of using baited purple traps for adult EABs in these areas is being studied. The interested student is encouraged to use LCPL from a handheld green laser pointer to possibly attract EABs. It is possible to purchase a green diode laser which has ∼ 120 mW power for ∼ $150 which might be capable of attracting and killing beetles of this type. These handheld lasers can also be pulsed at ∼10 Hz which produces higher peak power. One can find many YouTube™ demonstrations of He-Ne lasers killing spiders. Special safety precautions are required for the 120 mW laser.

Photon energies and molecular properties

In the field of spectroscopy, we study the wavelengths or frequencies of light absorbed by atoms and molecules. We find experimentally that microwave or millimeter wavelengths of light correspond to the excitation of molecular rotation. Infrared wavelengths of light correspond to the simultaneous excitation of both rotations and molecular vibrations. Finally, UV-visible light corresponds to the excitation of valence electronic states of atoms and molecules, as well as their vibrations and rotations. Light in the vacuum ultraviolet or X-ray regions can excite higher energy transitions of valence electrons or those of core electrons. A complete coverage of atomic and molecular spectroscopy is beyond the scope of this book. We present review and introductory material where appropriate for specific experiments. However, because several experiments employ either infrared or Raman measurements, we present in Chapter 14 a more complete introduction to rovibrational spectroscopy. The interested reader is referred to several specialized texts on atomic and molecular spectroscopy cited there for more details.

We are interested not only in the spectral regions in which atoms and molecules absorb, but also what changes can be made to molecular properties following light absorption. For this we need to consider the energies of light in different spectral regions and how these compare to the energies needed to cause physical or chemical changes in atoms and molecules. As stated in the introduction, the energy of a photon is related to Planck's constant, h, and the radiation frequency, ν, through $E = h\nu$. Using $h = 6.626 \times 10^{-34}$ J/sec and ν in cycles per second (Hz) the photon energy is given in joules. The standard joule energy units, or their calorie equivalents, are employed in thermochemistry, but in spectroscopy more convenient units are used. With the relation $c = \lambda\nu$, other units such as wavenumbers $(1/\lambda = cm^{-1})$ can be easily derived. Also since a volt is equal to a joule per coulomb, we can relate the electron volt (eV) to the joule and then obtain eV in wavenumbers. Likewise using the mechanical equivalent of heat as 1 kcal = 4.184 kJ, we can also relate the cal to the electron volt and wavenumbers (cm^{-1}). The result is: 1 eV = 8065.5 cm^{-1} = 23.06 kcal/mol. In the microwave region of the spectrum, frequency units are generally employed (e.g., GHz). In the case of infrared and Raman spectroscopy, cm^{-1} or micron $(\mu = 10^{-6}$ meter) units are used. In reporting bond dissociation energies, ionization potentials or electron affinities, it is common to use eV, kcal/mol, or cm^{-1}. Many of the experiments to be discussed in this book involve these energy units. Table 1.1 shows examples of the conversion of wavelengths, frequencies, and energy units for different regions of the spectrum.

The most important physico-chemical change that can be induced by light absorption is bond breaking, also known as *photochemistry*. This is an extremely important aspect of spectroscopy and it impacts strongly on the field of kinetics. Photochemistry is widespread, occurring in exotic environments such as the atmosphere or interstellar space. It is also employed by synthetic organic or inorganic chemists to induce reactions, by microelectronics engineers to produce metal films from volatile inorganic compounds for integrated circuits, and by a range of medical specialists for skin or other tissue treatment or cutting. It is therefore important to understand what colors of light can induce photochemistry, and what molecular level energetics are involved. We therefore present a collection of representative bond dissociation energies, indicated as D_0 values, in Table 1.2. As shown, bond energies are typically 100–200 kcal/mol (400–800 kJ/mol). Referring to Table 1.1, it can be seen that this corresponds to light in the ultraviolet or vacuum-ultraviolet wavelength regions. Visible light, which is the main component of the solar spectrum, does not usually cause photochemistry. It is the small amount of ultraviolet light in the solar spectrum penetrating through our atmosphere that causes photochemistry on human skin, i.e., sunburn. The delicate balance of UV-absorbing molecules in the atmosphere, such as ozone, is therefore extremely important for human life on Earth.

Table 1.2 *Bond dissociation energies for some common molecules. The student should convert the kcal/mol into eV. Taken from [11].*

Molecule/bond	D_0 (kcal/mol)	Molecule/bond	D_0 (kcal/mol)
H_2	104.2	CH_3CH_2-H	101.1
CO	192.3	HCC-H	133.3
N_2	225.0	C_6H_5-H	112.9
HCl	103.2	H-CH_2OH	96.1
H_2O	118.8	CH_3O-H	104.6
NH_3	107.6	HCOO-H	112
CH_3-H	105.0	H-CHO	88.1

Light absorption can also ionize atoms or molecules, exciting their outermost electron until it is no longer attached to the core system. In the case of neutral atoms or molecules, the minimum energy required to remove an electron is known as the ionization energy (IE) or ionization potential (IP). Ionization of such a neutral produces the corresponding singly charged cation:

$$M + \text{light (energy} = h\nu) \rightarrow M^+ + e^- \tag{1.14}$$

Multiply charged ions can also be produced, but this requires much greater energies. Table 1.3 provides examples of the ionization energies for selected small molecules. As shown, these values for stable small molecules generally fall in the range of 10–15 eV. This is in the vacuum-UV region of the spectrum. Consistent with everyday experience, sunlight in the visible wavelength region does not cause ionization. Open shell species, i.e., radicals, generally have lower ionization energies. Similar trends apply for atoms, as shown in Table 1.4. The closed shell rare gas atoms have the highest IP values, while these are much lower for open shell species such as the alkali metals. Within a group of elements in the periodic table, larger atoms have lower IP values because the outermost electrons are in higher shells farther removed from the positive nuclear charge.

Light absorption can also induce the elimination of excess electrons from anions, in a process known as photodetachment:

$$M^- + \text{light (energy} = h\nu) \rightarrow M + e^- \tag{1.15}$$

The minimum energy required for this process is the electron affinity (EA). Because such anions have an extra electron beyond the exact number needed to neutralize the positive nuclear charge(s) in the system, the binding energies of

Table 1.3 *Ionization potentials for some common molecules. The student should convert eV to kcal/mol or cm^{-1}. Taken from [12].*

Molecule	IP (eV)	λ (nm)	Molecule	IP (eV)	λ (nm)
H_2	15.4	80.5	H_2O	12.6	98.4
N_2	15.6	79.5	NH_3	10.2	121.6
O_2	12.1	102.5	CO_2	13.8	89.8
CO	14.0	88.6	C_2H_2	11.4	108.8
NO	9.25	134.0	C_6H_6 (benzene)	9.24	134.2
HCl	12.7	97.6	CH_3OH (methanol)	10.8	114.8
I_2	9.28	133.6	CH_3COCH_3 (acetone)	9.70	127.8

Table 1.4 *Ionization potentials for selected atoms. The student should convert eV to kcal/mol or cm^{-1}. Taken from [12].*

Atom	IP (eV)	λ (nm)	Atom	IP (eV)	λ (nm)
H	13.60	91.2	He	24.6	50.4
C	11.26	110.1	Ne	21.56	57.5
O	13.62	91.0	Ar	15.76	78.7
N	14.53	85.3	Xe	12.13	102.2
F	17.42	71.2	Na	5.14	241.2
Cl	12.97	100.8	K	4.34	285.7
I	10.45	118.6	Cs	3.89	318.7
Al	5.99	207.0	Fe	7.87	157.5
Mg	7.65	162.1	Co	7.86	157.7
Ca	6.11	202.9	Ni	7.64	162.3

these electrons are relatively low—much lower than the ionization energies of neutral atoms and molecules. Table 1.5 provides examples of EA values for selected atoms and molecules. Unlike the photoionization of neutrals, which generally requires vacuum-ultraviolet light, anion photodetachment usually requires such low energies that it can be accomplished with infrared or visible light. Because of this,

Table 1.5 *Electron affinities for a number of atoms and molecules given in eV as well as the equivalent nm and cm⁻¹. Taken from [12].*

Atom/molecule	EA (eV)	λ (nm) or $1/\lambda$ cm⁻¹	Atom/molecule	EA (eV)	λ (nm) or $1/\lambda$ cm⁻¹
H	0.756	6098 cm⁻¹	SF_6	1.49	832.1 nm
N_2	−1.9	-	Na	0.55	4436 cm⁻¹
O_2	0.44	3549 cm⁻¹	Al	0.44	3549 cm⁻¹
NO	0.02	161.3 cm⁻¹	C	1.2	10,330 nm
OH	1.83	677.5 nm	Si	1.39	892 nm
H_2O	negative	-	Ag	1.3	953.7 nm
CN	3.82	324.6 nm	Au	2.3	539.1 nm

anions are not expected to survive in the atmosphere during the daytime. CN is a notable exception, since attachment of an electron to this open-shell radical produces a closed-shell anion isoelectronic to neutral N_2. Some neutral species are so reluctant to attach an excess electron that they simply do not do it. The electron affinity of such a species (nitrogen and water are examples) is by definition negative. The exact value of the negative EA for molecules such as these can be probed as resonances in electron-scattering experiments.

The experiments in this book provide several examples of how lasers can be used for light absorption, photochemistry, and photoionization. The energetic properties of atoms and molecules mentioned here can all be calculated using computational quantum chemistry and programs such as Gaussian or GAMESS, as discussed in Chapter 15 (Quantum Chemistry Calculations). Once the student has become proficient in such methods, calculations of the bond energies, ionization potentials, and electron affinities for such small molecules can be easily carried out. Since these are small molecules the calculation times are short. Similar calculations can also predict the spectroscopic constants and absorption spectra for these systems.

Photon sources and detectors

Chapter 4 provides an introduction to the various types of lasers employed today in basic and applied research. Also Chapters 5, 26, and 27 describe how laser light of a given frequency can be shifted to higher or lower frequency using various nonlinear optical techniques. Chapters 16, 18, and 26 demonstrate how laser light can be used to access energy levels much higher than that of the laser photon through multiphoton absorption processes. For completeness we briefly describe some of the non-laser light sources available for research, especially those used in conjunction with lasers. Such light sources are also employed in the design and construction of lasers.

Light sources

The most common light sources known throughout everyday experience are based on the principle of incandescence, i.e., materials glow when they get hot [13]. The glowing is based on black-body emission, as discussed above. Incandescence explains candle- and fire-light, as well as the operation of filament-based light bulbs. In candles and fire, exothermic oxidation reactions of the fuel provide the energy and small aerosol grains of soot provide the hot glowing material. In light bulbs, the current running through a resistive wire causes it to get hot and glow. Incandescence is simple to implement in the lab, but is not a very efficient form of light generation. Another source of incandescent light is of course that from the Sun. Although its heat is generated from nuclear plasma processes at its core, its outer gases are heated by this and radiate, producing an emission spectrum characteristic of a black-body at about 5800 K. The outermost atmosphere of the Sun contains volatile species such as metal atoms, which absorb some of this continuous white light. The resulting dark line spectrum of missing colors was studied by Wallaston [14] and later documented by Fraunhofer,[15] who first labeled the atomic transitions. These lines can be seen in sunlight with a prism- or grating-based handheld spectroscope.

The next most common types of light sources are based on various forms of electrical discharges, which can be used in either continuous or pulsed modes of operation. Discharges are ignited when a high voltage is applied to electrodes mounted across the gas contained in an insulating tube (usually glass or quartz). Figure 1.9 shows a diagram of such a device. The applied voltage generates a small amount of ionization of the gas, and then the resulting positive ions and electrons are accelerated through the gas in opposite directions by the field present between the electrodes. Energetic collisions cause additional ionization, and the formation of excited electronic states of atomic and molecular neutrals and ions present, as well as other related chemical and physical processes. Some of the typical discharge processes are indicated below, using as examples the gases argon and mercury, which are present in ordinary fluorescence light bulbs:

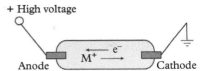

Figure 1.9 *Schematic diagram of a gas discharge lamp.*

Electron impact ionization: $e^- + Ar \rightarrow Ar^+ + 2e^-$

$\qquad e^- + Hg \rightarrow Hg^+ + 2e^-$

Electron excitation of excited states: $e^- + Ar \rightarrow Ar^* + e^-$

$\qquad e^- + Hg \rightarrow Hg^* + e^-$

Electronic energy transfer: $Ar^* + Hg \rightarrow Hg^* + Ar$

Penning ionization: $Ar^* + Hg \rightarrow Hg^+ + Ar + e^-$

Charge transfer: $Ar^+ + Hg \rightarrow Hg^+ + Ar$

Recombination: $Ar^+ + e^- + M \rightarrow Ar + M$

In overhead fluorescence lights used for room illumination, mercury is typically added to obtain a light spectrum pleasing to human eyes. The emission generated from several atomic transitions of excited mercury atoms, together with a white phosphor painted on the inside of the lamp surface, adds to produce an effect closely approximating the "whiteness" of sunlight. This is why mercury continues to be used in these devices in spite of its well-known toxicity and environmental risks. Other less noxious gases do not produce the same desirable white emission spectrum. Figure 1.10 shows this mercury emission spectrum as observed with a simple handheld prism-based spectroscope (e.g., Krüss model 1501). Such a device is convenient for observing emission spectra of gas discharges of all types, and also for looking at the spectrum of sunlight, where the Fraunhofer lines can be seen. The emission lines of mercury here are superimposed on the rainbow of emission from the white phosphor.

The xenon lamp is another discharge light source of great significance to spectroscopy and to the laser itself. Figure 1.11 shows a photograph of such a lamp and its emission spectrum. This device can produce ~100 W of continuous radiation over the wavelength range shown. Pulsed xenon lamps were used in the past to pump dye lasers, and are now used to pump Nd:YAG lasers, which provide strong laser emission at 1064 nm (see Chapter 4). To achieve high efficiency, all xenon lamps operate at high pressure (up to 35 atmospheres) and should be handled with face guards for protection in case of breakage. The bulbs consist of fused quartz with tungsten electrodes. Electron collisions with the anode produce high temperatures and water cooling is necessary in all laser pumping applications. The lamp shown in Figure 1.11 is cooled by forced air.

Similar in operation to the xenon lamp is the deuterium lamp. Deuterium discharge lamps provide radiation extending from 112 to 900 nm and are especially popular for short-wavelength applications as the light source in UV-visible spectrometers. The spectrum shows a fairly flat continuum from 180 to 370 nm with strong Balmer lines at 486 and 656 nm. Deuterium provides greater emission in

Figure 1.10 *Emission spectrum of an overhead fluorescence light measured with a handheld spectroscope and photographed with a digital camera. The emission lines of mercury appear on a broad background.*

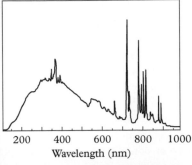

Wavelength (nm)

Figure 1.11 *An example of a xenon discharge lamp made by Oriel mounted in a cabinet with an output focusing lens extending to its right. The spectrum of emission produced (right) contains a black-body continuum along with some of the xenon emission lines.*

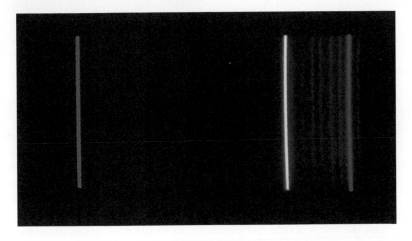

Figure 1.12 *The emission lines at 656.3, 486.1, and 434.2 nm from the Balmer series, produced with a hydrogen discharge lamp, detected with a hand-held spectroscope, and photographed with a digital camera.*

the molecular band compared to hydrogen. Figure 1.12 shows an example of the spectrum of a hydrogen discharge lamp with the prominent Balmer series lines at 656.3 nm (red), 486.1 (cyan), and 434.2 nm (blue).

Discharge lamps employing the hollow cathode design are used to provide atomic emission lines for spectrometers and for the wavelength calibration of lasers. Figure 1.13 shows a typical example. Lamps containing samples of different atomic and molecular gases (frequently rare gases) are used to provide reference spectra for comparison to measured flame spectra in Atomic Absorption (AA) spectrometers used for elemental analysis. This kind of lamp can also be used to calibrate a tunable laser using the method of "optogalvanic spectroscopy." In this technique, the laser is directed into the discharge region

Figure 1.13 *Picture of a hollow cathode discharge lamp (left) such as those used for optogalvanic spectroscopy.*

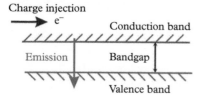

Figure 1.14 *Schematic of a light-emitting diode.*

of the lamp and the current is measured as the laser tunes through atomic or ionic energy transitions of gaseous species. When transitions between any pair of ground and/or excited states are resonant with the laser, an increase or decrease in the current through the discharge is produced by changing the population of the excited states. Chapter 24 provides more information on optogalvanic spectroscopy.

A final kind of light source becoming important in both everyday life as well as in laser science is a light emitting diode (LED). Figure 1.14 shows a diagram of how such a device works. A semiconducting material has populated valence electronic states known as the valence band, and unpopulated excited states known as the conduction band, analogous to the pattern of HOMO-LUMO molecular orbitals in aromatic molecules. These states in a semiconductor are separated in energy by a "band gap." In some semiconducting materials (but not all), the optical transition across the band gap is optically allowed. Injection of charge into the conduction band can occur when a voltage is applied to such a device, resulting in light emission at a wavelength corresponding to the band gap energy. LEDs can be used as ordinary light sources with a single specific color, or they can be modified to produce laser light (see Chapter 4). For home lighting, LED-based light bulbs have been developed which combine multiple diode colors to achieve the hue of white light closely approximating sunlight. Such devices are quite efficient in their power consumption, are long-lasting, and do not generate toxic waste, but they are more costly to produce than incandescent or fluorescent bulbs.

Light detectors

The detection of light works the same way for ordinary sources and for lasers, and so we introduce detectors here that can function for both applications. The kind of detector required depends on the power level of the light to be detected and the sensitivity required. The simplest methods use heat generated by light absorption. Using the mechanical equivalent of this heat (4.186 J/cal), one can determine the laser power. This is especially useful for higher power lasers. In calorimeter-type detectors, the light hits a piece of metal, usually dark in color so that it absorbs efficiently, whose heat capacity is known (see Figure 1.15). A thermocouple records the temperature increase of this absorber, indicated as the "volume absorbing disk" in the figure. Using the standard relation $\Delta E = C_p \Delta t$, the laser power can be determined. Such devices are calibrated at the factory and use either an analog or digital display to provide the average power in watts. The temperature change is rather slow in such devices, with time constants of a few seconds, and so they are only useful for average power measurements and not shot-by-shot measurements for pulsed lasers. Calorimeters are not particularly sensitive, and at low power levels their signal can drift from room temperature variations. Closely related to calorimeters are bolometers, which have heat deposited into a material whose resistance changes with temperature. Bolometers are commonly cooled to liquid nitrogen or even liquid helium temperatures to eliminate the effects of thermal

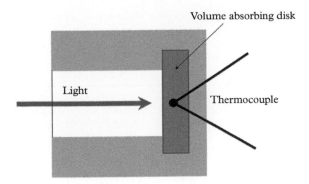

Volume absorbing disk

Light

Thermocouple

Figure 1.15 *Illustration of the simple calorimeter light detector.*

noise. In this configuration they become extremely sensitive, and can detect low levels of light with small photon energies in the IR or microwave regions.

Semiconductor photodiodes represent another common and relatively inexpensive method of light detection. As shown in Figure 1.16, these detectors work in the reverse direction compared to light-emitting diodes. In these detectors, photons having energies above the band gap can be absorbed, transferring an electron from the valence band to the conduction band. As in LED operation, the semiconducting material must have an appropriate band gap for a p–n junction and have an allowed transition for light absorption across the band gap. The photocurrent produced can be detected directly or amplified in an external circuit to give a signal proportional to the laser power. Photodiodes for visible or UV light detection are relatively inexpensive. However, those with small band gaps designed for infrared light detection must be cooled with liquid nitrogen or liquid helium to reduce band-gap absorption induced by thermal background. In some photodiodes an interstitial material, "I," is placed between the p and n junction materials. The resulting "PIN" diode detector is often employed for photon detection. Unlike the p and n regions the "I" region is lightly doped and charge carriers from the p–n junctions can spill over into this region allowing for faster response. A thorough description of photodiodes and PIN photodiodes can be found in [15]. Table 1.6 provides a list of common semiconductors used for photodiodes and their band gaps. The right side of Figure 1.17 shows three examples of common photodiode detectors like those used in many of the experiments in this book. The second detector is a diode mounted directly on a BNC cable and is ideally suited to detect the IR light in many of the experiments described. These are relatively inexpensive detectors and require no external power supplies.

The most sensitive light detectors are photomultiplier tubes (PMTs). These allow detection of low light levels even down to the single-photon limit. PMTs operate via the photoelectric effect, where light incident on a metal can eject electrons if the photon energy is great enough. The binding energy of electrons to atoms and molecules is expressed as the ionization energy or ionization potential,

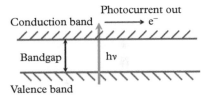

Conduction band

Photocurrent out

e^-

Bandgap

$h\nu$

Valence band

Figure 1.16 *Diagram showing the basic operation of the p–n semiconductor photodiode detector.*

Table 1.6 *Summary of the band gap and characteristics of a number of photodiode detectors.*

Semiconductor	Band gap (eV)	Wavelength range (nm)	Energy range (eV)
silicon	1.2	190–1100	1.13–6.5
germanium	0.67	400–1700	0.73–3.1
indium gallium arsenide	0.75	800–2600	0.477–1.55
lead sulfide	0.37	1000–3500	0.354–1.24
mercury cadmium telluride	0–1.5	400–14000	0.089–3.1

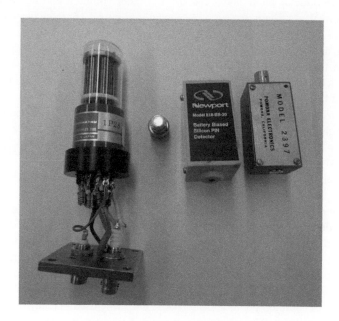

Figure 1.17 *Photographs of four light detectors commonly employed in experimental chemical physics research. The photomultiplier tube (left) has attached resistors that form the voltage divider chain. The three devices on the right are photodiodes mounted in different ways.*

as noted above. In the case of solid metal, this binding energy is known as the *work function*. Because metals have delocalized electrons, the energy required to emit an electron is lower than it is for atoms and molecules. Typical work functions are 3–5 eV. Referring to Table 1.1, it can be seen that this corresponds to ultraviolet light. Therefore, a plate of copper metal (work function = 4.7 eV) can emit electrons, generating a photocurrent that can be detected, when light in the ultraviolet region strikes it. A photomultiplier tube constructed with such a copper plate is called a "solar-blind" tube, because it cannot detect visible light. To generate a photoelectric signal with lower energy visible light, metals with lower work functions are required, such as alkali metals like cesium (work function = 2.1 eV). Unfortunately, alkali metals are too soft for mechanical devices, and so alloys containing these are used instead. Alloy plates containing cesium can generate a photoelectric signal down to red visible wavelengths, but photon energies below this (i.e., far-red or near-IR) do not produce a signal. The plate used to produce a photoelectric signal from light is called a photocathode.

A photomultiplier tube uses a photocathode together with other components to amplify the signal, as shown in Figure 1.18. The photocathode is biased with negative high voltage, and it is connected electrically to a stack of 10–20 "dynode" plates, each with a progressively lower negative voltage determined by its connection to a voltage divider chain of resistors. A lower negative voltage is by definition more positive, so the first dynode plate attracts the photoelectrons generated at the photocathode (one per photon). The photoelectrons are accelerated by the field between these plates to energies of 50–100 eV, and when these collide with the dynode plate this energy far exceeds the work function, allowing a "splash" of many secondary electrons to be ejected. The same process happens at each

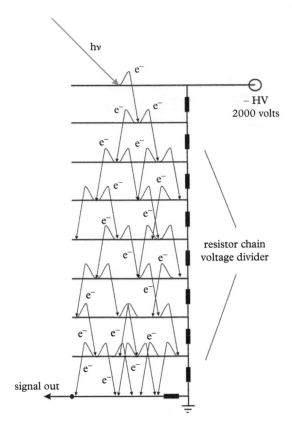

Figure 1.18 *The design of a PMT detector, showing the electron cascade process that leads to amplification of signal. Photoelectrons from the photocathode are accelerated down a dynode chain, producing more secondary electrons at each stage. A voltage divider applies progressively less negative voltage to each stage, making it more attractive to the electrons. The photocurrent is collected at the last plate at ground potential.*

subsequent dynode stage, producing more and more secondary electrons down the stack. This amplification process ultimately can produce a factor of up to 10^6 electrons at the output of the PMT for each photon that hits the photocathode. This is the "gain" factor for the tube. Because of this high gain, PMTs can detect much lower light levels than photodiodes. Background gas would tend to limit the mobility of the electrons through this device, and so this kind of detector is sealed under vacuum. For visible wavelengths the front window can be glass, but ultraviolet detection requires a quartz front window. An *electron* multiplier tube (EMT) has the same design as a photomultiplier tube, but with the front end open and an ordinary metal plate instead of a low work function photocathode. This is employed inside the vacuum system in mass spectrometers to detect ions. The left side of Figure 1.17 shows a photograph of an inexpensive model 1P-28 PMT.

Photomultipliers and electron multipliers use components (metal plates, wires, and resistors) that have been available for over 50 years. The same design concepts are used with more modern materials in "microchannel-plate" detectors. Microchannel plates are thin (1–2 mm) ceramic disks made with many micron diameter channels running through them, which can be biased electrically from front to back just like the dynode stack of plates in a PMT. Photons, electrons, or ions hitting the surface of these plates generate electrons which cascade through these channels, generating more secondary electrons as they hop along the way. The gain can be comparable to a PMT or EMT, but the path is much shorter and therefore the response time is much faster. A photo of a microchannel plate is shown in Figure 1.19, together with a mounting configuration for two plates in series, known as a dual channel-plate configuration.

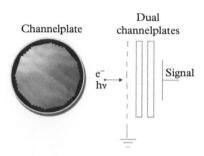

Figure 1.19 *Picture of a single channel-plate electron multiplier and the configuration for a dual channel-plate detector.*

REFERENCES

1. J. Heirtzler, "The longest electromagnetic waves," *Sci. Am.* **206**, 128 (1962). As an example, at one end of the spectrum, electromagnetic waves with frequencies of 10^{-2} Hz (which corresponds to a period of 100 seconds and to a wavelength of about 5000 Earth radii) have been detected at the Earth's center.

2. J. C. Mather, E. S. Cheng, D. A. Cottingham, R. E. Eplee, Jr., D. J. Fixsen, T. Hewagama, R. B. Isaacman, K. A. Jensen, S. S. Meyer, P. D. Noerdlinger, S. M. Read, L. P. Rosen, R. A. Shafer, E. L. Wright, C. L. Bennett, N. W. Boggess, M. G. Hauser, T. Kelsall, S. H. Moseley, Jr., R. F. Silverberg, G. F. Smoot, R. Weiss, and D. T. Wilkinson, "Measurement of the cosmic microwave background spectrum by the COBE FIRAS instrument," *Astrophys. J.* Part 1 **420**, 439 (1994).

3. S. J. Williamson and H. Z. Cummins, *Light and Color in Nature and Art*, John Wiley & Sons, 1983.

4. D. Sadeh, "Experimental evidence for the constancy of the velocity of gamma rays, using annihilation in flight," *Phys. Rev. Lett.* **10**, 271 (1963).

5. J. Dryzek and D. Singleton, "Test of the second postulate of special relativity using positron annihilation," *Am. J. Phys.* **75**, 713 (2007).

6. R. P. Feynman, *QED: The Strange Theory of Light and Matter*, Princeton University Press, Princeton, NJ, 1985.

7. E. Hecht, *Optics*, fourth edition, Addison-Wesley Publishing Co., San Francisco, CA, 2002.

8. R. Guenther, *Modern Optics*, John Wiley & Sons, Hoboken, NJ, 1990.

9. R. A. Beth, "Mechanical detection and measurement of the angular momentum of light," *Phys. Rev.* **50**, 115 (1936).

10. V. Sharma, M. Crne, J. Ok. Park, and M. Srinivasaro, "Structural origin of circularly polarized iridescence in jeweled beetles," *Science* **325**, 449 (2009).

11. S. J. Blanksby and G. B. Ellison, "Bond dissociation energies of organic molecules," *Acc. Chem. Res.* **36**, 255 (2003).

12. E. P. Hunter and S. G. Lias, NIST Chemistry WebBook, NIST Standard Reference Database Number 69, Eds. P. J. Linstrom and W. G. Mallard, National Institute of Standards and Technology, Gaithersburg, MD, 20899, http://webbook.nist.gov.

13. B. E. A. Saleh and M. C. Teich, *Fundamentals of Photonics*, John Wiley & Sons, Hoboken, NJ, 1991.

14. W. H. Wollaston, "A method of examining refractive and dispersive powers, by prismatic reflection," *Philos. Trans. Royal Soc.* **92**, 365 (1802).

15. J. Fraunhofer, "Bestimmung des Brechungs-und des Farben-Zerstreuungs-Vermögens verschiedener Glasarten, in Bezug auf die Vervollkommnung achromatischer Fernröhre" (Determination of the refractive and color-dispersing power of different types of glass, in relation to the improvement of achromatic telescopes), *Denkschriften der Königlichen Akademie der Wissenschaften zu München* (Memoirs of the Royal Academy of Sciences in Munich) **5**, 193 (1814–1815).

2

Basic Optics

Introduction

Table 2.1 *Refractive index values for selected materials at 20 °C and 589 nm.[5]*

Material	n_D
air	1.00045
water	1.333
ethanol	1.361
methanol	1.328
acetone	1.359
glycerol	1.473
benzene	1.501
crown glass	1.50–1.54
quartz	1.458
calcium fluoride	1.434
zinc selenide	2.624
sapphire	1.76–1.78
diamond	2.419

The use of lasers for measurements in physics or chemistry requires a basic knowledge of optics, which explains how light propagates through materials and how its properties can be manipulated. A full course in optics is well beyond the scope of this book. The interested reader is referred to many excellent texts on this subject.[1–3] However, we review here selected topics that should be generally familiar from undergraduate physics courses, and emphasize those aspects that are particularly useful for the understanding of how lasers work and how their light may be employed effectively for various laboratory experiments.

The first important concept of optics is that light travels at different velocities as it propagates through different materials. Energy must be conserved, and since it is proportional to the frequency through $E = h\nu$, the frequency of a specific light beam must be the same in different materials. Because $c = \lambda\nu$, and the frequency is constant, if the speed changes the wavelength must change, and it does; it gets shorter in materials than it is in vacuum. The ratio of the speed of light in vacuum to that in a material is the refractive index, n.

$$n = c/v_{material} \tag{2.1}$$

Because light moves slower in materials than it does in vacuum, n is always a number ≥ 1. The refractive index turns out to be an especially important parameter governing many aspects of how light interacts with materials. Some examples of refractive index values for selected materials are given in Table 2.1. The refractive index in these materials also varies with temperature and with the wavelength of light. Therefore, refractive index values are usually tabulated at 20 °C and at the standard wavelength of 589 nm, the wavelength of the sodium "D" line (corresponding to the 3s → 3p, $^2S → ^2P$, atomic transition in neutral sodium atoms). In the solar spectrum first documented by Fraunhofer,[4] in the early nineteenth century when the assignments were not yet known, this line was the fourth in order counting from the red and was thus called the "D" line. The index value at this wavelength is therefore often written as n_D. This yellow light is also seen in the common flame tests with sodium salt solutions used in qualitative chemical analysis.

Laser Experiments for Chemistry and Physics. First Edition. Robert N. Compton and Michael A. Duncan.
© Robert N. Compton and Michael A. Duncan 2016. Published in 2016 by Oxford University Press.

Reflection, absorption, refraction, and transmission

When light traveling through one material comes into contact with an interface with another material, it can be reflected, refracted, absorbed, or transmitted through the material. The path that light follows is defined relative to the normal to the surface as defined by the interface between the two materials. To demonstrate these ideas, it is simplest to consider light traveling through air as it hits the surface of a solid material, which necessarily has a higher index of refraction. If the incoming path is along the surface normal, the polarization of the light does not matter and there is no refraction. In this situation, the light can either be transmitted or reflected at the initial interface. The amount of light reflected depends on the refractive index of the material, n_2 compared to that of air, n_1. We define the ratio, $\rho = n_2/n_1$, and then the amount of light reflected at such an interface is

$$I_r = \left(\frac{1-\rho}{1+\rho}\right)^2 \tag{2.2}$$

For the example of an air/glass interface, using the refractive index values above for the 589 nm wavelength, $\rho = 1.5$ and $I_r = 0.04$. This means that 4% of the light will be reflected and 96% will be transmitted through the interface. The refractive index of quartz is similar to that of glass, and the value of about 4% reflectance at a surface is typical for both materials near normal incidence at visible wavelengths.

The light passing through an initial interface can then be transmitted through the material or absorbed within it, depending on the absorption spectrum and the wavelength of the light. Every material has its own characteristic absorption/transmission spectrum, which can be measured with UV-visible or infrared spectrometers. The transmission spectra of quartz and glass, measured with a standard UV-visible spectrometer, are compared in Figure 2.1. Similar transmission curves are available online for many optical materials from the manufacturing companies that sell these materials. Such companies include CVI Laser Optics (http://cvilaseroptics.com/), Melles Griot (http://mellesgriot.com/), and Thorlabs (https://www.thorlabs.com/). From such measurements, the useful wavelength regions for optical materials can be determined. Glass transmits in the visible, but not well in the UV or IR; quartz transmits in the visible and UV, but not in the mid-IR (its UV transmission varies substantially with the grade of quartz); calcium fluoride and magnesium fluoride transmit in the visible, UV and near-IR; zinc selenide transmits well in the IR; etc.

When light encounters an interface between two materials at an angle other than normal to the surface, it can be reflected or transmitted, but also *refracted*, or bent in its path through the material, as shown in Figure 2.2. Absorption in the material is neglected in the remaining discussion. Refraction is described by Snell's Law:

$$n_1 \sin \theta_1 = n_2 \sin \theta_2 \tag{2.3}$$

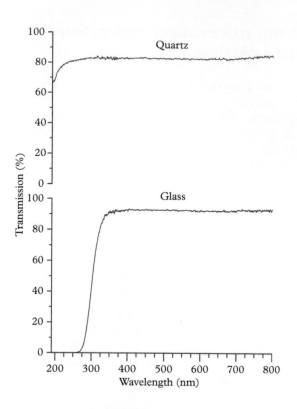

Figure 2.1 *The transmission spectra of light through plates of boro-silicate glass and quartz, as measured with an ordinary UV-visible spectrometer. The maximum transmission is about 92% because of reflective losses at the incoming and outgoing surfaces.*

Figure 2.2 *Examples of reflection, transmission, refraction, internal reflection, and transmission, when a beam of light encounters an interface at an angle away from normal. R1 is the beam resulting from reflection at the initial interface, R2 is that resulting from reflection off the back interface, and R3 is that reflected off the front interface as the R2 beam exists. Higher order reflections are also possible, with diminishing intensities.*

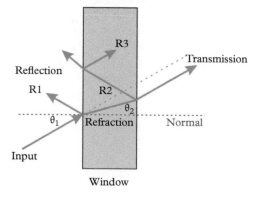

where θ_1 is the angle away from normal incidence for a beam in material of index n_1 and θ_2 is the refracted angle after the interface in a material with index n_2. Figure 2.2 shows an example such as that for an air–solid interface, where $n_2 > n_1$, and the refraction at the initial interface causes light to bend closer to the normal angle. The reverse is true for an interface in which $n_2 < n_1$, such as that in Figure 2.2 where the light comes out of the material back into air. The refraction going in is exactly offset by that coming out of such a material, and the transmitted beam is parallel to the hypothetical continuation of the input beam (dashed blue line) but slightly offset from it.

Refraction is particularly important in optics because it is the basis for the formation of lenses, when light is transmitted through materials having curved surfaces. Depending on the curvature of the surfaces involved, a lens can be either positive (focusing) or negative (defocusing). The bending of light for a particular geometry of lens depends on the surface curvature and also on the refractive index of the material. For example, the availability of new high-index plastic materials is absolutely crucial in the design of eyeglasses and contact lenses in which the lens material is thin and light-weight. Spherical lenses, with a circular profile and thickness varying radially outward from the center, focus (or defocus) a round spot to a smaller (or diverging) round spot. These are perhaps the most familiar, but are by no means the only shapes of lenses. For example, cylindrical lenses have a rectangular profile, with thickness varying from center to edge in either the horizontal or vertical direction, but not both. These can focus a round spot from a laser to a line whose length is the same as the original spot diameter. The variety of lens shapes, sizes, lens combinations, and their applications is almost unlimited. We refer the interested reader to more comprehensive textbooks on optics for formulas covering lens types and combinations.[1,2] Various software programs are available for designing/modeling lens systems (http://www.optics-lab.com/; http://arachnoid.com/OpticalRayTracer/; http://www.stellarsoftware.com/).

As noted above, the refractive index of materials is wavelength dependent, which results in an extremely important aspect of lens performance. Table 2.2 shows the wavelength variation of the refractive index for selected materials. Because of this, lenses have different focal lengths for different colors of light. As shown in Figure 2.3, the focal length for blue light is shorter than that for red. To eliminate confusion, the focal lengths of lenses are quoted in optics catalogs at the standard wavelength of 589 nm. Also because of this wavelength dependence, images containing objects with mixed colors cannot be focused simultaneously with a simple lens, an effect known as *chromatic aberration*. The human eye contains a lens and also suffers from this. However, sunglasses that filter out certain colors can sometimes cause images to appear sharper because fewer wavelengths are present to be focused. In laser applications, lenses are often used to focus light

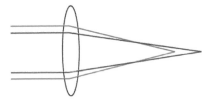

Figure 2.3 *Diagram showing the wavelength dependence of the focal length for a positive lens.*

Table 2.2 *The variation of refractive index with wavelength for selected materials.[5]*

Material	400 nm	500 nm	600 nm
boro-silicate glass (BK7)	1.5302	1.5213	1.5160
synthetic fused silica (quartz)	1.4696	1.4625	1.4580
sapphire	1.7858	1.7746	1.7680

on a sample. It is therefore essential to remember that the focal length of a lens changes if a different colored laser is employed.

Although there are many possible combinations of lenses, one particularly important configuration is the combination of two lenses to form a telescope. Such a device was of course produced by Galileo and used for his celestial observations, but these devices are also important in laser science. As shown in Figure 2.4, a simple telescope can be produced with a positive–positive or negative–positive lens combination. In either case, collimated light of a small diameter can be converted to a larger diameter (or vice versa). This allows small images of distant objects to be magnified in size. However, it also allows a laser beam with a small spot size to be magnified, or beam expanded, to a larger size (or vice versa). As shown in Figure 2.4, situating the lenses with spacings corresponding to their respective focal lengths causes the outgoing beam to be collimated. In this configuration, the beam expansion, or the ratio of the beam waist after (w_f) relative to before (w_i) expansion, is given by $w_f/w_i = f2/f1$, or $w_f/w_i = f_p/f_n$.

However, the lens spacing can also be varied smoothly by mounting one lens on a small platform whose position can be translated with a drive screw (known as a translation stage) relative to the other (fixed position) lens. If this is done, the telescope becomes adjustable, allowing the beam size in far-field to be either gradually focusing or diverging, so that the size of a laser spot or its focal point can be controlled at any point downstream. This is of course how handheld telescopes are focused for different observers; a screw adjustment in a tube varies the lens spacing. Two-lens telescopes based on transmission of light through lenses are called "refractors." These devices are subject to chromatic aberration, as noted above, and cannot focus different colors simultaneously. Telescopes can also be made with two or more curved mirrors, as shown in Figure 2.5. Such a "reflecting" telescope does not suffer from chromatic aberration because the light does not pass through any material. These devices are used for astronomy, but they can also be used for laser beams. They are particularly useful for lasers in the far-IR or vacuum-UV, where highly transmitting materials are not readily available.

The wavelength dependence of the refractive index, also known as dispersion, explains how prisms separate white light into its different component colors. Each color is bent to a different angle by refraction at the incoming and outgoing surfaces, resulting in different colors being bent to different angles. As shown

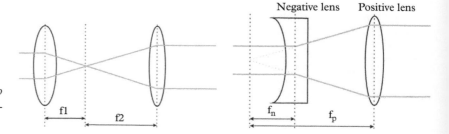

Figure 2.4 *Telescopes formed from two positive lenses (left) or from a negative–positive combination (right).*

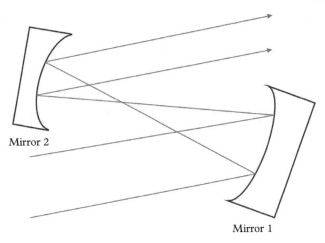

Mirror 2

Mirror 1

Figure 2.5 *A reflecting telescope using curved mirrors instead of lenses.*

in Table 2.2, blue light has a higher index than red light, and so the blue color going through a prism is bent to a larger angle than red. The wavelength dependence of the refractive index in water vapor causes different color components of sunlight in the atmosphere to bend to different angles, producing a rainbow. Rainbows actually involve a combination of refraction and diffraction processes. White light from the sun behind an observer is diffracted upon passing through water droplets, producing a refracted beam back to the observer which makes an arc of ~ 41° with the order red/yellow/green/blue (the 41° is a result of the index of refraction of water being 1.333). If the sunlight is bright enough a second reflection and refraction within the rain drops will produce an observable secondary rainbow with an inverted color pattern blue/green/yellow/red. The area between the primary and secondary rainbow is somewhat dim and is called Alexander's "Dark Band." It should also be noted that the rainbows are a cumulative scattering from many raindrops which is an example of Babinet's Principle (see Hecht [1] page 508, and Chapter 30).

Light that is incident on a material at angles off-normal can also be reflected, but the amount of reflection depends on both the angle of incidence and the polarization of the light. However, it is not the vertical or horizontal polarization in the lab frame that is important; it is the polarization *relative to the interface*. As shown in Figure 2.6, the angle of incidence θ_i is defined as the angle between the incident ray and the normal to the surface whereas the angle of reflection θ_r is defined as that angle between the reflected ray and the normal ($\theta_i = \theta_r$). The *plane of incidence* is then defined by the plane containing both the incident and reflected rays and the normal to the surface. If an electric vector component of the E&M ray is in this plane of incidence, it is defined as having "*p*" polarization. If the electric vector is perpendicular to this plane, the light has "*s*" polarization. The angular dependence of the reflected fraction of light can be determined by solutions of Maxwell's equations for polarized light, giving the reflected light intensity as a

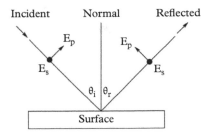

Incident   Normal   Reflected

E_p E_p

E_s E_s

$\theta_i \mid \theta_r$

Surface

Figure 2.6 *s- and p-polarized light incident on a surface. The plane of incidence is defined by the incident ray and the normal to the surface. The E_s-vector for s-polarized light is perpendicular to the plane of incidence whereas the E_p-vector for p-polarized light is in the plane of incidence.*

function of incident angle for *s*- and *p*-polarized light. If we define R_s and R_p as the percent of the intensity of reflected light for *s*- and *p*-polarization, respectively, then the angular dependence of the reflectivity for so-called *external* reflectance (at an interface going from low index to high) is given by the Fresnel equations:

$$R_s = \left[\sin^2(\theta_i - \theta_t)/\sin^2(\theta_i + \theta_t) \right] \tag{2.4}$$

$$R_p = \left[\tan^2(\theta_i - \theta_t)/\tan^2(\theta_i + \theta_t) \right] \tag{2.5}$$

where θ_i is the incident angle and θ_t is the transmitted (refracted) angle. Figure 2.7 shows a graph of R_s and R_p versus the angle of incidence for the case in which light from a vacuum is incident upon glass ($n_i = 1$ and $n_t = 1.5$).

As noted above, the reflectance for light at normal incidence to the glass surface for both *s*- and *p*-polarized light is ~ 4%, and near grazing incidence (90°) the reflectivity is near 100%; but the reflectivity as a function of angle varies significantly for *s*- versus *p*-polarization. At θ_i ~ 56%, a surprising result is that the reflectivity for *p*-polarization is approximately zero. This special angle is known as *Brewster's angle* and corresponds to complete transmission of *p*-polarized light. Brewster's angle, θ_B, is given by:

$$\theta_B = \tan^{-1}(n_t/n_i) \tag{2.6}$$

For an air–glass interface $\theta_B = 56.3°$ (index of refraction of glass = 1.5), whereas for an air–water interface $\theta_B = 53°$ (index of refraction of water = 1.33). A useful way to remember which plane of polarization is transmitted and which is reflected is to note that *s*-polarized light "skids" off the surface and *p*-polarized light "plunges" through the interface.

Figure 2.7 *Intensity of externally reflected light for s- and p-polarized light incident from a vacuum and transmitted through glass (n = 1.5). Note that the s-polarized light is totally transmitted at θ_i ~56°, indicated by the dashed red vertical line, which is known as Brewster's angle.*

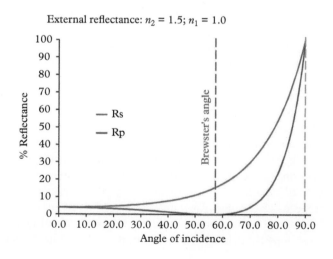

Brewster's angle is often used to minimize reflective losses in a light beam. One common design is shown in Figure 2.8, where a window bringing light into a container such as a vacuum system is mounted at Brewster's angle. The vertical polarization on the page is *p*-polarization relative to the surface, and this light passes through the window without any reflective loss. Brewster's angle can also be used to sort the polarization of a light beam. If a beam of mixed polarization hits a window mounted at this angle, all of the *p*-polarized light will be transmitted, as well as a fraction of the *s*-polarized light. However, the light reflected from such a surface will be purely *s*-polarized. If multiple windows are used at this angle, in a "pile of plates" configuration, the transmitted beam becomes more nearly pure *p*-polarized and a greater fraction of the *s*-polarized light can be separated out by reflection. A stack of plates adjusted smoothly at angles other than Brewster's can be used to systematically attenuate a laser beam by reflecting a greater and greater fraction of it at steeper angles. This method can be used to avoid saturation of a signal, which can cause line broadening, or to determine the order of a multiphoton process. As described in Chapter 16, in a multiphoton process the level of the signal depends on the laser intensity to the *n*th power. Thus, a plot of the log of the signal as a function of the laser power (measured with a calorimeter or diode) will give the order of nonlinearity.

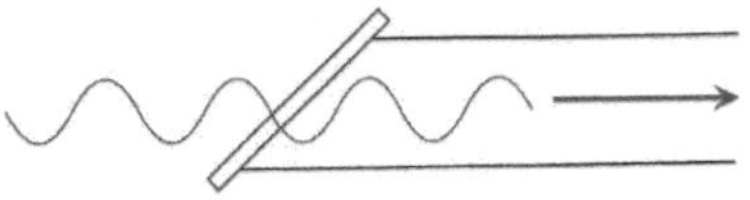

When light is moving inside a material such as glass or quartz, and hits the rear surface to emerge back into air, *internal* reflectance can occur. Fresnel's equations again describe the reflectance versus angle in this situation, but the incident and transmitted angles are reversed. The angular dependence of this internal reflectance is plotted in Figure 2.9. As shown, both polarizations are reflected by about 4% at normal incidence, as noted earlier. Again, both polarizations differ in their angular dependence for small incident angles. At the complement of Brewster's angle coming into the material, there is also a Brewster's angle for light emerging from it which occurs at 33.7° for glass. At this angle, *p*-polarized light is again totally transmitted. However, both polarizations undergo another surprising

Figure 2.8 *Brewster's angle window designed to transmit polarized light from the lab into a sealed chamber without any reflective losses. The wave shown has p-polarization and will be completely transmitted.*

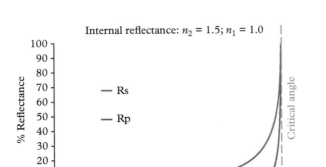

Figure 2.9 *Intensity of internal reflected light for s- and p-polarized light incident from glass (n = 1.5) and transmitted outward into vacuum (n = 1.0). θ = 41.8° is the critical angle, beyond which total internal reflectance occurs.*

result, shown for this example of refractive indices at 41.8°, where the reflectivity becomes 100%, i.e., all the light undergoes *total internal reflectance*. The angle at which this occurs is known as the *critical angle*, θ_c. It depends on the refractive index ratio of the two materials, as shown below:

$$\theta_c = \arcsin(n_2/n_1) \tag{2.7}$$

At angles greater than this, light is reflected completely within the material. A common application of this effect is fiber-optic cable, in which light can be transferred long distances without significant losses in its intensity. In laser science, total internal reflection allows the use of prisms for various applications in the lab. Figure 2.10 shows a right-angle turning prism as it is used to reflect a high-powered pulsed laser. Ordinary mirrors with metal film coatings of aluminum, silver, or gold are used to reflect non-laser light beams or low-power laser beams, but pulsed lasers with microsecond, nanosecond, or shorter pulse widths have such high peak powers that they burn the metal coatings, thus destroying mirrors. Dielectric coatings can produce mirrors with much higher damage thresholds, but these rely on multilayer coatings whose thickness is a multiple of the wavelength to be reflected. Such dielectric mirrors are therefore wavelength specific and relatively expensive. They are used for Nd:YAG laser harmonics or for excimer lasers, but they do not work for dye lasers that scan over many visible and UV wavelengths. However, a right-angle prism employing total internal reflectance has no coatings to burn and is not wavelength specific. It needs to be used at angles normal to the input faces, as shown in the figure, but is otherwise convenient for a wide variety of wavelengths and pulse energies. Figure 2.11 shows a different prism configuration that can reflect high intensity laser beams back along the same direction from which they came. This also uses total internal reflectance. This kind of prism can be mounted on a movable stage to make a delay line for pump-probe experiments using ultrafast laser systems.

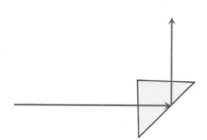

Figure 2.10 *Right-angle prism using total internal reflectance to reflect a laser beam at a right angle. The beam enters and exits normal to those faces, resulting in a 45° with the back surface, which is greater than the critical angle for materials like glass and quartz.*

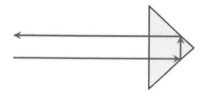

Figure 2.11 *Prism retro-reflector using total internal reflectance.*

Polarization of light and its manipulation

As discussed in Chapter 1, light waves are composed of oscillating electric and magnetic fields oriented perpendicular to each other. In laser optics, we are primarily concerned with the electric field, which oscillates in a plane. If all the light waves from a source oscillate in the same plane, the polarization is linear. If these waves have random orientations, the light is unpolarized. Linearly polarized light can be produced in several different ways, and its polarization can be modified, as described below. Plane-polarized light can be viewed as equal portions of right- and left-circularly polarized light, as described in Chapter 1. Alternatively, circularly polarized light can be viewed as containing equal amounts of plane-polarized light whose waves are at right angles to each other and $\frac{1}{4}$ cycle out of phase with each other. The vector sum of such waves traces out a corkscrew as it propagates;

depending on the phase relationship, the screw direction can be clockwise or counter-clockwise. These are defined from the standpoint of the receiver as *left-* or *right-*circularly polarized. If the amplitudes of the interacting waves are not equal, elliptical polarization is also possible, but this is not important in laser applications in this text.

The most common way to produce polarized light from unpolarized light is through the use of polarizing, or polaroid, filters. Edwin Land invented the polarizing filter by doping a long chain polymer with iodine and stretching out the material making a long so-called "picket fence" pattern of the aligned polymer molecules. These polymer molecules absorb light preferentially when it oscillates along their length, parallel to the picket fence pattern, thus blocking its transmission. They do not absorb light that oscillates perpendicular to the picket fence, which is thus transmitted. Polaroid sunglasses are usually oriented to block *s*-polarized light which is reflected more from common surfaces. The simplest way to determine the direction of polarization of polarizing film is to observe unpolarized light reflected from a surface. The intensity of the reflected light will appear greatest when the polarizer is oriented along the electric vector of the reflected (*s*-polarized) light. As an interesting note, astronomers have found that star light passing through interstellar elongated dust grains, which are aligned over large regions of space by magnetic fields, is partially polarized due to absorption of the light polarized along the length of the grains. Thus interstellar dust grains provide cosmic sunglasses!

The transmission of polarized light through a polaroid film is not a sharp function of angle, but rather falls off more gradually. Specifically, light whose E vector is at an angle θ relative to the polarization direction of the polarizer will be diminished in intensity by a factor of $\cos^2 \theta$ since only the $\cos \theta$ component will pass through the polarizer and the intensity is proportional to E^2. This is called the Law of Malus (Etienne Louis Malus, 1775–1812) and can be easily verified using two polarizers. This represents an excellent way to accurately control or to vary the intensity of a laser beam and the attenuation versus angle is independent of the wavelength.

Polarized light can also be produced using reflection from a Brewster's angle window. As discussed above, *s*-polarized light is preferentially reflected, but only partially. So multiple plates like this would be needed to isolate a large fraction of the available *s*-polarized light.

The polarization of light can also be manipulated with so-called "optically active" materials, such as calcite. In such materials, light transmission depends in an unusual way on the polarization, and the polarization of light can be changed from linear to circular, or rotated in its plane of linear polarization. The full details of optical activity are beyond the scope of this discussion, but practical aspects of this phenomena are not too difficult to understand. Optically active crystals have molecular lattices without inversion symmetry. Because of this, light passing through these materials exhibits an effect known as "double refraction," or "birefringence." An incident beam of light separates into an *ordinary* beam with refractive behavior following Snell's Law, and an *extraordinary* beam (often written as *o*- and *e*-beams), whose polarization is rotated 90° from the ordinary beam, with a different refractive index. Such crystals have either one or two special optical axes, and the behavior of transmitted beams depends on whether the propagation is along an axis, perpendicular to it, or at an angle to this axis. Along the optical axis, there is no extraordinary wave and no birefringence. At an angle to the optical axis, the ordinary and extraordinary beams become physically displaced from each other. This can be seen with a laser pointer, whose beam will separate into two beams. It can also be seen by holding such a crystal over written text on a typed page, which results in a double image of the text when viewed through the crystal. If the incident beam propagates perpendicular to the optical axis, the ordinary and extraordinary beams follow the same path, but because they are moving at different speeds through the material, their perpendicular waves move in and out of phase with each other, depending on the length of the path. The refractive index of the extraordinary beam actually varies as a function of the angle away from normal to the optical axis, so that tilting the crystal changes the phase relationship between the two beams. These various behaviors of light going through optically active crystals turn out to be incredibly useful for several applications in optics and lasers.

A quarter-wave plate uses incident light perpendicular to the optical axis, so that the ordinary and extraordinary beams follow the same path. This is produced by cutting a plate of optically active material such that the optical axis is parallel to the face where light enters. These beams travel at different speeds and their waves come into and out of phase with each other. The thickness of the plate is chosen so that the beam of light exits when the beams are exactly $\pi/2$ or $\lambda/4$ out of phase with each other. The resultant of the two plane-polarized beams is then a beam that is circularly polarized. This is a common way to produce circularly polarized light. If this circularly polarized light passes through another identical quarter-wave plate, it shifts back to linearly polarized, but the plane of polarization is rotated 90° from that of the initial beam. This is illustrated in Figure 2.12.

In the above description, the quarter-wave plate has to be selected for a particular wavelength. A very useful (but somewhat expensive) device is a Soleil–Babinet compensator which consists of two wedges cut at angles so that they can slide against each other making a variable thickness. A screw mechanism can adjust the thickness to give a quarter-wave plate for a range of wavelengths. Finally,

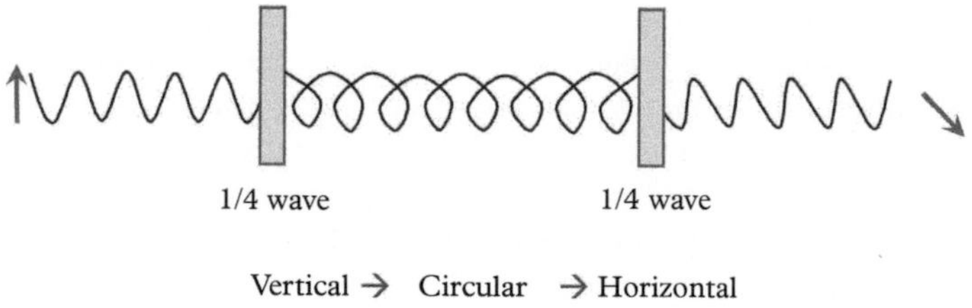

Figure 2.12 *The action of two quarter-wave plates in sequence, which first produces circularly polarized light and then plane-polarized light rotated by 90° from the initial wave.*

1/4 wave   1/4 wave

Vertical → Circular → Horizontal

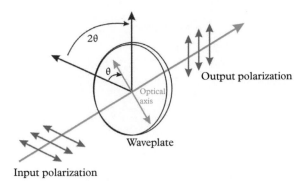

Figure 2.13 *The design of a half-wave plate, which rotates the linear polarization of a light beam.*

perhaps the most precise method to produce circularly polarized light is a single Fresnel rhomb whose face can be rotated at ± 45° relative to a linearly polarized light beam. The exiting light will be right- or left-circularly polarized. Of course the light beam is displaced from its original direction. This circular polarizer will work for any wavelength that the material transmits.

A half-wave plate is cut like a quarter-wave plate, but its thickness is chosen to shift the *o*- and *e*-waves out of phase by π or $\lambda/2$. The plate involved is usually cut in the shape of a disk. As shown in Figure 2.13, transmission of light through this results in a rotation of the plane of polarization of an input wave that varies with the orientation of the angular disk. If the incident plane of linear polarization of the light is at an angle of θ from the optical axis, the transmitted wave will be linearly polarized but rotated by an angle of 2θ from its original orientation. Half-wave plates are therefore convenient devices for the systematic adjustment or rotation of polarization.

Another useful polarization rotation device is a combination of two Fresnel rhombs. Incident polarized light into either rhomb will exit as a rotated polarized beam depending on the orientation of the incident polarization to the optical axis of the rhomb. Rotating the rhomb by 2π radians will continuously rotate the incident light phase by 4π radians. Commercial double-Fresnel rhombs have a calibrated angle readout to high precision and accuracy.

Some crystals that are not normally optically active become active when an electric field is applied across them, thus polarizing the molecular species in the lattice in an asymmetric way. There are two variations of this effect. In the *Pockels* effect the induced birefringence varies linearly with the applied field, whereas in the *Kerr* effect it varies quadratically with the field. A *Pockels cell*, which employs this effect, is therefore a voltage-controlled waveplate. As demonstrated later in the discussion of Nd:YAG lasers (Chapter 4), such devices are particularly useful for fast switching of light transmission, which can operate much faster than any shutter based on mechanically moving parts.

Another common device whose operation depends on optically active crystals is a Glan–Taylor (also called Glan–air) prism, as shown in Figure 2.14, which

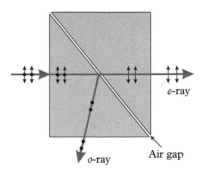

Figure 2.14 *Glan–Taylor prism used to obtain linearly polarized light.*

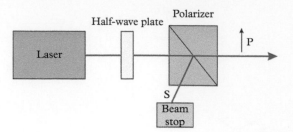

Figure 2.15 *Waveplate/polarizer combination used as a variable attenuator.*

is actually two prisms joined via an air space. Unpolarized light passing through such a prism separates into *o*- and *e*-beams with orthogonal polarization. Because these beams have different refractive indices, their respective Brewster's angles and critical angles are different. Although there are subtle details about how this works, the *o*-ray is totally reflected at the interface while the *e*-ray is transmitted, thus separating the polarizations.

Another useful device for the controlled attenuation of laser beams employs a combination of a half-wave plate and a polarizer (usually a Glan–Taylor prism). Such a device is shown in Figure 2.15. Like the simple polaroid polarizer, the Glan–Taylor prism transmits polarized light with a $\cos^2 \theta$ angular dependence. When the plane of polarization is rotated with a half-wave plate, therefore, its transmission is smoothly attenuated. Such waveplate/polarizer combinations are sold commercially as "variable attenuators."

The same polarization rotation effects achieved with quarter-wave or half-wave plates can also be accomplished with materials that are not optically active, using a special kind of prism known as a Fresnel rhomb. In this device, light undergoing total internal reflectance at two (four) carefully selected angles gets phase shifted in the same way that it does in quarter (half) waveplates. A Fresnel rhomb has an advantage over waveplates in that it works over a wide range of wavelengths.

The polarization behavior of light in optically active crystals is especially important in nonlinear optics for frequency doubling, sum-frequency generation, and optical parametric oscillation. These topics are discussed in Chapter 5.

Interference and optical cavities

The wave nature of light leads to behavior typical of other waves, i.e., interference. Depending on their phase relationship, the amplitude of waves can add constructively to reinforce each other, or destructively to cancel each other. In laser science, such wave behavior is employed and manipulated in *optical cavities*, which are essential in many applications. An optical cavity is formed whenever two reflecting surfaces (i.e., mirrors) are mounted in aligned fashion to reflect light beams back and forth between them. The reflectivity can be partial or nearly

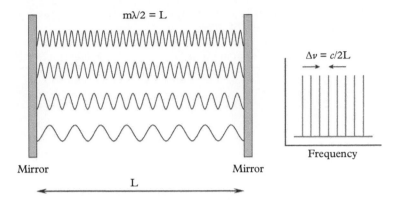

Figure 2.16 *Longitudinal modes with different wavelengths fitting into an optical cavity (left), and the spectrum of modes (right), where* Δv *is the mode spacing.*

total, depending on the quality of the mirrors and the application. Figure 2.16 shows an example of such an optical cavity, with parallel flat mirrors. As shown in Chapter 3, optical cavities are essential in the design and construction of virtually all kinds of lasers.

As shown in Figure 2.16, waves "fit" into an optical cavity when a multiple of half of their wavelength has the same dimension as the length of the cavity, defined as the spacing between the mirrors. Such waves can reflect back and forth with their peaks and valleys overlapping, or in phase, so that constructive interference occurs. Other waves with slightly different wavelengths can reflect back and forth, but their peaks and valleys become out-of-phase with each other and destructive interference occurs, damping out their intensity. Wavelengths, and their corresponding frequencies, which fit into a cavity in this lengthwise fashion and oscillate, are known as *longitudinal cavity modes*. As is evident in the figure, many waves with different wavelengths can fit into such a cavity, beginning with the longest wavelength, where $\lambda/2$ is the cavity length, and extending to infinitely many wavelengths shorter than this. However, the spectrum of wavelengths that fit into a cavity of a given length is not continuous, as there are many frequencies that do not oscillate in a stable fashion. The spectrum of stable mode frequencies for a cavity therefore contains a set of discrete frequencies with a constant mode spacing, as shown on the right side of Figure 2.16. The mode spacing is given by $\Delta v = c/2L$, and is greater for shorter cavity lengths and smaller for longer cavity lengths. Table 2.3 shows some examples of mode spacings in cm^{-1} for different cavity lengths.

Figure 2.16 shows the waves that fit into a cavity along the center line normal to the parallel end mirrors. However, waves at angles slightly away from this center line can also oscillate for several round trips before they walk out of the cavity. Such waves can contribute to lasing action in a laser cavity, and they produce a more complex spatial pattern due to their angular interference. These give rise to *transverse electrostatic modes*, indicated with the abbreviation TEM. The transverse mode pattern is simplest when the oscillation is along the cavity center

Table 2.3 *Longitudinal mode spacings for typical optical cavity (air spaced) lengths, in cm^{-1} and GHz (1 cm^{-1} = 29.979 GHz).*

L (cm)	Mode spacing (cm^{-1})	Mode spacing (GHz)
0.1	5.0	150.0
1.0	0.5	15.0
5.0	0.1	3.0
8.0	0.06	1.87
10	0.05	1.5
20	0.025	0.75

line, as shown in Figure 2.16. This gives rise to a circular cross-sectional spot of intensity, indicated as the TEM$_{00}$ mode. Other paths produce more complex spatial intensity patterns. Optical cavities are also sometimes constructed of curved rather than flat mirrors of various shapes. The interested reader is referred to more advanced texts on optics or laser design to get a sense of the variety of cavities employed for different applications.

A special kind of optical cavity known as an *etalon* is used for several applications such as narrowing the output linewidth of tunable lasers or calibrating their wavelength. A solid etalon consists of a quartz or glass window with precisely flat sides that are both mirror-coated to be partially (90–95%) reflective. An air-spaced etalon has two partially reflective mirrors separated by an air gap. In either case, the cavity length, defined by the thickness of the solid piece or the width of the gap, is only a few millimeters, resulting in longitudinal mode spacings that are on the order of a cm^{-1} (see Table 2.3). Incident light is transmitted through the partially reflective coatings of an etalon only when the waves fit as resonant longitudinal waves in its cavity. In an air-spaced etalon, the wavelength must be a multiple of twice the cavity length, as noted above, i.e., $m\lambda/2 = L$, or $m\lambda = 2L$. In a solid etalon, the refractive index of the material causes the wavelength to change (see Figure 2.17), and thus the resonant condition changes to $m\lambda = 2nd$.

If a tunable laser is directed into an etalon, the transmitted light intensity as the laser tunes can be measured with a detector such as a photodiode, producing a pattern, or spectrum, of transmitted frequencies, as shown in Figure 2.18. The resulting peaks, known as *transmission fringes*, appear at regular spacings of the longitudinal modes. In the case of etalons, this spacing is called the *free spectral range* (FSR). The linewidth of each transmitted mode, known as the *bandwidth* (BW), depends on the quality of the etalon, including the flatness and percent reflectivity of the surfaces. The quality of the etalon is expressed as the *finesse*, f, where f = FSR/BW. For a high-finesse etalon, the bandwidth is usually

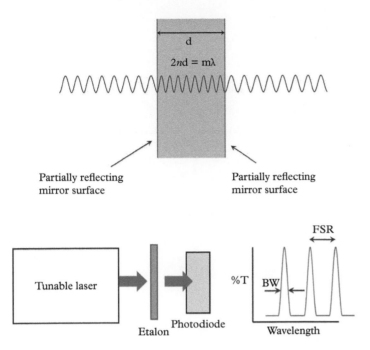

Figure 2.17 *Solid etalon, showing the change in wavelength for a longitudinal mode as light passes through the etalon material.*

Figure 2.18 *The pattern of transmission fringes resulting when a tunable laser passes into and out of resonance with a solid etalon. The free spectral range (FSR) and bandwidth (BW) are indicated.*

sharper than the linewidth of the laser employed, and the experimental width of transmitted lines provides a measure of the linewidth of the laser.

If the light enters the etalon exactly normal to the surfaces, the free spectral range and bandwidth are reproducible constants of the etalon, and can thus be used for wavelength calibration of the tunable laser, or of spectroscopic features measured with it. This is done by recording the series of transmission fringes at the same time that the spectrum for a frequency standard is recorded (such as the electronic spectrum of iodine vapor or atomic transitions in a gas discharge lamp) together with the desired spectrum to be calibrated, as shown in Figure 2.19. The etalon fringes provide a frequency marker scale to define relative spacings in the new spectrum and to connect these to absolute frequencies in the standard spectrum, as shown in Figure 2.20.

Etalons can also be employed to provide absolute frequency calibration without having to tune the laser. This can be done by sending a diverging beam of laser light through the etalon, such as that which has passed through the focus of a lens. If this is done, the path lengths of rays at certain angles different from the normal path can also be transmitted through the device. Only certain angles fit the resonance condition:

$$m\lambda = 2nd\,\cos\theta \qquad (2.8)$$

The transmitted rays can be projected onto a viewing surface, or recorded with a spatially sensitive CCD detector, where they form the Airy disc (bull's eye)

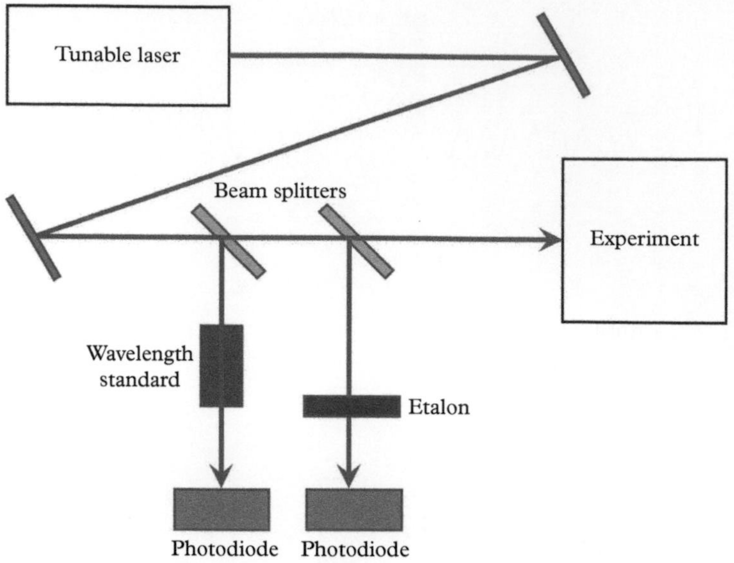

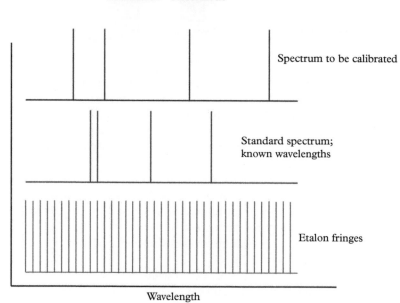

Figure 2.19 *The experimental config-uration used to calibrate spectra meas-ured with a tunable laser by simulta-neously recording etalon fringes and a standard spectrum.*

Figure 2.20 *The spectra resulting from a tunable laser scan including etalon fringes and a standard spectrum. The known FSR of the etalon defines the fre-quency scale, including all the spacings between standard spectral lines and those for a new molecule.*

pattern, as shown in Figure 2.21. The ring spacing in this pattern for an un-known wavelength can be compared to that for a known wavelength, e.g. from a He-Ne laser, if both are passed through the same etalon, to determine the absolute frequency of the unknown laser. This kind of device is called a *Fabry–Pérot* inter-ferometer. Commercial devices using this concept are also called *wavemeters*, and are sold by companies such as Bristol Instruments (http://www.bristol-inst.com/).

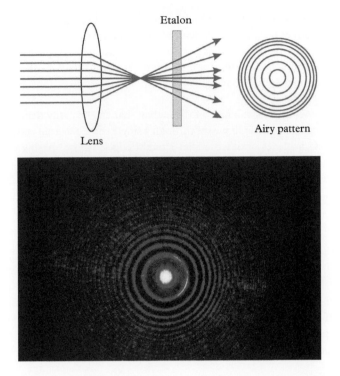

Figure 2.21 *A diagram of how diverging light after passing through a lens can be sent through an etalon to produce an Airy pattern (left), and a photo of such a pattern generated with a helium-neon laser, lens, and etalon in this configuration.*

In addition to the etalon devices discussed so far, it is also possible to produce another kind of etalon device with a time-variable spacing. This is done with an air-spaced etalon, with one end mirror mounted to a piezoelectric crystal. When a sawtooth voltage ramp is applied to the piezo, its motion generates a small reproducible variation in the length of the cavity. The transmission through such a *spectrum analyzer* varies in time as the fixed laser wavelength comes into and out of resonance with the changing cavity length. This transmission can be displayed on an oscilloscope to generate a pattern of time-varying transmission fringes like those shown for frequency variation in Figure 2.18. This pattern can be viewed in real time as the laser optics are adjusted to produce the sharpest linewidth, or it can be compared to the same pattern generated from a He-Ne or other known laser wavelength to determine the absolute wavelength of the tunable laser.

Diffraction

Closely related to the interference of light waves in optical cavities is their interference when encountering small apertures or objects close in size to their wavelength. In these cases, diffraction occurs, producing many interesting characteristic patterns depending on the details of the objects or apertures encountered.

Diffraction can occur when light passes through a pinhole, a single slit, a double slit, or a pattern of lines or slits drawn onto or etched into some appropriate substrate. Diffraction demonstrations provide colorful examples of the wave behavior of light, but diffraction gratings are also practical and useful devices employed throughout optics and laser science. We provide a selection of topics here that is far from complete.

Perhaps the simplest example of diffraction that can be easily demonstrated is that which occurs when light passes through a single slit. Such a slit can be formed from two razor blades or other similar straight edges, whose spacing can be adjusted to form an extremely small aperture. In this case, the intensity pattern of transmitted light resulting from diffraction is given by the Fraunhofer diffraction formula as,

$$I_\theta = I_0 \left[\frac{\sin\left(\frac{b\pi\theta}{\lambda}\right)}{\frac{b\pi\theta}{\lambda}} \right]^2 \tag{2.9}$$

Figure 2.22 shows a plot of this function above the experimental pattern, generated with a green laser pointer and a small slit, displayed on a wall, and photographed. Diffraction through a slit can also be demonstrated by observing the light passing between one's middle and index finger when these are held close together.

More complex diffraction patterns are produced for arrangements of two or more slits. A pinhole with appropriately small dimension produces a bull's eye pattern very much like that depicted in Figure 2.21. There are many examples of such phenomena available on the Internet and elsewhere, and we therefore limit this discussion here.

The most common application of diffraction in optical instrumentation is in the form of diffraction gratings, which consist of many closely spaced ruled parallel grooves or lines on an appropriate optical substrate. Diffraction gratings can be used in transmission, in which case the substrate must be a transparent material, or they can be used in reflection, in which case surface reflection is required. The reflection mode is more common in modern instrumentation. When light is incident on a reflection grating, the diffraction pattern from the multiple lines on the substrate gives rise to an angular pattern of spots, each one of which has its own pattern of angular dispersion of the wavelengths present. Each of the major spots is known as an *order* of the grating, and the angular dispersion within each order is described by the grating equation:

$$m\lambda = d \left(\sin\alpha + \sin\beta\right) \tag{2.10}$$

where m = 0, 1, 2 ... is the order of the spot, d is the grating constant (spacing between lines), and α and β are the incident and outgoing angles from the grating, measured from the surface normal. It can be shown readily that a smaller line

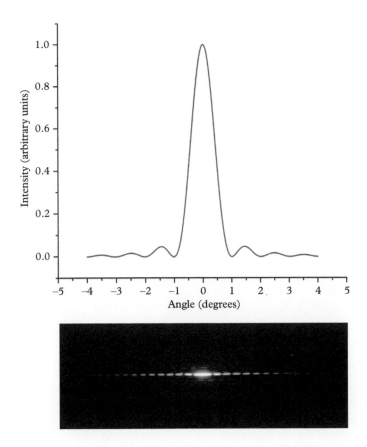

Figure 2.22 *A plot of the intensity versus angle for Fraunhofer diffraction from a single slit compared to the intensity measured experimentally with a green laser pointer passing through a small slit.*

spacing produces greater angular dispersion of different wavelengths. Some of the first diffraction gratings made for spectroscopy were produced by Professor Henry Augustus Rowland at Johns Hopkins University in about 1882, using diamond to produce scratches on glass substrates. He used these gratings to measure the best spectra up to that time for the atomic lines present in the solar spectrum. Later gratings were fabricated using milling machines to cut lines into substrates. However, modern technology employs laser holography for photographic printing of the microscopic line patterns.

One of the fathers of the development of the laser, Professor Arthur Schawlow, described a simple demonstration of a grating-like device using a ruler, which can be used to determine the wavelength of a laser.[6] If one shines a He-Ne laser on the 1/64 inch scale of a shiny metal ruler at an oblique angle, as illustrated in Figure 2.23, the reflected beam will exhibit a number of bright dots on a wall a distance away. The first spot up the wall is the central bright one and is seen with no interference from reflections due to all other shiny areas. It is the same spot that would be seen from reflection from the rest of the unruled area of the ruler.

48   *Basic Optics*

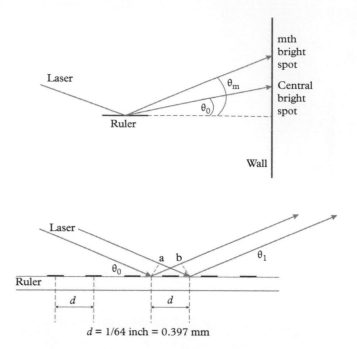

Figure 2.23 *Geometry of a laser beam reflected from the part of a metal ruler containing 1/64 inch divisions. The reflected beam shows the interference pattern on a wall resulting from reflections by the shiny parts between the 1/64 divisions.*

Figure 2.24 *Illustration showing two laser rays reflected from adjacent shiny spots on the ruler. The two rays travel different distances and under the proper phase conditions they interfere constructively or destructively.*

If the laser beam is directed so that the beam subtends many of the shiny divisions, an interference pattern is seen on the wall. The resulting pattern can be understood by referring to Figure 2.24. In this figure one can see that the difference in path length of the two beams will be a – b. These two rays interfere constructively if a – b is an integral multiple of the wavelength of light or $m\lambda = a - b = d \cos \theta_0 - d \cos \theta_m$. The number m again represents the order of the grading. Thus, the laser wavelength $((d/m) (\cos \theta_0 - \cos \theta_m))$ can be calculated from measuring $\cos \theta_m$ and $\cos \theta_0$ for any order m given that 1/64 inch = 0.397 mm.

Diffraction gratings are used as the wavelength dispersion device in monochromators, which are used in ordinary UV-visible spectrometers. These devices were also used in the past for infrared spectroscopy, but they have now been replaced by FT-IR devices based on interferometry. An example of the layout for a UV-visible monochromator is shown in Figure 2.25. The device works equally well with the input and output directions reversed. Slits are employed to define input and output angles more precisely, and an input lens is employed in combination with a curved mirror to expand the light on the grating. The specific lens focal length and placement is required for correct collimation and beam expansion of the light on the grating. This increases the resolution, which depends on the number of grating lines illuminated through the relation

$$R = \lambda/\Delta\lambda = Nm \tag{2.11}$$

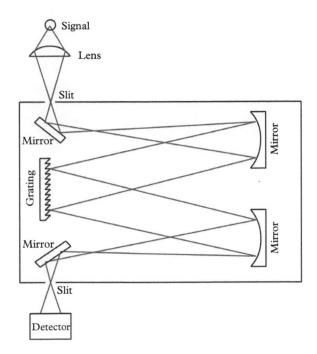

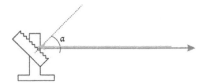

Figure 2.25 *A schematic layout of a UV-visible monochromator using a diffraction grating in reflection mode for wavelength dispersion. Curved mirrors are employed together with a lens, producing beam expansion, so that more lines on the grating are illuminated for better resolution.*

where N is the number of lines illuminated and m is the grating order. The resolution is greater when the light from higher grating orders is employed and when the beam is expanded more on the grating.

Although the most familiar application of diffraction gratings is in spectrometers for wavelength *analysis*, gratings are also used in lasers for wavelength *selection*. In particular, a diffraction grating can be used in the so-called "Littrow" configuration, in which the input and outgoing angles are the same, as shown in Figure 2.26. The grating equation is then modified to the following form:

$$2d \left(\sin \alpha\right) = m\lambda \qquad (2.12)$$

In this configuration, the grating acts essentially as a wavelength-specific, or wavelength-selective, mirror. It can be substituted for one of the mirrors in an optical cavity to further select, beyond the allowed longitudinal modes, those which fall in a narrow wavelength range. This concept is employed in laser cavities, as discussed in Chapters 3 and 4, for selection of certain specific lasing wavelengths when several are produced by the lasing material. Tilting the grating to an appropriate angle allows only a selected wavelength to be reflected back into the cavity. When a range of output wavelengths is produced by the lasing material, as in the broad fluorescence spectrum of a dye laser, tilting the grating makes it possible to scan the laser wavelength within the available output range.

Tuning the wavelengths of light by tilting the angle of a diffraction grating is important in both grating-based spectrometers and when gratings are

Figure 2.26 *A diffraction grating in the Littrow configuration, which can be employed as a wavelength-selective mirror.*

used for wavelength selection or tuning in a laser cavity. In both situations, it is important to notice in the grating equation (equations 2.10 and 2.12) that the wavelength selected depends on the sine of the angle at which light comes into or out of the grating, rather than directly on the angle. Therefore, if a motorized scanning system is employed to rotate the grating directly, the resulting scan of the output wavelength will not be linear. A common solution to this well-known problem is to employ a *sine-bar drive* mechanism, as shown in Figure 2.27. In this device, the action of a motor is employed to turn a fine-pitched threaded rod, which causes a plate to push against the sine bar, varying its angle. Turning the drive shaft varies the length of the section of the threaded rod corresponding to the side of a triangle, defined as shown in the figure. If the length of the sine bar is l, the length of the varied section is l sin α, and scanning the drive screw smoothly varies the sine of the input and output angle on the grating. The action of the motorized scan therefore produces a linear variation of the wavelength, as desired for most spectroscopy and laser tuning purposes.

As discussed above, diffraction occurs when a laser is directed through a pinhole aperture, where the waves interfere. This aspect of diffraction also affects how lasers are focused with lenses. Because of the effects of diffraction, lasers (or other light sources) cannot be focused to an arbitrarily small spot. Instead, the spot size is limited by diffraction to a minimum waist, ω_0. As shown in Figure 2.28, focusing produces this "diffraction-limited" minimum beam waist, which extends over a *confocal length*, l, during which the spot size is relatively constant, before expanding outward again. The minimum beam waist depends on

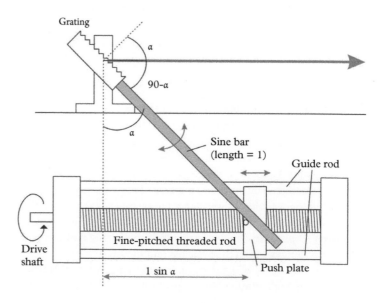

Figure 2.27 *The sine-bar drive used for linear wavelength scanning of a diffraction grating.*

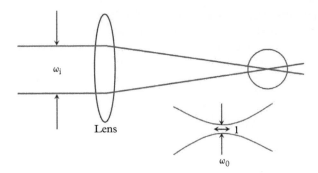

Figure 2.28 *An expanded view of the focus region of a laser, showing the confocal waist, ω_0, and confocal length, l.*

the initial spot size of the laser, ω_i, on the focal length of the lens employed, f, and on the wavelength, λ, through the following relationship:

$$\omega_0 = \frac{\lambda f}{\pi \omega_i} \tag{2.13}$$

This equation is not completely accurate, as it assumes an infinitely thin lens, but it serves to illustrate the qualitative behavior seen. As indicated, the focal spot size can be smaller if the focal length of the lens is shorter, if the initial beam waist is larger, or if a shorter wavelength is employed. The focal length of the lens can be adjusted over some reasonable ranges, and the initial beam waist can be increased with a beam-expanding telescope, but the wavelength has the greatest effect.

These issues are extremely important in several aspects of practical technology. The same optical issues seen for focusing a light source apply to focusing an image of an object. Thus, this diffraction effect limits the detail that can be resolved with an optical microscope. This aspect of light focusing in microscopes was first investigated by Ernst Abbe in 1873,[7] who derived a different formula for microscope resolution incorporating similar ideas. In practice, optical microscopes have difficulty resolving objects smaller than microns in dimensions. This is why electron microscopes, using electron beams which can be focused more tightly as a result of employing shorter wavelengths ($\lambda=h/p$), and other related instruments are used for more precise imaging.

The limited ability of laser light to be sharply focused also affects how integrated circuits are manufactured, such as those used in personal computers, tablets, cell phones, etc. Many circuits are produced with the technique of *photolithography*, in which a focused laser beam is employed to burn patterns into a substrate or to decompose volatile inorganic compounds over a surface, leaving behind a patterned deposit of metal atoms. These photo-produced patterns of metal form the microscopic wires that connect elements in a micro-circuit. The size of such circuits is thus limited by the resolution at which such patterns can be produced. The diffraction-limited focusing of the laser is the limiting factor in this technology. To make the sharpest patterns, and thus smallest circuits, ultraviolet and vacuum-ultraviolet lasers are employed for these processes, such as excimer lasers. Similar optical processes are employed with laser writing and reading of DVDs used for video recording. The sharper definition at shorter wavelengths explains the improved resolution obtained with blue diode lasers in Blu-Ray disc technology.

....

REFERENCES

1. E. Hecht, *Optics*, fourth edition, Pearson/Addison Wesley, San Francisco, 2001.
2. J. R. Meyer-Arendt, *Introduction to Classical and Modern Optics*, third edition, Prentice-Hall, Englewood Cliffs, NJ, 1972.
3. W. Demtröder, *Laser Spectroscopy: Basic Concepts and Instrumentation*, third edition, Springer-Verlag, Berlin, 2003.
4. J. Fraunhofer, "Bestimmung des Brechungs-und des Farben-Zerstreuungs-Vermögens verschiedener Glasarten, in Bezug auf die Vervollkommnung achromatischer Fernröhre" (Determination of the refractive and color-dispersing power of different types of glass, in relation to the improvement of achromatic telescopes), *Denkschriften der Königlichen Akademie der Wissenschaften zu München* (Memoirs of the Royal Academy of Sciences in Munich) **5**, 193 (1814–15).
5. *Optics Guide* 5, Melles-Griot, Irvine, CA, 1990.
6. A. L. Schawlow, "Measuring the wavelength of light with a ruler," *Am. J. Phys.* **33**, 922 (1965).
7. E. Abbe, "Beiträge zur theorie des mikroskops und der mikroskopischen wahrnehmung," *Archiv für Mikroskopische Anatomie* **9**, 413 (1873).

General Characteristics of Lasers

3

Introduction

Lasers are light sources with special properties and performance characteristics. Certain atomic and molecular materials can be used to produce a laser device, but others are not suitable for this. In this chapter we discuss these issues to give students a better perspective on why we have the laser devices that are presently available, and why some desirable kinds of lasers are not available. The light that comes from lasers is similar to that from ordinary light sources in some respects, but quite different in others. We also discuss the general characteristics of laser light here so that students can appreciate what makes lasers special and why this is important for chemical physics research. Specific examples of common laboratory lasers and how they work are presented in Chapter 4.

Requirements for a laser

Atomic or molecular quantum states for emission

The first requirement for a laser is a liquid, solid, or gaseous material with quantum states having energy differences corresponding to the desired color of light to be generated by emission. Lasing occurs through photon emission from an excited state to a lower state. Selection of appropriate materials requires a detailed knowledge of spectroscopy; a full coverage of this is beyond the scope of this work, but we provide some brief comments here. At the lowest energy range, quantum states corresponding to molecular rotational levels occur in the microwave region of the spectrum. Specific energies and units for different regions of the spectrum were presented earlier in Table 1.1. Moving upward to the infrared region, quantum state differences here correspond to molecular vibrations. Finally, visible and ultraviolet absorption and emission corresponds to electronic state differences in atoms and molecules. The properties of light absorption and emission are not the same in these different regions, as explained below, which greatly affects the possible laser systems that can be developed.

Laser Experiments for Chemistry and Physics. First Edition. Robert N. Compton and Michael A. Duncan.
© Robert N. Compton and Michael A. Duncan 2016. Published in 2016 by Oxford University Press.

Pumping mechanism

The second requirement for the production of a laser is a mechanism for excitation, known as "pumping," molecules to produce excited states. The pumping mechanism is different for different kinds of lasers, particularly depending on the material to be pumped. Liquid and solid materials are best pumped using light from a lamp, such as a flashlamp (see Chapter 1), or by using another laser. Flashlamp optical pumping is employed for crystal-based systems, such as Nd:YAG lasers. Optical pumping with another laser is common for dye lasers, where the lasing material is a liquid solution. In the case of solid crystals or liquid solutions, the concentration of the absorber must be adjusted to make the material transparent enough for illumination throughout its volume.

In gas lasers, pumping is generally accomplished with electrical discharges, using fast electron collisions in the discharge for excitation of atoms or molecules. The individual processes involved are similar to those discussed in Chapter 1 for the operation of discharge lamps. Electrical discharge pumping is employed for nitrogen, CO_2, He-Ne, argon-ion, krypton-ion, and excimer lasers. In every one of these systems, high voltage power is present and special precautions should be used if the laser power supply is opened. In pulsed lasers, high voltage capacitors are also present, which are a danger for shock even when the power to the laser is turned off.

A final kind of pumping that is less common in laboratory lasers involves excitation via exothermic chemical reactions. An example is shown below for the HCl "chemical" laser:[1]

$$H\cdot + Cl_2 \rightarrow HCl\,(v \leq 5) + Cl\cdot \quad \Delta H = 45 \text{ kcal/mol}$$

Gaseous hydrogen or a hydrogen/nitrogen mixture is discharged initially, producing hydrogen atoms, which then react with Cl_2 gas. As in all gas lasers, partial pressures and additives are adjusted to optimize performance. The reaction is exothermic, and the dynamics of the process direct this exothermicity into excited vibrations of the HCl product, which then emits in the infrared.

Another example of a chemical laser is the so-called chemical-oxygen-iodine laser or COIL, which has been under development for many years by the military. It involves a chemical reaction between Cl_2, I_2, and an aqueous mixture of hydrogen peroxide and potassium hydroxide, which results in heat and the production of the singlet delta oxygen molecule, $O_2(^1\Delta_g)$. This metastable form of oxygen lives for ~45 minutes (see Chapter 14). Reactions of this excited molecule with I_2 give a spin-orbit excited I atom, $I\,(^2P_{1/2})$ which undergoes emission to the lower $I\,(^2P_{3/2})$ state, lasing at 1.315 μm (see Figure 17.1, p. 274).

Efficient emission

Regardless of the pumping mechanism that produces excited states, atoms and molecules eventually relax to their more stable ground-state configurations. The primary mechanism for relaxation of isolated atoms and small molecules is fluorescence, or re-emission of radiation of the same energy which was initially absorbed to excite the system. For a laser system, fluorescence emission must be *efficient*. However, depending on the dynamics of the system, the energy in an excited state may also be lost by processes other than fluorescence. These include collisional transfer to another species, by breaking a bond (i.e., photochemistry), or by other so-called "radiationless" decay processes (e.g., intersystem crossing, internal conversion). In all cases, however, the average time required for relaxation of the excited state is referred to as the lifetime, τ, of that state. Excited-state lifetimes of atoms and molecules are important diagnostics with which to investigate their dynamics.

The usual framework employed to investigate the efficiency and intensity of light emission is that developed in the early work of Einstein.[2] Building on concepts developed by Planck, Einstein assumed a two-level quantum state system involving a ground state (0) and an excited state (1), as shown in Figure 3.1, and derived expressions for the kinetics of light absorption and emission.

The relative populations in the ground and excited states are given by the usual Boltzmann relationship, $N_1/N_0 = e^{-\Delta E/kT}$, and at typical conditions of temperature and energy spacing the population of the ground state is much greater than that of the excited state, i.e., $N_0 \gg N_1$. Einstein then treated the kinetics of light absorption in the same way as a bimolecular reaction, with photon density $\mu(\nu)$ and ground-state population as "reactants" and a rate coefficient B_{01}:

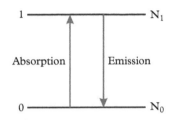

Figure 3.1 *Two-level system used by Einstein to describe absorption and emission kinetics.*

$$\text{absorption rate} = -\frac{d[N_0]}{dt} = B_{01}\,[N_0]\,[\mu(\nu)] \qquad (3.1)$$

The similarity to a bimolecular reaction rate equation should be evident to students who have studied kinetics. A similar rate equation applies to emission from the excited state back down to the ground state induced by the presence of a radiation field, known as "stimulated emission." Here the excited-state population is relevant and the downward coefficient is given as B_{10}.

$$\text{stimulated emission rate} = -\frac{d[N_1]}{dt} = B_{10}\,[N_1]\,[\mu(\nu)] \qquad (3.2)$$

These equations put the ideas of light absorption and emission into the framework of kinetics. At this point, however, Einstein made a conceptual advance, arguing that the coefficients for absorption and stimulated emission should be equivalent, i.e., that $B_{01} = B_{10}$. This makes sense because the physical mechanism of an oscillating electric field coupling to the electrostatic charge distribution about the atom or molecule is the same for both processes.

In a second advance, Einstein included the process of "spontaneous emission," by which excited molecules fluoresce and return to the ground state in the absence of any external radiation field. The kinetic equation for this includes the excited-state population, but no photon density, and a new rate coefficient, the Einstein "A" coefficient.

$$\text{spontaneous emission rate} = -\frac{d[N_1]}{dt} = A_{10}\,[N_1] \qquad (3.3)$$

In a situation in which a continuous light field is present, the excited state would decay as the sum of the stimulated and spontaneous rates. If a light is pulsed on and then turned off quickly, the excited state would decay in the absence of light via only the spontaneous rate, as indicated in equation 3.3. Equation 3.3 is in the standard form for the rate of a first-order reaction, which is well known in kinetics. Its integration gives the population of the excited state as a function of time,

$$[N_1]_t = [N_1]_{t=0} \, e^{-A_{10}t} \tag{3.4}$$

where $[N_1]_{t=0}$ is the number of molecules initially excited. Experimentally, we usually cannot detect this population directly. However, the fluorescence yield is directly proportional to the number of excited molecules, and this can be detected as a function of time. A standard experiment in photochemistry and spectroscopy is to excite atoms or molecules with a pulse of light and to measure the resulting fluorescence with a photodiode or PMT detector as it decays in time (see Chapter 28). As indicated in equation 3.4, the population follows an exponential decay with time, and so does the fluorescence. Exponential decays all go to $t = $ infinity when their intensity goes to zero, but the rates at which they approach zero are not the same. The exponential decay rate is expressed as the "lifetime" of the decay, τ, which is the point in time when the intensity of the signal has fallen to $1/e$ of its initial value. In the case of equation 3.4, where the decay in fluorescence is governed only by the spontaneous lifetime, the lifetime is known as the "radiative lifetime," τ_{rad} and this is given by the inverse of the A_{10} coefficient, $\tau_{rad} = 1/A_{10}$. It should be evident by now that A_{10} has units of sec^{-1} and τ_{rad} has units of sec.

Building on Planck's treatment of black-body radiation, Einstein was able to make yet another conceptual advance, relating the A and B coefficients to each other and to the absorption strength, another experimentally observable quantity.

$$B_{01} = \frac{8\pi^3}{3h^2} |{<}\Psi_m |\hat{u}| \Psi_n{>}|^2 \tag{3.5}$$

$$A_{10} = \frac{8\pi h\nu^3}{c^3} B_{01} \tag{3.6}$$

$$A_{10} = \frac{64\pi^4\nu^3}{3hc^3} |{<}\Psi_m |\hat{u}| \Psi_n{>}|^2 \tag{3.7}$$

In these equations, the bracket notation is used to indicate the dipole moment matrix element, which is the integral connecting initial and final states via the dipole moment operator, $\hat{u}$. The numerical value of this matrix element can be calculated, but the value of A_{10} can also be obtained experimentally from the absorption spectrum via the relationship

$$A_{10} = \frac{\nu^2 \int \varepsilon \, d\nu}{3.5 \times 10^8} \tag{3.8}$$

where $\int \varepsilon \, dv$ is the "integrated extinction coefficient" and the numerical values for the constants have been included.

This treatment provides a convenient framework for describing the relationship between absorption strengths and radiative lifetimes in different regions of the spectrum. The integrated extinction coefficient comes directly from the experimentally measured absorption spectrum, and then the radiative lifetime can be predicted. If the absorption strength is large, the radiative lifetime will be short, and vice versa. Forbidden transitions have very weak absorption strengths, and therefore long lifetimes. The frequency cubed dependence of the Einstein A_{10} coefficient is also important to notice. When the radiative rate is high, the radiative lifetime is short. This indicates that radiative lifetimes at low frequencies (e.g., microwave region) will be relatively long, those at higher frequency (e.g., infrared) will be shorter, and those at the highest frequencies (e.g., UV-visible) will be the shortest, consistent with experimental observations. Table 3.1 provides examples of radiative rates for selected atoms and small molecules illustrating this trend. This frequency dependence also applies to the absorption strength. Those systems having greater radiative rates and shorter radiative lifetimes have stronger absorption strengths, and therefore absorption strengths are relatively weak in the microwave region, stronger in the infrared, and strongest of all in the UV-visible. Likewise, because the Einstein A and B coefficients are related to each other, this same trend applies for stimulated emission strengths, which is most relevant for laser action.

The discussion so far has focused on fluorescence, or radiative decay, as the only process involved in relaxing the excited state. However, as noted earlier, other processes such as collisional energy transfer (aka "quenching"), bond-breaking, intersystem crossing, internal conversion, etc., may also occur in some systems. If other decay channels are active, then the overall decay rate is the sum of all the rates for processes at work and is greater than that due to fluorescence alone, i.e.

$$k_{total} = k_{rad} + k_c + k_d + k_{isc} + k_{ic} \ldots \tag{3.9}$$

where the various k values are the rate coefficients for collisional energy transfer, dissociation, intersystem crossing, internal conversion, etc., respectively.

Table 3.1 *Examples of Einstein A_{10} coefficients for selected systems.[3,4,5]*

Atom/molecule	A_{10} (sec^{-1})	Lifetime (sec)	Comment
HCl (v = 0 → v = 1)	40.2	2.5×10^{-2}	infrared
HF (v = 0 → v = 1)	194.5	5.1×10^{-3}	infrared
Na (^{2}S → ^{2}P, "D" line)	6.1×10^7	1.64×10^{-8}	visible, allowed
O$_2$ ($^3\Sigma_g^- \to {}^1\Delta_g$)	1.47×10^{-4}	6.8×10^3	visible, forbidden

Additional decay processes increase the overall rate of relaxation of excited molecules, and the excited-state lifetime, as observed by the decay of fluorescence, will then be shorter than the radiative lifetime. Specifically, the actually observed fluorescence lifetime, $\tau_f = 1/k_{total}$. If the expected radiative lifetime τ_{rad} is known from the absorption strength, and a shorter lifetime is observed experimentally, this can be used to identify the presence of non-radiative decay processes and to determine their rate. If these other processes are taking place, the overall amount of fluorescence emission that can be used to make a laser will be less. This is expressed in the "quantum yield" for fluorescence, which is given by $\Phi_f = k_{rad}/k_{total}$. If fluorescence is efficient, a quantum yield near 1.0 is possible, but if other decay processes are efficient, the yield can be much less.

The likelihood of these other non-radiative processes has been studied extensively in many experiments. Collisional quenching depends on the pressure of gases present and their per-collision efficiency of quenching. A reference point for this is the collision frequency or "hard-sphere" collision rate from elementary gas laws,

$$Z_{ab} = N_a N_b (r_a + r_b)^2 \sqrt{\frac{8kT}{\pi \mu_{ab}}} \qquad (3.10)$$

where N_a and N_b here are number densities of the two gases, r_a and r_b are their atomic radii, and μ is the reduced mass. Quenching can happen no faster than collisions, but it does not necessarily occur on every collision, and so Z_{ab} gives an upper limit on k_c. Collisional quenching has been studied extensively in the context of gas laser systems,[6] and is an important consideration in their operating conditions. The rates of other intramolecular non-radiative processes, i.e., those that involve energy transfer within or between quantum states of isolated molecules, are often estimated using Fermi's Golden Rule,

$$k_{nr} = \left| <\Psi_i \left| \hat{H} \right| \Psi_f> \right|^2 \rho(v) \qquad (3.11)$$

This indicates that the rate of energy flow from one set of quantum states prepared initially, e.g., by light absorption, into another set of final quantum states at the same energy depends on a coupling matrix element that connects these states with a quantum mechanical operator that describes the process. Such coupling can be evaluated computationally, and varies substantially from one system to another. More important here is that the rate also depends on the "density of states," $\rho(v)$. The state density, usually expressed as the number of states per cm^{-1}, depends on the number of vibrational and electronic states of the system that are possible at that energy. The density of electronic states is much higher for open-shell species, such as a transition metal atom with a partially filled d shell, where many electronic and spin configurations lie close in energy. The number of vibrational modes for nonlinear polyatomic molecules is 3N–6, where N is the number of atoms. At any energy above the ground state, all the possible combinations or overtones of quanta in different modes that result at that energy provide the density of states. For both vibrational and electronic states, the densities increase dramatically with system size. Non-radiative processes therefore become much more important, and quantum yields much smaller, on average, for larger molecules. Thus, although there are exceptions to this (e.g., dye lasers), most laser systems are based on atoms and small molecules, which have fewer and less rapid non-radiative decay channels and thus higher quantum yields.

Population inversion

According to the Einstein treatment above, the rate of excitation in a two-level system is given by equation 3.1, and the overall rate of relaxation is given by the sum of equations 3.2 and 3.3. Because the population of ground-state molecules is greater than excited-state molecules, the rate up is greater than the rate down and there is initially net excitation. However, upon continued pumping, the populations in the ground and excited states could eventually become the same. When this happens, since $B_{01} = B_{10}$, the rates up and down would be the same. Further pumping just causes population to cycle repeatedly between the ground and excited states. This corresponds to a temperature of infinity, according to the Boltzmann relationship, since $N_1/N_0 = 1 = e^{-\Delta E/kT}$, and this can only be true if the argument of the exponential is zero. This situation is not useful for making a laser, since there is no *net* emission. Laser emission requires a greater excited-state population than that in the ground state, which cannot be achieved in this two-level system. Another requirement for lasing, therefore, is a pattern of quantum states and pumping that makes it possible to produce an excess population in an excited state, known as a "population inversion." As a side note, an inversion such that $N_1 > N_0$ inserted into a Boltzmann relationship leads to the unphysical result of a *negative* temperature.

The first laser (MASER) obtained a population inversion by spatially separating out the rotationally excited ammonia molecules from the ground-state molecules with an inhomogeneous electric field. This procedure worked for the special case of ammonia, but is not generally applicable.

Population inversions in other systems are possible when more than two states are available (usually true in real systems) and when these states fall into a convenient pattern. Figure 3.2 shows the pattern for a so-called three-level system. Pumping in this system can populate level 2, and this level has less population than level 1. However, level 3 lies at high enough energy that it is not populated thermally and is therefore essentially empty. As soon as there is population in level 2, it is inverted with respect to (has more population than) level 3 and 2 → 3 laser emission is possible.

Another situation which can produce a population inversion is a four-level system, such as that shown in Figure 3.3. In this case, pumping produces an excited state of species A which can transfer energy, e.g. through a collision, to

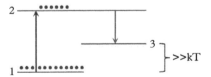

Figure 3.2 *The three-level system that makes it possible to produce a population inversion. In this case, level 2 can be inverted with respect to level 3.*

Figure 3.3 *A four-level system that makes it possible to produce a population inversion. Level 3 is populated collisionally following pumping of level 2, and this is inverted with respect to level 4, so that laser emission 3 → 4 becomes possible.*

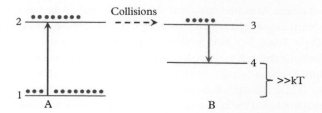

another species B with an energy state at the same level. The species B has a lower level that is high enough to have no thermal population. State 3 is therefore inverted with respect to 4, and laser emission is possible, as shown. There are many variations on these three- and four-level concepts which are employed in different laser systems.

Optical feedback and gain

In spontaneous emission, fluorescence is emitted at the average rate given by the A_{10} coefficient and light comes out of the sample in random directions. This light is not as useful as laser light. In stimulated emission, fluorescence is generated at the exact same frequency, at the same time, and in the same direction as the stimulating light field. This allows light to be concentrated and controlled to form a more intense beam. Therefore, production of a laser requires that stimulated emission be much greater than both absorption (which would diminish its intensity) and spontaneous (uncontrolled) emission. Mathematically, this is expressed as:

$$\frac{B_{10}[N_1][\mu(\nu)] - B_{01}[N_0][\mu(\nu)]}{A_{10}N_1} \ggg 0 \qquad (3.12)$$

To enhance emission rather than absorption, we need a population inversion, as noted above. To enhance stimulated emission over spontaneous emission, we need "optical feedback," which takes the emission generated initially and uses it to stimulate other molecules to emit in the same way. As shown in Figure 3.4, this is

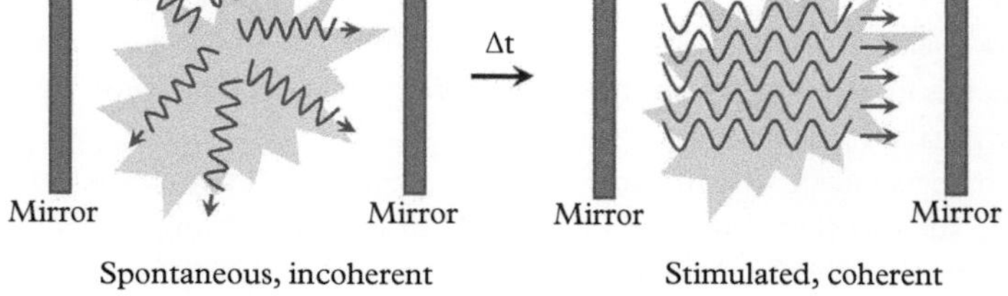

Figure 3.4 *Conversion of random, incoherent spontaneous emission into stimulated, coherent emission in an optical cavity.*

accomplished by placing the lasing material in an optical cavity (see Chapter 2 for more details about cavities). In this cavity, the first few spontaneous emission processes generate light that goes out in all directions and with waves in random phases with respect to each other (i.e., incoherent). However, in an appropriate cavity, some fraction of the initially emitted waves hits the cavity end mirrors and is reflected back into the lasing material. If excited species are still present, stimulated emission can cause these to emit in the same direction and in phase with the stimulating wave. If there is enough stimulation, the emission becomes more stimulated and less spontaneous as the intensity builds. "Gain" occurs when the increase in emission from this build-up process for each round-trip oscillation in the cavity becomes greater than the losses in the cavity from imperfect mirror reflections or from absorption and scattering processes. This is expressed as

$$G = R_1 R_2 \, e^{2(\beta - \alpha)L} \qquad (3.13)$$

where G is the round-trip gain, R_1 and R_2 are the reflectivity of the mirrors, L is the cavity length, α is the combined losses per unit length from absorption and scattering, and β is the small-signal gain coefficient, defined as:

$$\beta = B_{10} \, (N_1 - N_0) \, \frac{h\nu}{4\pi c} \qquad (3.14)$$

B_{10}, N_1, and N_0 here are defined as in equations 3.1–3.3 above. For efficient lasing G must be greater than 1. The reflectivity of one of the end mirrors must be less than 1.0 in order to couple light out of the cavity for use. This reflectivity, and the cavity design (shape of mirrors, sample volume and concentration, etc.), are variables that are engineered for each different laser system to optimize its performance.

Characteristics of laser emission

Following the discussion above, some of the special characteristics of laser emission become evident. Laser emission is more intense than spontaneous emission. As discussed in Chapter 1, the intensity can be measured in power per unit area (e.g., W/cm^2) for continuous wave (CW) systems or in energy per unit area for pulsed systems (e.g., J/cm^2). Because of the highly directional beam, the light from a one watt laser is far more powerful than a 1-W light bulb, which sends its light out in all directions. Laser emission is *coherent*, meaning that the waves are in phase. This is a consequence of stimulated emission. Likewise, laser beams are generally plane polarized, another consequence of stimulated emission. Laser beams are collimated, but not perfectly. Depending on the design of the optical cavity, the number of round-trip oscillations on average before light is coupled out, the wavelength employed, etc., laser beams have more or less divergence, which is measured in milliradians (mrad) or degrees. Lasers are generally monochromatic, i.e., a single color, but of course this is imprecise language. The actual color can be expressed as a range of wavelengths present, which is known as the "bandwidth" or "linewidth." CW lasers tend to have lower gain, thus requiring more highly reflective mirrors, and less light is coupled out of their cavities. This results in a greater number of round-trip cavity transits on average, and narrower linewidth. In certain designs, the linewidth can be far less than a cm^{-1}, even down into the megahertz range (1 cm^{-1} = 29.979 GHz). Pulsed lasers have higher gain and use lower reflectivity mirrors, but produce higher peak powers in their output. With fewer round-trip oscillations, they have broader linewidths, usually in the 0.1–1.0 cm^{-1} range. More information about cavity designs and other characteristics of lasers are given in subsequent chapters and in more specialized texts on this subject.

....

REFERENCES

1. J. V. V. Kasper and G. C. Pimentel, "HCl chemical laser," *Phys. Rev. Lett.* **14**, 352 (1965).
2. A. Einstein, "Zur Quantentheorie der Strahlung (On the quantum mechanics of radiation)," *Physikalische Zeitschrift* **18**, 121 (1917).
3. E. Arunan, D. W. Setser, and J. F. Ogilvie, "Vibration-rotation Einstein coefficients for HF/DF and HCl/DCl," *J. Chem. Phys.* **97**, 1734 (1992).
4. A. Kramida, Y. Ralchenko, J. Reader, and NIST ASD Team. *NIST Atomic Spectra Database* (ver. 5.2). http://physics.nist.gov/asd. National Institute of Standards and Technology, Gaithersburg, MD, 2014.
5. M. G. Mlynczak and D. J. Nesbitt, "The Einstein coefficient for spontaneous emission of the $O_2(a\,^1A_g)$ state," *Geophys. Res. Lett.* **22**, 1381 (1995).
6. J. T. Yardley, *Introduction to Molecular Energy Transfer*, Academic Press, New York, 1980.

Laboratory Lasers

4

Introduction

The lasers that we have today in the laboratory are the result of many years of research and development, which began in the late 1950s and early 1960s. Building on developments with radar during World War II, which required bright microwave light sources, there was a boom in the study of microwave spectroscopy beginning just after the war. This foundation of work led Charles Townes to develop the first laser-like device, known as the MASER (microwave amplification by stimulated emission of radiation), in his research program at Columbia University.[1] An autobiography by Townes describing this early work is highly recommended.[2]

Soon after the MASER was developed, Townes and others pointed out how the concepts learned could be extended into the optical wavelength region. For this work, Townes shared the Nobel Prize in Physics in 1964 (with N. G. Basov and A. M. Prokhorov) the same year that Martin Luther King received the Nobel Peace Prize. In the early 1960s, several new laser devices were demonstrated for the first time, and the field was off to a rapid start. The first optical device demonstrated was the ruby laser, in which the lasing material is a ruby crystal (aluminum oxide with chromium ions doped into it) excited by a flashlamp. This was reported by Ted Maiman and coworkers at Hughes Aircraft Corporation in California.[3] This initial report was almost immediately reproduced by a group at Bell Labs in New Jersey.[4] Soon after this, the helium-neon laser was reported by Ali Javan and coworkers, also at Bell Labs.[5] This was the first example of a gas laser, pumped by a continuous electrical discharge. Although the ruby laser was eventually replaced by more efficient crystal materials like Nd:YAG (neodymium ions doped into a yttrium-aluminum garnet crystal), the He-Ne laser is still used widely today in the research laboratory and for many practical applications in the everyday world (e.g., barcode readers in the grocery store). Other early developments included pulsed nitrogen lasers, developed by H. G. Heard at Energy Systems Inc.,[6] and CO_2 gas lasers, developed by C. K. N. Patel at Bell Labs,[7] the latter of which can operate in either pulsed or continuous wave (CW) modes. The argon ion laser was invented in 1964 by William Bridges at Hughes Aircraft,[8] and is also still commonly used today. Other important lasers used in many research labs include the Nd:YAG laser, developed by J. E. Geusic,[9] and the excimer laser, first reported by Nikolai Basov et al. at the Lebedev Physical Institute in Moscow.[10] Basov first described a xenon dimer-based excimer

Laser Experiments for Chemistry and Physics. First Edition. Robert N. Compton and Michael A. Duncan.
© Robert N. Compton and Michael A. Duncan 2016. Published in 2016 by Oxford University Press.

laser, but this concept was rapidly extended to rare gas halides by several groups in the early 1970s. Semiconductor diode lasers are extremely important in communications and audio/video technology (e.g., CD and DVD players). Infrared versions of these were first developed in 1962 by two US groups led by Robert N. Hall at the General Electric Research Center [11] and by Marshall Nathan at the IBM Watson Research Center.[12] The General Electric group submitted their results earlier and they also made a resonant cavity for their diode. Soon after this, the first visible wavelength laser diode was demonstrated by Nick Holonyak, Jr., in 1962.[13] The lasers mentioned so far provide fixed wavelengths of radiation, but tunable laser light is needed for studies of atomic and molecular spectroscopy. The first tunable visible lasers were organic dye lasers, developed independently at about the same time by Sorokin's group [14] and that of Schaefer.[15] Many other types of lasers are known and these are discussed in other texts devoted to these topics.[16–21] Here, we focus on a few selected kinds of systems that are more important in current research and technology and that are employed in the laboratory experiments described in this book.

Helium-neon lasers

Perhaps the most common laser employed in laboratory research and in practical applications is the helium-neon (He-Ne) laser. The continuous beam of red light provided by this laser is convenient for alignment of optics in research labs or for numerous applications in industry, such as providing a straight line in the construction of buildings or pipelines. He-Ne lasers are also extremely convenient for optics demonstrations or undergraduate laboratory experiments because they are inexpensive and relatively safe to operate (see Chapter 6 on laser safety).

The He-Ne laser consists of a discharge tube with a moderate pressure of a mixture of helium and neon gases. A high voltage applied to electrodes in the tube generates a continuous discharge. The atomic states involved are shown in Figure 4.1. Energetic collisions between electrons accelerated in the discharge and

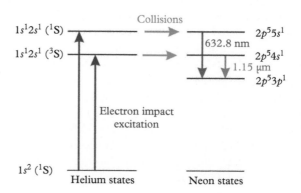

Figure 4.1 *The atomic energy levels involved in the He-Ne laser.*

helium atoms produce the so-called metastable excited states having the $1s^1 2s^1$ electronic configuration. When the electrons have opposite spins, the excited state has the configuration of ^1S, located at 20.6 eV, and when the spins have the same sign the configuration is ^3S, located at 19.8 eV. Both of these excited states have extremely long lifetimes because their relaxation by the emission of light is highly forbidden by angular momentum (both states) and spin (^3S state) selection rules. Relaxation by electron collisions is possible, but unlikely under the discharge conditions. However, the energy of these excited states can be transferred efficiently in collisions with neon, producing neon excited states. Because the ionization potential of neon (21.6 eV) is lower than that of helium (24.6 eV), its excited states also lie at lower energies. The $2p^5 5s^1$ and $2p^5 4s^1$ configurations are populated by collisions, and these levels are therefore inverted with respect to lower levels of neon that are not populated thermally at room temperature. The familiar laser radiation at 632.8 nm occurs when the $2p^5 5s^1$ level emits down to the $2p^5 3p^1$ level. Other transitions are also possible, but this particular one provides the most convenient and efficient lasing action. The operation of the laser produces virtually no by-products, and therefore the gases are sealed in the laser tube and do not need to be replaced. With a good seal, the laser tube can last for many years of operation.

The He-Ne laser usually provides milliwatt levels of radiation. This level is bright enough to be easily seen, even at a distance, but low enough in power so as to not be particularly dangerous to use in the lab. It is a class IIIb laser that can be used without goggles as long as there is no direct viewing of the main laser beam. In addition to the several experiments described here, He-Ne lasers are also used as wavelength standards in instruments such as Fourier-transform infrared spectrometers or wavemeters, and they can be used to calibrate grating-based UV-visible spectrometers.

Argon ion lasers

Another discharge-pumped continuous output laser providing visible wavelengths is the argon ion laser. This laser is based on transitions in the argon singly positive cation. The ionization energy of argon is 15.76 eV, and therefore the discharge in neutral argon gas must first form ions and then excite these. The atomic levels involved are those of the excited argon cation. The excitation occurs via electron collisions, as in the He-Ne laser, but the emitting level is the $3s^2 3p^4 4p^1$ ($^2D_{5/2}$) state at 158,730 cm^{-1} (19.7 eV) above the ground state of Ar$^+$. Emission occurs to the $3s^2 3p^4 4s^1$ state at 138,244 cm^{-1}, providing the main lasing line at 488.0 nm. Another nearby transition lands on the same lower level, providing another option for light at 514.5 nm. These two lines are the strongest, but several other weaker transitions are also possible (351.1, 363.8, 454.6, 457.9, 465.8, 476.5, 496.5, 501.7, 528.7, 1092.3 nm). For situations in which maximum laser output is desired, the sum of all these lasing transitions can be employed; this is common for pumping a dye laser. When a specific wavelength is desired, an angle-adjusted prism can be inserted into the cavity to allow gain at only one wavelength along the lasing axis. The laser output can be quite substantial, with typical power levels of up to 30 watts or more.

Because of the ionization and highly excited states involved, the discharge for an argon ion laser requires more energetic conditions than those used for the He-Ne laser. Much higher currents are required and much more substantial power supplies, making the argon ion laser much more expensive than a He-Ne. The discharge tube often has a solenoid coil around it to magnetically confine the ions near its center, where lasing is more intense. The discharge tube walls are usually coated with a high-temperature ceramic, such as beryllium oxide, to avoid sputtering of the glass walls by energetic ions. The tube must also be cooled by circulating water. Overall, a tube for such a laser is quite expensive (>$20,000) and must be replaced every year or so, making the operation of such a laser inconvenient and costly. For this reason, newer lasers such as diode-pumped YAG lasers are now commonly replacing argon ion lasers, but many such lasers are still actively used in chemistry and physics labs around the world. Krypton ion lasers operate under similar conditions to those used for argon, and provide other visible wavelengths, the most intense of which is a red line at 647 nm.

The most common application of argon ion lasers is in Raman spectroscopy, and these lasers have been the standard in commercial Raman instruments for many years. Another common application is in pumping dye lasers (see later discussion). Less common, but quite important for modern chemical physics is the application of these lasers for anion photoelectron spectroscopy. Many common free radicals have electron affinities in the 1–2 eV range, such that the 488 nm radiation (2.54 eV) is enough to induce photodetachment. Energy analysis of ejected electrons provides electron affinities and vibrational structure for the ground states of radicals that are difficult to study by other methods.[22] Argon ion and krypton ion lasers are sold commercially by companies such as Coherent and Newport/Spectra Physics.

Pulsed nitrogen lasers

The nitrogen laser is one of the earliest examples employing *pulsed* electrical discharges for excitation of the laser material. In this device, excited states of the N_2 molecule are produced by electron collisions in a fast (few nsec) pulsed electrical discharge. The ground state of N_2 is closed shell, with a $^1\Sigma_g^+$ configuration. Various excited singlet states exist (numbered *a, b, c* states for historical reasons), but they lie at high energies and excitation of these is not efficient. However, excited triplet states (numbered *A, B, C,* etc.) are found at lower energies. The *C* state is excited efficiently with electron collisions, producing greater population here than in the *A* and *B* states, leading to the population inversion. Emission back to the singlet ground state is forbidden, but is allowed between the excited triplet states.

The C state emits efficiently back to the B state, with a transition at 29,671 cm^{-1} (337 nm), producing the UV laser emission. The lifetime of the excited C state is a few nsec, producing a short pulse of laser emission. Since the B and C states have vibrational and rotational structure, the main lasing line centered at 337 nm may also have side-band structure.

Because near-instantaneous pulsing with a short timewidth is desirable, the discharge circuit is designed to store charge at a high voltage and then dump it rapidly into the gas through suitable metal electrodes. Figure 4.2 shows a typical example of a pulsed laser discharge circuit, such as those used for nitrogen, CO_2, or excimer lasers. A high-voltage power supply charges up a capacitor, which serves as a reservoir for charge. Solid-state switches cannot handle the high voltages and currents necessary for these devices. Therefore, the circuit is triggered by a special high-voltage switch known as a "thyratron." The thyratron is a ceramic canister containing hydrogen gas biased between two internal electrodes at a level very close to its breakdown voltage. When a trigger pulse of a few hundred volts (this can be generated with solid-state transistors) is applied, adding to the bias voltage, the hydrogen breaks down, conducting through its discharged vapor. This completes the circuit, allowing the capacitor to dump its charge into the gas in the laser cavity. In early versions of gas discharge lasers like this, an automobile spark plug was employed, but the time "jitter" (shot-to-shot reproducibility of firing) for these circuits was extremely bad (microsecond variations). Thyratrons have very low jitter of about 1 nsec, justifying the use of this extremely expensive and exotic component. When the circuit triggers, voltage is applied to the electrodes and the nitrogen gas in the cavity discharges along a rectangular volume between them, producing an output that is rectangular in cross-section. The inductor and diode in the circuit provide a filter against a reflected spike of high voltage that might go back into the power supply, potentially damaging it. In a nitrogen laser, the high voltage used is 3–5 kV, and the thyratron is relatively small.

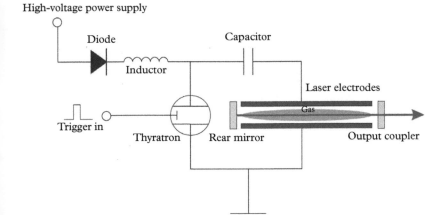

Figure 4.2 *Typical pulsed-discharge circuit for gas lasers such as nitrogen, CO_2, or excimer systems.*

However, in excimer lasers, pulses in the range of 40 kV are used, and a much larger, more expensive, thyratron is required. Because the lifetime of the nitrogen excited state is short, the pulsewidth of a nitrogen laser is essentially determined by the transit time of light in the discharge tube. A large laser with a 1-meter tube would have a pulsewidth of roughly 3.3 nsec, whereas a small laser (such as one used for matrix-assisted laser desorption ionization [MALDI]) might have a pulsewidth of about 500 psec.

The typical output of a nitrogen laser is in the microjoule/pulse to milli-joule/pulse range. The only wavelength produced is that at 337 nm. In the 1970s this was essentially the only pulsed laser source available for photochemistry or dye laser pumping. In recent years, nitrogen lasers have been replaced in these applications by excimer and Nd:YAG lasers, but these systems are still used in inexpensive undergraduate lab experiments. The most common application of nitrogen lasers in present technology is for MALDI mass spectrometry of biomolecules (see experiment in Chapter 12).[23,24] In this experiment, a 20–100 µJ/pulse is required, which is easily provided by a nitrogen laser. Many of the experiments in this text employ nitrogen lasers or nitrogen-pumped dye lasers.

CO_2 lasers

CO_2 lasers operate via either pulsed or continuous discharge pumping, providing fixed frequencies of infrared light at 10.6 and/or 9.6 µm. Pulsed operation is far more common. In this case the discharge circuit is schematically just like that in Figure 4.2, but with the gas mixture, voltage level, and optics chosen for CO_2 operation in the infrared. CO_2 lasers are based on the molecular vibrational transitions, as shown in Figure 4.3. The vibrational levels are labeled with the number

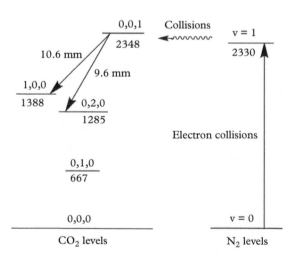

Figure 4.3 *The vibrational levels of N_2 and CO_2 involved in the action of the CO_2 laser.*

of quanta in the three vibrations of CO_2 (symmetric stretch, bend, asymmetric stretch) with a three-number sequence indicating v_1, v_2, v_3, respectively. Excitation in a pulsed discharge produces vibrationally excited N_2 molecules, whose $v = 1$ lifetime is long because emission from this state is dipole-forbidden. Collisions of N_2 with CO_2, aided by a small amount of translational energy, produce the 0,0,1 vibrational level of the triatomic, which lies at just slightly higher energy. Allowed emission from this state of CO_2 to the 1,0,0 and 0,2,0 levels proceeds efficiently. Further emission and collisions relax these levels so that the process can be repeated.

The rotational constant of CO_2 is 0.39 cm^{-1}. Because of this relatively high value, rotational fine structure is present and easily resolved in its vibrational bands (see Chapter 14). The laser output therefore has the "picket-fence" appearance of these individual rovibrational transitions, corresponding to P-, Q-, and R-type branches, extending across a bandwidth of several cm^{-1}. In versions of the laser with just a mirror in the optical cavity, all of these different wavelengths are present in the output. When additional spectral resolution is desired, a grating can be substituted for the rear mirror and individual rotational lines can be selected for the lasing action.

The CO_2 laser is extremely efficient, producing pulse energies up to 1 J or continuous power levels up to 100 W. The wall-plug efficiency can be as high as 20%. The timewidth of pulses is usually a few (20–50) nsec up to several μsec. The 10.6 and 9.6 μm wavelengths correspond to energies of about 940 and 1040 cm^{-1} respectively, which fall in the region of several vibrational fundamentals for medium-sized organic and inorganic molecules. If vibrations are slightly off resonance, they can be excited anyway because the high pulse energy of this laser causes bands to be power broadened. Once excited, molecules can continue to absorb to higher and higher levels in the intense radiation field, eventually dissociating. Many of the early applications of these lasers were therefore in the area of multiphoton absorption and infrared multiphoton photodissociation (IR-MPD). Now that other tunable lasers are available in the infrared, these applications are much less common. However, continuous CO_2 lasers are used industrially to cut metal, and pulsed CO_2 lasers are used in cosmetic skin-removal treatments.

Rare gas-halide excimer lasers

The term *excimer* is a contraction of "excited dimer," and refers to a whole class of molecules and molecular complexes that are bound only in an excited electronic state and not in the ground state. While there are many different kinds of excimers, this term has become synonymous with the particular kind of rare gas-halide molecules that form efficient excimer lasers. Examples of the wavelengths available from these excimer lasers are shown in Table 4.1.

Although other rare gas-halide combinations can also produce some laser emission, these excimer species emit efficiently enough for common usage.

Table 4.1 *Rare gas-halide excimer lasers and their wavelengths and photon energies.*

Excimer	Wavelength (nm)	Photon energy (eV)
XeF	350	3.54
XeCl	308	4.02
KrF	248	5.0
KrCl	222	5.58
ArF	193	6.42

In each case, excitation of a gas mixture containing the rare gas atom and a halide precursor (F_2 or HCl) is accomplished with a pulsed discharge, using a circuit like that shown in Figure 4.2. Electron collisions in the discharge produce metastable excited states of the rare gas (RG) atoms, like those mentioned above for the He-Ne laser, and they dissociate the halide molecules present, producing halide atoms. In the metastable excited state, the rare gas atom has a vacancy, or "hole," in its p shell, with an extra electron trapped in a higher orbital. This RG* species with the p-hole configuration resembles a halogen atom, and then the RG*-X species can form a halide-like diatomic in its excited state. The forbidden transition of the rare gas atom becomes allowed in the reduced symmetry of the diatomic, and emission occurs. However, the ground state is unbound, with a repulsive potential energy surface (see Figure 4.4), guaranteeing that there will be a population inversion. The molecule dissociates as soon as it emits down to this state, producing reagents which can be recycled. For different output wavelengths, the excimer laser cavity is filled with a different recipe of gases at specific partial pressures, with helium or neon added as a non-reactive buffer gas. Rapid collisions are required for the excimer to form and for reagents to recombine before the next gas pulse. This requires a total pressure of about 5 atm in the discharge tube. Discharges at such high pressures are difficult to achieve with precise timing, and so the circuit for an excimer employs much higher voltages (40 kV) than those used for other discharge lasers. These voltage levels are lethal, and only specially trained personnel should open the enclosure to work on these lasers.

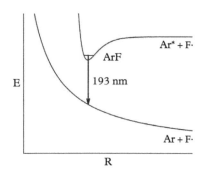

Figure 4.4 *The form of the potential energy curves for the ground and excited states of rare gas-halide excimer lasers, with ArF as an example.*

Because of the special conditions required, the operation of an excimer laser is much more complicated than that of other lasers. The halide gases used are extremely corrosive (and explosive in the case of fluorine) when used in pure form, and so mixes of these at low concentration (5% in helium) are generally used. Special corrosion-resistant regulators and vented bottle enclosures are required for these gases. Reactant gas partial pressures, buffer gas pressures, and discharge voltages must be adjusted to specific levels for each system. Cooling water running through radiator-like heat exchangers is required to prevent gas over-heating. Optics must be changed for different UV wavelengths. The resulting pulsewidths are comparable (10–20 nsec) for different mixes, but the pulse energies possible and the gas fill lifetimes vary significantly from one system to another. The fluorine gas mixtures are more corrosive than the HCl mixes, and generate more volatile by-products from reactions with metal surfaces (electrodes, capacitors, wires, etc.) in the cavity that can quench lasing action. Therefore, the chloride mixes (e.g., XeCl) generally have much longer gas lifetimes than the fluoride mixtures. In early versions of excimer lasers, gas lifetimes were as short as a few hours for mixes such as ArF, but newer gas tanks are made from inert ceramics, making gas lifetimes much longer now (several weeks). When a gas mixture is depleted, it can be pumped out and replaced with a fresh mixture of the same gases in just a few minutes time. However, if the cavity has to be vented to replace or clean optics, it must be evacuated thoroughly and its surfaces have to be re-passivated with halogen gas before continuing operation.

Considering all of this, the operation of excimer lasers can be quite challenging, and special training is required for operators. In spite of this, these lasers are still used quite frequently for research applications. They provide the only source of high pulse energy ultraviolet light, which can be used for photochemistry or for photoionization studies. Likewise, the pulse rate for such gas discharge lasers can be quite high (up to kHz repetition rates), providing high average power. These systems can be employed to pump dye lasers (see below), where they are especially efficient for blue–green dyes. Industrial applications include photochemical vapor deposition or photolithography, widely used in microelectronic chip fabrication for cell phones or computers. Because of the short wavelengths provided by mixes such as ArF, the diffraction-limited resolution for pattern definition is the best generally available at present. However, new methods using laser generation of metal vapor plasmas have made it possible to produce extreme ultraviolet (XUV) light at 13 nm, and this is the basis of emerging lithography technology.[25] Commercial excimer lasers are sold by several companies, including Lambda Physik and GAM.

Nd:YAG lasers

The first solid-state laser was that made from ruby, as noted above. Like many other gemstones, ruby has a transition metal present as an impurity in a stable lattice such as aluminum oxide, silicon oxide, or their mixtures. Over the years, many other such crystal-based laser systems were investigated to find those with the best emission properties, mechanical properties, and thermal stability. Eventually, Nd:YAG (neodymium-doped yttrium-aluminum garnet; commonly abbreviated as the "YAG" laser) became the most commonly used system for this. Its main output wavelength from an atomic transition of the Nd^{3+} ion is at 9391 cm^{-1} (1064 nm) in the near infrared region. This wavelength itself is not particularly useful for spectroscopy because it lies higher in energy than molecular vibrations and lower than most molecular electronic transitions. However, because its output is so intense, its operation so convenient, and its beam quality so uniform, this fundamental wavelength can be converted to more useful wavelengths using nonlinear optical techniques. These methods include frequency doubling or sum-frequency mixing (see Chapter 5). The 1064 nm fundamental can be converted to the 532 nm (bright green) "second harmonic" via frequency doubling,[26] sum-frequency mixing of these two wavelengths produces the "third harmonic" at 355 nm, and then frequency doubling of the 532 nm output produces the "fourth harmonic" at 266 nm. The latter two wavelengths lie in the ultraviolet region. These nonlinear (two-photon) processes are efficient in this system because of the high peak power (up to 1–2 J/pulse) and short pulsewidth (about 5 nsec) of this laser.

The design of this solid-state laser is quite different from that of the pulsed-discharge gas lasers discussed above. In particular, excitation of chromophores doped into a solid crystal lattice requires bright white light rather than an electrical discharge. However, this light can be generated with a pulsed-discharge "flashlamp," which resembles a bright strobe light. The light from such flashlamps is confined and directed to the laser rod via a mirrored enclosure, with water cooling to avoid thermal expansion and possible cracking of the laser crystal. In a so-called "long-pulse" mode of operation, a low level of laser emission is produced throughout the time width of the flashlamp, which lasts 200–300 μsec. A similar low level can be produced with continuous lamps to produce a continuous YAG laser. However, for the most efficient nonlinear conversion processes, the light should have the highest possible peak power, and this requires "Q-switching." In this process, the laser oscillation in the cavity is blocked until a large population is produced in the excited state by the full length of the flashlamp pulse, and then the entire population is dumped suddenly to produce an intense pulse of light lasting about 5 nsec. Conceptually, this is like blocking the mirrors in the laser cavity with a shutter and then suddenly removing it. However, because this has to be done on the nanosecond timescale, no mechanical shutter is fast enough to be used. Instead, Q-switching is achieved with optical components consisting of a polarizer, a quarter-wave plate, and a Pockels cell, as shown in Figure 4.5. The output of the YAG crystal is polarized linearly when it passes through the polarizer before it reaches the quarter-wave plate. A single pass through this device

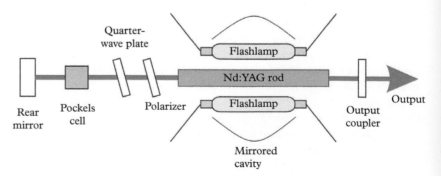

Figure 4.5 *Nd:YAG laser oscillator design.*

converts the polarization from linear into circular. If the Pockels cell is not activated, the light hits the rear mirror, gets reflected back through the quarter-wave plate, and the second pass through this converts the circularly polarized light back to linear polarization that is rotated 90° from the initial orientation. This light cannot pass through the polarizer, and therefore oscillation in the cavity is blocked. This is the configuration applied when the flashlamp initially fires, so that population builds in the excited state, but lasing is prevented. If a voltage is applied to the Pockels cell, it acts as a second quarter-wave plate, and two passes through this rotates the light another 90°, bringing it back to its original plane, allowing it to pass through the polarizer and back to the YAG rod. In actual operation, a sudden high-voltage pulse applied near the end of the flashlamp pulse switches the Pockels cell on to activate it, and oscillation back and forth through the cavity can occur. This is how the built-up excited state population is suddenly allowed to lase in a fast pulse.

Figure 4.6 shows a schematic of the optical configuration used for frequency doubling of a YAG laser. The 1064 nm (infrared) output from the laser is directed into a frequency doubling crystal, usually made from potassium dihydrogen phosphate (KDP). The angle of the crystal must be adjusted properly to obtain the phase matching between the input beam and the second harmonic output beam (see Chapter 5). The refractive index of the crystal, which affects the velocities of both beams and their phase matching, depends on temperature, and so this crystal is usually enclosed in a temperature-stabilized box. The second harmonic beam at 532 nm (green) is collinear with the residual 1064 nm input beam (about 50% of the 1064 nm beam gets converted to 532 nm), and these must be separated if the green light is to be used. This is accomplished with a dielectric mirror that reflects only the green and transmits the IR, which is usually dumped into an enclosed container for safety reasons. If the third harmonic (355 nm; near UV) is to be generated, the collinear 1064 and 532 nm beams are directed into a second KDP crystal, where they combine via sum-frequency generation to produce this beam. An appropriate dielectric mirror can then be used to separate this UV beam from the residual IR and green beams. Another KDP crystal (cut at the correct crystal angle) and dielectric mirror can also be used to generate the second harmonic of the 532 nm beam, producing 266 nm (UV), but this is less

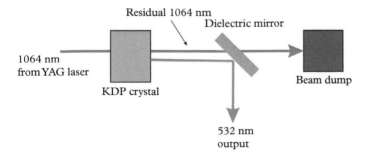

Residual 1064 nm
Dielectric mirror
1064 nm from YAG laser
KDP crystal
Beam dump
532 nm output

Figure 4.6 *The optical configuration for frequency doubling a Nd:YAG laser.*

common. The theory of nonlinear optics that makes frequency doubling possible is discussed in Chapter 5.

The various harmonic beams from YAG lasers have many applications in chemistry and physics, and some of these are demonstrated in the experiments in this book. These colors can be absorbed directly by many materials for the study of fluorescence or photochemistry. The second or third harmonic beams can be used to pump dye lasers or either visible or infrared optical parametric oscillators (OPOs) to generate tunable visible or infrared light. These lasers are used for several experiments in this book, such as that on multiphoton ionization (Chapter 13, 16, and 18). They can also be used for laser ablation of solid metal targets to produce gas-phase metal atom clusters (Chapter 12).[27] Larger YAG lasers for research applications are sold by several companies, including New-port/Spectra Physics, Continuum, and Quantel. Smaller YAG lasers appropriate for undergraduate lab courses are sold by these same companies and also by New Wave Research.

Dye lasers

Some of the lasers mentioned above generate visible and near-UV wavelengths of light, but the wavelengths provided are fixed and not tunable. For studies of UV-visible spectroscopy, broadly tunable light in these regions with narrow spectral linewidth is needed. This can be provided by dye lasers. As noted above, dye lasers were developed early in the chronology of lasers, but these systems are still used widely today. Dye lasers provide visible light based on the fluorescence spectra of large conjugated aromatic molecules. Dye molecules like these have been used for many years to color materials such as fabric for clothing. Figure 4.7 provides the molecular structures of two typical dyes.

Because dye molecules are highly conjugated, they absorb strongly at visible wavelengths and fluoresce brightly. A dye laser controls this fluorescence, making it into an intense directional laser beam and adjusting or "tuning" its wavelength. A liquid solution containing the dye molecule at a specific optimized

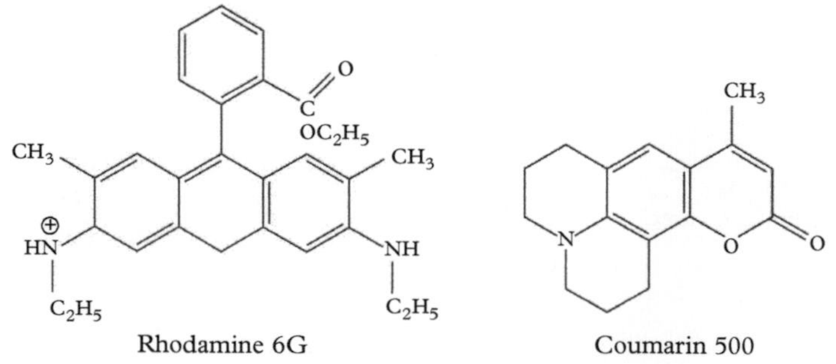

Figure 4.7 *The structures of two representative dye molecules used in dye lasers.*

Rhodamine 6G

Coumarin 500

concentration is prepared in one or more dye cells. These may be sealed cuvettes such as those used for samples in UV-visible spectroscopy if the laser is to be used at low power, or they may be flow-through containers with circulation pumps to avoid over-heating the solution in higher power systems. In either case, the dye solution is excited, or "pumped," to generate fluorescence with another laser such as an argon ion (488 nm), nitrogen (337 nm), excimer (XeCl; 308 nm), or Nd:YAG (532 or 355 nm). Continuous or pulsed operation is possible depending on the pump laser used, and the optical design is different for each. A schematic diagram of a pulsed dye laser, such as that used with a Nd:YAG as the pump source, is shown in Figure 4.8. Optics in the oscillator section include a beam expander and a diffraction grating. In much the same way that the grating in a spectrometer selects wavelength, the grating here selects a specific wavelength from within the wide spectral range of the fluorescence. This wavelength selection is shown in Figure 4.9. In a typical system, the frequency linewidth can be 0.1 cm^{-1}, and improvements over this are possible using intracavity etalons (see Chapter 2). Modern commercial dye lasers employ various schemes to improve beam expansion on the grating, and thus the linewidth, and to minimize the amplification of stray broadband fluorescence (known as amplified spontaneous emission, or "ASE"). The specific wavelength selected is reflected back through the dye cell and oscillates in the cavity formed from the grating (rear reflector) and the front mirror. Because the emission is very efficient, the front mirror is usually just a quartz window, which reflects roughly 4% of the light that hits its front surface. The remainder of the light exits the oscillator cavity and stimulates the dye solution in the amplifier cell at the selected wavelength, and amplification to higher power occurs here. The laser can be tuned across the fluorescence range of the dye by

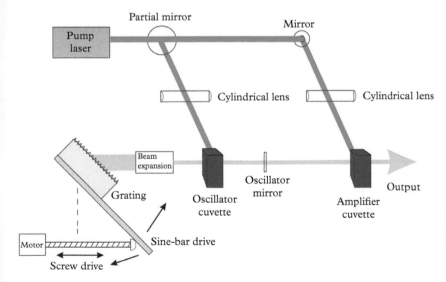

Figure 4.8 *A schematic diagram of a pulsed dye laser. The dye concentration in the amplifier cuvette is usually lower than that in the oscillator.*

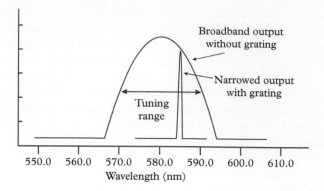

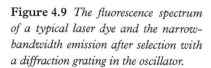

Figure 4.9 *The fluorescence spectrum of a typical laser dye and the narrow-bandwidth emission after selection with a diffraction grating in the oscillator.*

tilting the grating angle, as is done in ordinary spectrometers. However, as in normal spectrometers, the wavelength varies with the sine of the angle of the light hitting the grating (measured from the normal). To get a linear wavelength scan, a sine-bar drive (see Figure 2.27) is employed on the grating, as shown in Figure 4.8, which converts the linear motion of a drive motor into a smooth variation in the sine of the incident angle. Scanning is usually accomplished under computer control.

Each dye solution can only be used within its fluorescence spectrum. Typical dyes have a fluorescence that spans 15–20 nm, although some have ranges as large as 40–50 nm. An interesting aspect of these fluorescence spectra is that they are approximately the same regardless of the excitation wavelength used to excite the dye. This observation was explained many years ago by Michael Kasha, and is known as "Kasha's Rule."[28] Higher excited states that might be produced in organic molecules in solution are rapidly relaxed by solvent collisions to the lowest excited state, and this is the one that emits. To generate other wavelengths, the solution must be changed to another dye with fluorescence in another segment of the visible spectrum. Covering the full visible spectrum is possible, but this requires the use of 15–20 different dye solutions (see Figure 17.2 for an example where many dyes were used to record the full spectrum of the iodine molecule). The tuning ranges for different laser dyes in a PDL-2 dye laser (sold formerly by Spectra Physics) pumped by a Nd:YAG laser are shown in Figure 4.10. Dyes can be reused, but when they are changed the residue from each one must be rinsed out of the system, generating much solvent waste. Long wavelength scans covering the whole visible spectrum are tedious and time consuming, and therefore rarely done. Just like other grating-based spectrometers, dye laser wavelengths must be calibrated. This can be done with known wavelengths from a He-Ne laser, from atomic lines in gas discharge lamps like those used in atomic absorption instruments, or with wavemeter devices using one or more etalons calibrated with a He-Ne laser.[21]

Much attention has been given over the years to the optimized design of dye laser oscillators, which are the key element in wavelength selection.

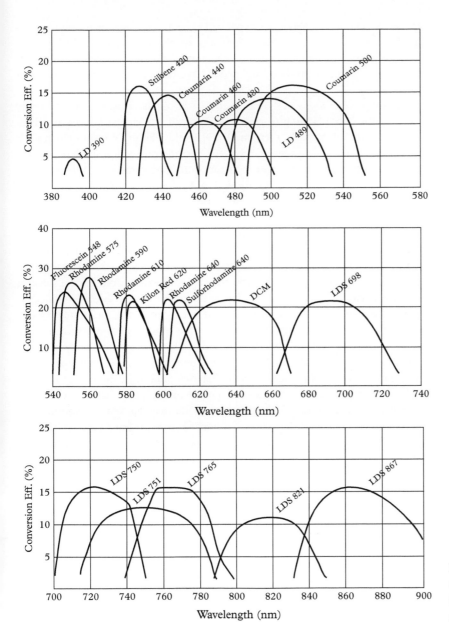

Figure 4.10 *Tuning curves for differ-ent laser dyes. (Permission to use granted by Newport Corporation. All rights reserved.)*

In particular, it is desirable to have the narrowest spectral width, or resolu-tion, for the output wavelength. This activity has focused on pulsed dye lasers pumped by Nd:YAG or excimer lasers, as these are used more for ultra-sensitive spectroscopy. As described in Chapter 3, diffraction gratings provide greater res-olution when more lines are illuminated on their surface. This suggests that the dye laser oscillator spot should be expanded to be as large as possible on the grating. However, it is also desirable to have a relatively small spot size for the laser output for more efficient excitation of the sample. These considerations have resulted in two main design concepts which have been used for pulsed dye laser oscillators.

Figure 4.11 shows one of the earliest dye laser oscillator designs developed by Professor Theodore Hänsch (formerly at Stanford University; now at the Technical University of Munich, Germany).[29] In this design, a beam expander is inserted in the cavity between the dye cell and the diffraction grating (Littrow configuration) that forms the rear reflector of the cavity, increasing the spot size on the grating and improving the resolution. In the original Hänsch design, the beam expander was formed with a standard two-lens telescope. However, this proved to be difficult to align, as any slight displacement of the beam from the center of either lens causes deflection of the beam away from the center line of the grating. More recent designs employ a four-prism beam expander that only expands the beam in the vertical direction as shown. This provides equally good resolution and is easier to align. However, it requires multiple optical surfaces inside the cavity. Anti-reflection coating of the prisms avoids excessive reflective losses. The resolution can be increased further by inserting an intracavity etalon into the oscillator. The mode spacings in this etalon cavity are broader than those for the full oscillator cavity. Tilting the etalon to a specific angle allows it to act as a wavelength-selective filter, passing only one of the several longitudinal cavity modes that would otherwise be lasing. Scanning with such an etalon filter in place is difficult, as it requires simultaneous tilting of the grating and tilting of the etalon angle to track with the cavity mode selected, but this can be done with computer control of the angles. Professor Hänsch received the Nobel Prize in Physics in 2005 for his studies of high-resolution spectroscopy using lasers like this.

A second design concept came along somewhat later in the development of dye lasers, made possible by new technology for making diffraction gratings that

Figure 4.11 *Hänsch design for pulsed dye laser oscillator, with a two-lens telescope for beam expansion and intracavity etalon. The dye concentration is optimized to absorb the pump laser just inside the front face of the dye cell, establishing the fluorescence volume and optical axis for lasing. The wedged output coupler prevents reflection from the outer surface from returning to the cavity, which would generate an unwanted second cavity length and set of cavity modes.*

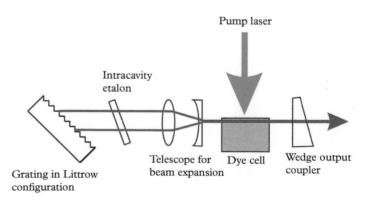

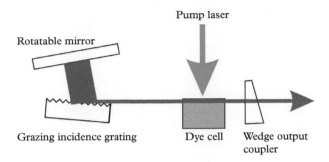

Pump laser

Rotatable mirror

Grazing incidence grating   Dye cell   Wedge output
                                                     coupler

Figure 4.12 *Littman design for pulsed dye laser oscillator.*

could be used at a steep input angle. The so-called "grazing incidence" design shown in Figure 4.12 was introduced by Professor Michael Littman (formerly at MIT; now at Princeton University).[30] By tilting the diffraction grating at a steep angle, more lines are illuminated without any beam expansion optics, reducing reflective losses. A mirror reflects one of the orders off this grating back through the system; tilting this mirror provides the tuning. Because fewer optics are present, the laser cavity in the Littman design can be shorter, resulting in wider longitudinal mode spacing, making it easier to select and lase on a single cavity mode. A variation on this design used a second diffraction grating in the Littrow configuration instead of the mirror. This provided even higher spectral resolution, but much lower efficiency for the cavity. Modern dye lasers implement one or the other of the Hänsch or Littman design concepts, as well as many hybrid variations of these ideas, in their design.

The laser dye solutions that fluoresce efficiently produce mainly far-IR and visible light. Ultraviolet light can also be generated, however, using frequency doubling of the visible wavelength in nonlinear optical crystals like those used in YAG lasers. Typical crystals used for this include KDP and more recently beta-barium borate (BBO). Crystals such as these are cut so that the optical axes are at specific angles to their input face, thus optimizing the phase matching. This causes each crystal to function over a limited segment of the spectrum. Several crystals must be employed to cover the full visible wavelength range.

Dye lasers are used for many techniques in spectroscopy, including direct absorption measurements, fluorescence excitation or dispersed fluorescence spectra, fluorescence lifetimes, photoionization, and photodissociation measurements. In recent work, dye lasers are sometimes replaced by visible OPO systems (see later), which offer similar tuning ranges and linewidth without the need to use or dispose of solvents and their waste. Unfortunately, the crystals used in visible OPO systems are expensive and are often subject to burn damage at the high pump laser powers used, and so many labs still prefer dye laser systems for UV-visible spectroscopy. Dye lasers are sold by many of the same companies that sell excimer, Nd:YAG, or argon ion lasers (e.g., Coherent, Spectra Physics, Lambda Physik, Continuum).

Optical parametric oscillators

Dye lasers provide robust sources of tunable visible and far-IR radiation, but the limited tuning range of each dye and the associated solvent changes and waste disposal issues make working with these systems inconvenient. Therefore, a long-standing goal of laser technology has been a solid-state device that would be convenient to operate and would cover all visible wavelengths. Optical parametric oscillators (OPOs) succeed in this goal to some degree.

An OPO works conceptually as the reverse of frequency doubling or sum-frequency generation (SFG). While doubling or SFG involve the combination of two lower energy photons to produce a higher energy one, an OPO takes a higher energy photon and "splits" it into two smaller ones, as shown in Figure 4.13. In this way, a high-power ultraviolet laser can be used to generate visible light, or a high-energy visible laser can be used to produce infrared light. In frequency doubling, the two photons that combine have the same wavelength, whereas in SFG the two photons have different wavelengths; an OPO is more like the reverse of SFG since the two wavelengths produced are not usually the same. The specific wavelengths produced and the wavelength tuning are controlled by the phase-matching angle of the crystal, and this can be controlled further by making a cavity around the crystal with a diffraction grating for the rear reflector. The higher frequency generated is called the "signal" beam and the lower frequency is called the "idler." At the "degeneracy point" the signal and idler wavelengths are the same, and this is exactly the reverse of frequency doubling. The theory of nonlinear optics that makes OPO operation possible is discussed in Chapter 5.

In the usual design of visible OPOs, the pump laser is a Nd:YAG operating on its third harmonic at 355 nm. This pump laser must have exceptional beam quality (uniform spatial intensity, i.e., no "hot spots") to avoid burning the expensive crystal. This wavelength is employed with a BBO crystal, producing signal wavelengths throughout the visible spectrum. The signal and idler beams are separated with a prism; this works well except near the degeneracy point where the wavelengths are too close and the generated intensity is low. Frequency doubling of this output can be used to produce ultraviolet wavelengths. Therefore, visible OPOs effectively replace dye lasers in their frequency coverage. Pulse energies and linewidths are comparable when grating oscillators are employed. In applications where higher frequency resolution is not required, lower cost/lower resolution (5 cm^{-1}) models are also available. The disadvantage of OPO systems is that

Figure 4.13 *The schematic diagram of a visible OPO laser.*

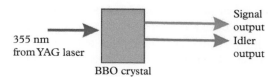

the BBO crystals used are expensive and they are subject to burning at the high pump powers employed (200–500 mJ/pulse). High quality YAG lasers with optics designed to produce uniform spatial and temporal beam profiles are required for this, which are also more expensive than ordinary lasers. With these caveats, visible OPOs provide a convenient source of tunable visible and UV radiation, covering all the same applications as dye lasers, but without the inconveniences of solvent and waste handling.

OPOs can also be used to convert visible light into tunable infrared wavelengths. A commercial device for this application is produced by LaserVision.[31] The schematic layout of this device is shown in Figure 4.14. It is pumped by the 1064 nm fundamental of a Nd:YAG laser, which is split into two beams. The first part of this beam is frequency doubled to produce 532 nm, and this is employed to pump the oscillator section of the OPO. Here one or two potassium titanyl phospate (KTP) crystals are employed to split the 532 nm beam into signal and idler beams. The signal beam is usually blocked and the idler beam goes forward to the amplifier (OPA) stage of the system, which consists of four potassium titanyl arsenate (KTA) crystals. Here, the idler beam is combined with the remaining 1064 nm beam using difference-frequency generation. If the oscillator beam is adjusted to produce an idler beam at 6000 cm^{-1}, the signal beam would be at 12,782 cm^{-1} to conserve energy from the 532 nm beam (18,782 cm^{-1}). Difference-frequency generation with the 1064 nm beam (9391 cm^{-1}) produces mid-IR at 3391 cm^{-1}. This is in the region of C–H and O–H stretch fundamentals for many molecules, and this laser can be used for infrared spectroscopy. In this configuration, the tuning range is 2000–4500 cm^{-1}. However, the range can be extended to lower frequencies (600–2000 cm^{-1}) with another stage of difference-frequency generation in a silver gallium selenide crystal.

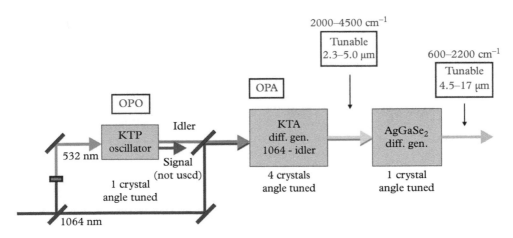

Figure 4.14 *The schematic of an infrared OPO/OPA laser system.*

Diode lasers

Another large class of lasers that are extremely important in industry, and are gaining applications in research, includes semiconductor diode lasers. These lasers are most well known in laser pointers, barcode readers, and in CD or DVD players, but they also provide relatively inexpensive and reliable sources of narrow-band (<1 MHz) tunable laser light, in visible or infrared wavelength regions, which can be used effectively for high-resolution spectroscopy of atoms and molecules. The lasing material in these systems is a small piece of solid semiconductor material whose band gap lies at the energy where emission is desired. Direct band-gap emission is forbidden for pure component semiconductors like silicon or germanium, and requires instead that compound semiconductors such as gallium arsenide, indium phosphide, or gallium nitride be used. A p–n junction of this material contains a depletion region near the interface, in which there are no charge carriers. Forward electrical bias introduces holes from the p region and electrons from the n region, which recombine in the depletion zone, resulting in emission of light across the band gap. Depending on the band-gap energy, the emission can be infrared or visible radiation. The emission from just such a simple p–n junction is spontaneous, and must be amplified in an optical cavity to achieve stimulated emission and amplification. This is accomplished by cutting the semiconductor crystal with flat parallel surfaces at its ends; the interfacial reflection here effectively produces the end mirrors that form the cavity. The output of diode lasers coming from a small crystal is highly divergent because of diffraction and must be collimated with lenses or other optics to provide a useful beam. The wavelength is tunable over narrow ranges by variation of the injection voltage.

The first diode lasers produced suffered from limited wavelength ranges and unstable temperature and voltage tuning characteristics. Mode hopping, in which the laser switches the longitudinal mode that is lasing, producing an unwanted change in wavelength, was also a common problem during tuning. The introduction of stabilization methods employing a grating to feed the diode laser light back into the diode has largely overcome most of these problems.[32–35] The two common external-cavity diode laser (ECDL) designs are the Littrow and the Littman–Metcalf configurations. These two geometries are illustrated in Figure 4.15. In the Littrow geometry the light from the diffraction grating is reflected back into the laser in first order while the light from the grating in zeroth order is coupled out. For the Littman–Metcalf geometry, the light diffracted in the first order is returned back to the grating by a mirror or another grating and then back to the laser as shown in the figure.

There are several commercial sources of ECDLs available. However, over the past few years a number of low-cost external-cavity diode lasers built for use in undergraduate laboratories have appeared.[36, 37] Hawthorn et al. [38] describe in some detail an ECDL Littrow configuration tunable external-cavity diode laser resulting in a fixed direction of the output beam upon tuning. They report tuning

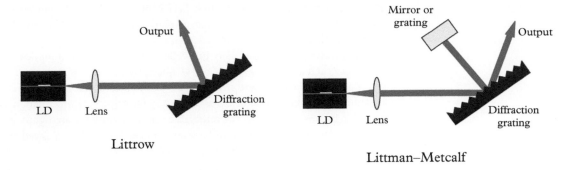

Figure 4.15 *Geometry of the ECDL designs: Littrow and the Littman–Metcalf configurations.*

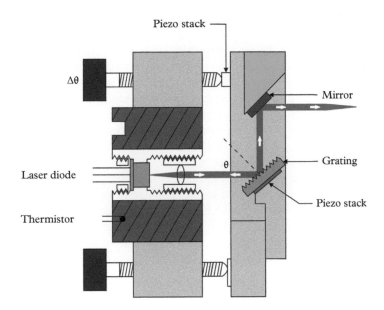

Figure 4.16 *Design of the external-cavity diode laser (ECDL) using the Littrow configuration allowing fixed direction of the laser beam during tuning (see [32] and [38]).*

the laser over a wavelength range greater than 10 nm without measurable movement of the output beam. Figure 4.16 shows how this is accomplished based upon the design of Arnold et al.[33]

Using the output mirror in the Littrow geometry, this laser provided a resolution of ~0.5 MHz and the lateral displacement of the output beam was found to be less than 0.4 mm over a tuning range of 10.5 nm. A complete description for the construction of this ECDL is found in [34] and online (http://optics.ph.unimelb.edu.au/atomopt/diodes.html). Also, Saliba and Scholten have described a similar laser having 0.1 MHz resolution.[39] Tunable diode lasers have contributed greatly to optical trapping of atoms [40] and to

Bose–Einstein condensation research.[41] More specialized "quantum cascade" diode lasers [42,43] are used for high-resolution infrared spectroscopy, but these are not employed in the experiments in this book.

Diode lasers are commercially important in fiber-optic communications and in devices such as CD or DVD players. The early versions of these devices used red diode lasers, which were much easier to produce at the time. However, the introduction of blue diode lasers significantly improved the information density available because the shorter wavelength could be focused more tightly (see discussion on the diffraction limit of focusing in Chapter 2). This allowed "Blu-ray" devices to be produced. Isamu Akasaki, Hiroshi Amano, and Shuji Nakamura shared the Nobel Prize in Physics in 2014 for their development of blue diode lasers.

Free electron lasers

The experiments in this book use only those laser sources readily available to college and university laboratories. However, an important laser light source which is not employed here is also worth mentioning. In the so-called free electron laser, or FEL,[44,45] a relativistic velocity (about 50 mega-eV) electron beam (the lasing medium) *undulates*, or wiggles, through a series of alternating magnetic fields, producing pulsed radiation whose polarization is perpendicular to the electron oscillation and along the direction of the electron beam. The schematic of such a laser is illustrated in Figure 4.17. The FEL can be configured to emit monochromatic (but incoherent) radiation over a wide frequency range from the microwave to terahertz to infrared to visible to ultraviolet to X-ray region of the electromagnetic spectrum (refer to Figure 1.1). The frequency of this device is tuned by either changing the energy of the electron beam or the strength of the magnetic field (by adjusting the gap between permanent magnets). Both of the authors and their students have made use of the Free Electron Laser for Infrared eXperiments (FELIX) in Radboud University, Institute for Molecules and

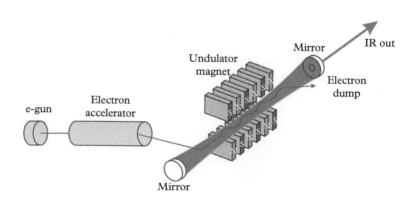

Figure 4.17 *Illustration of the operation of an infrared free electron laser, such as FELIX.*

Materials, Nijmegen, The Netherlands, in the study of the spectroscopy of molecular ions over the years. There are presently over 25 operating FELs in the world and others are under construction.

....

REFERENCES

1. J. P. Gordon, H. J. Zeiger, and C. H. Townes, "The MASER – New type of microwave amplifier, frequency standard and spectrometer," *Phys. Rev.* **99**, 1264 (1955).

2. C. H. Townes, *How the Laser Happened*, Oxford University Press, New York, 1999.

3. T. H. Maiman, "Stimulated optical radiation in ruby," *Nature* **187**, 493 (1960).

4. R. J. Collins, D. F. Nelson, A. L. Schawlow, W. Bond, C. G. B. Garrett, and W. Kaiser, "Coherence, narrowing, directionality, and relaxation oscillations in the light emission from ruby," *Phys. Rev. Lett.* **5**, 303 (1960).

5. A. Javan, W. R. Bennett, Jr., and D. R. Herriott, "Population inversion and continuous optical maser oscillation in a gas discharge containing a He-Ne mixture," *Phys. Rev. Lett.* **6**, 106 (1961).

6. H. G. Heard, "Ultraviolet gas laser at room temperature," *Nature* **200**, 667 (1963).

7. C. K. N. Patel, "Continuous-wave laser action on vibrational-rotational transitions of CO_2," *Phys. Rev. A* **136**, 1187 (1964).

8. W. B. Bridges, "Laser oscillation in singly ionized argon in the visible spectrum," *Appl. Phys. Lett.* **4**, 128 (1964).

9. J. E. Geusic, H. M. Marcos, and L. G. Van Uitert, "Laser oscillations in a Nd-doped yttrium aluminum, yttrium gallium and gadolinium garnets," *Appl. Phys. Lett.* **4**, 182 (1964).

10. N. G. Basov, V. A. Danilychev, Y. Popov, and D. D. Khodkevich, *Zh. Eksp. Fiz. i Tekh. Pis'ma. Red.* **12**, 473 (1970).

11. R. N. Hall, G. E. Fenner, J. D. Kingsley, T. J. Soltys, and R. O. Carlson "Coherent light emission from GaAs junctions," *Phys. Rev. Lett.* **9**, 366 (1962).

12. M. I. Nathan, W. P. Dumke, G. Burns, F. H. Dill, and G. Lasher, "Stimulated emission of radiation from GaAs p-n junctions," *Appl. Phys. Lett.* **1**, 62 (1962).

13. N. Holonyak, Jr. and S. F. Bevaqua, "Coherent (visible) light emission from Ga($As_{1-x}P_x$) junctions," *Appl. Phys. Lett.* **1**, 82 (1962).

14. P. P. Sorokin and J. R. Lankard, "Stimulated emission observed from an organic dye chloro-aluminum phthalocyanine," *IBM J. Res. & Develop.* **10**, 162 (1966).

15. F. P. Schäfer, W. Schmidt, and J. Volze, "Organic dye solution laser," *Appl. Phys. Lett.* **9**, 306 (1966).

16. O. Svelto, *Principles of Lasers*, Plenum Press, New York, 1989.

17. C. C. Davis, *Lasers and Electro-Optics, Fundamentals and Engineering*, Cambridge University Press, UK, 1996.

18. D. L. Andrews, *Lasers in Chemistry*, third edition, Springer-Verlag, Berlin, 1997.

19. G. R. Van Hecke and K. K. Karukstis, *A Guide to Lasers in Chemistry*, Jones and Bartlett Publishers, London, 1998.

20. W. T. Silfvast, *Laser Fundamentals*, second edition, Cambridge University Press, UK 2004.

21. W. Demtröder, *Laser Spectroscopy*, fourth edition, Volumes 1 & 2, Springer-Verlag, Berlin, 2008.

22. R. R. Corderman and W. C. Lineberger, "Negative ion spectroscopy," *Ann. Rev. Phys. Chem.* **30**, 347 (1979).

23. G. Hillenkamp and J. Peter-Katalinić, eds. *MALDI MS*, Wiley-VCH, Weinheim, Germany, 2007.

24. R. B. Cole, ed. *Electrospray and MALDI Mass Spectrometry*, John Wiley, Hoboken, NJ, 2010.

25. P. A. C. Jansson, B. A. M. Hansson, O. Hemberg, M. Otendal, A. Holmberg, J. de Groot, and H. M. Hertz, "Liquid-tin-jet laser plasma extreme ultraviolet generation," *Appl. Phys. Lett.* **84**, 2256 (2004).

26. P. A. Franken, A. E. Hill, C. W. Peters, and G. Weinreich, "Generation of optical harmonics," *Phys. Rev. Lett.* **7**, 118 (1961).

27. M. A. Duncan, "Laser vaporization cluster sources," *Rev. Sci. Instrum.* **83**, 041101 (2012).

28. M. Kasha, "Characterization of electronic transitions in complex molecules," *Disc. Faraday Soc.* **9**, 50 (1950).

29. T. W. Hänsch, "Repetitively pulsed tunable dye laser for high resolution spectroscopy," *Appl. Optics* **11**, 895 (1972).

30. M. G. Littman and H. J. Metcalf, "Spectrally narrow pulsed dye laser without beam expander," *Appl. Optics* **17**, 2224 (1978).

31. W. R. Bosenberg and D. R. Guyer, "Broadly tunable, single-frequency optical parametric frequency-conversion system," *J. Opt. Soc. Am. B* **10**, 1716 (1993).

32. M. W. Fleming and A. Mooradian, "Spectral characteristics of external-cavity controlled semiconductor lasers," *IEEE J. Quantum Electron.* **17**, 44 (1981).

33. A. S. Arnold, J. S. Wilson, and M. G. Boshier, "A simple extended-cavity diode laser," *Rev. Sci. Instrum.* **69**, 1236 (1998).

34. C. E. Wieman and L. Hollberg, "Using diode lasers for atomic physics," *Rev. Sci. Instrum.* **62**, 1 (1991).

35. K. B. MacAdam, A. Steinbach, and C. Wieman, "A narrow-band tunable diode laser system with grating feedback, and saturated absorption spectrometer for Cs and Rb," *Am. J. Phys.* **60**, 1098 (1992).

36. G. N. Rao, M. N. Reddy, and E. Hecht, "Atomic hyperfine structure studies using temperature/current tuning of diode lasers: An undergraduate experiment," *Am. J. Phys.* **66**, 702 (1998).

37. K. G. Libbrecht, R. A. Boyd, P. A. Willems, T. L. Gustavson, and D. K. Kim, "Teaching physics with 670 nm diode lasers – construction of stabilized lasers and lithium cells," *Am. J. Phys.* **64**, 1109 (1996).

38. C. J. Hawthorn, K. P. Weber, and R. E. Scholten, "Littrow configuration tunable external cavity diode laser with fixed direction output beam," *Rev. Sci. Instrum.* **72**, 4477 (2001).

39. S. D. Saliba and R. E. Scholten, "Littrow configuration tunable external cavity diode laser with fixed direction output beam," *Appl. Optics* **48**, 6965 (2009).

40. W. D. Phillips, "Laser cooling and trapping of atoms," *Rev. Mod. Phys.* **70**, 721 (1998).

41. A. J. Leggett, "Bose-Einstein condensation in the alkali gases: Some fundamental concepts," *Rev. Mod. Phys.* **73**, 307 (2001).

42. J. Faist, F. Capasso, D. L. Sivco, C. Sirtori, A. L. Hutchinson, and A. Y. Cho, "Quantum cascade laser," *Science* **264**, 553 (1994).

43. R. F. Curl, F. Capasso, C. Gmachl, A. A. Kosterev, B. McManus, R. Lewicki, M. Pusharsky, G. Wysocki, and F. K. Tittel, "Quantum cascade lasers in chemical physics," *Chem. Phys. Lett.* **487**, 1 (2010).

44. T. C. Marshall, *Free Electron Lasers*, Macmillan, New York, 1985.

45. C. A. Brau, *Free Electron Lasers*, Academic Press, Boston, 1990.

5

Nonlinear Optics

Introduction

Prior to the development of the pulsed laser, scientists were concerned mainly with light–matter interaction phenomena resulting from the excitation, scattering, or ionization of matter by "single photons." We call this "linear optics" in which the excitation or ionization, etc., is linearly proportional to the light intensity. However, when the electric field associated with the electromagnetic radiation approaches that of the electric field within an atom, the light–matter interaction can become "nonlinear." The simultaneous absorption of two photons was first described in the doctoral thesis of Maria Goeppert-Mayer in 1931.[1] The two-photon excitation is proportional to the square of the light intensity, hence it is a nonlinear process. Parenthetically, in 1960 Goeppert-Mayer, together with J. H. D. Jensen and Eugene Wigner, received the Nobel Prize in Physics for the development of the shell structure of the nucleus.

To further discuss nonlinear processes let us first estimate the electric field within an atom and compare this with the electric field intensity of a light wave. As a first approximation to the electric field within an atom, consider the ground state of a hydrogen atom. Taking the proton as a point charge with $Q = +1.6 \times 10^{19}$ C, the electronic charge is smeared out into a charge distribution around the center of the proton as

$$\rho(r) = -\frac{Q}{\pi a_0^3} e^{-\frac{2r}{a_0}} \tag{5.1}$$

where a_0 is the Bohr radius of the electron in a hydrogen atom and is equal to 0.529×10^{-10} m. To simplify the problem to obtain an order of magnitude for the electric field, let us assume a point charge for the electron as well as the proton. With this assumption the electric field halfway between the electron and proton can be calculated from Coulomb's Law ($F = Q^2/(4\pi\varepsilon_0 r^2)$) simply as

$$E = \left(9 \times 10^9 \, \text{Nm}^2/\text{C}\right)(2)\left(1.6 \times 10^{-19} \text{C}\right) \Big/ \left(0.529/\left(2 \times 10^{-10} \, \text{m}\right)^2\right) = 4.1 \times 10^{12} \, \text{V/m} \tag{5.2}$$

By convention the direction of the field is that of the path of a positive charge placed at the point of interest. If we use the fact that the electron charge is quantum mechanically "smeared out" around the Bohr radius, and the electric field

Laser Experiments for Chemistry and Physics. First Edition. Robert N. Compton and Michael A. Duncan.
© Robert N. Compton and Michael A. Duncan 2016. Published in 2016 by Oxford University Press.

extends radially outward from the proton, we can obtain a better estimate of the field between the electron and proton. From Gauss' law the integral of the electric field over an area A is equal to the charge enclosed within the area divided by ε_0, i.e. $\int E \cdot dA = Q_{enclosed}/\varepsilon_0$. Integrating the charge distribution contained within r in equation 5.1 one finds that the simple point charge approximation in equation 5.2 is reduced by the factor $[2(r/a_0)^2 + 2 (r/a_0) + 1]e^{-2r/a}$ which for $r = a_0/2$ equals 0.92. Thus we can take $\sim 10^{12}$ V/m as a qualitative estimate of the electric field within an atom or molecule.

The electromagnetic theory of light is used to estimate the electric field associated with a laser pulse. The rate of energy flow per unit area in a plane electromagnetic wave is given by the *Poynting Vector*, S, in terms of the electric and magnetic fields E and B, respectively, as

$$S = (E \times B)/\mu_0 \tag{5.3}$$

Where μ_0 is the magnetic permeability of free space defined by $\mu_0 \varepsilon_0 = c^2$. Using $E = c\,B$ we can write

$$S = E^2/\mu_0 c \tag{5.4}$$

If S is in watts/m^2 the maximum electric field of the light pulse is given by

$$E = \sqrt{(\mu_0 c\, S)} = \sqrt{\left((10^3 \text{ weber/amp-m})(3 \times 10^8 \text{ m/s})(S)\right)} = 5.48 \times 10^5 \sqrt{S} \text{ V/m} \tag{5.5}$$

To estimate the electric field in a laser pulse we must be able to first determine the power density, S, at say the focal point of a laser beam. We first consider a perfectly coherent Gaussian wave from a laser beam of diameter D and wavelength λ focused to the diffraction limit by a perfect lens (no aberration) of focal length f. Under these conditions the diameter, d, of the beam waist at the focus can be calculated from

$$d = f\lambda/\pi D \tag{5.6}$$

Most laser beams used in the experiments in this book are multimode and exhibit a finite beam divergence. Under these circumstances the approximate focal spot diameter reduces to

$$d = f\theta \tag{5.7}$$

where f is the focal length of the lens and θ is the laser beam divergence in radians. Generally, the laser manufacturer includes the divergence in their description of the laser. If not, it is easily estimated by projecting the laser on a movable card and measuring the increase in the area of the beam with distance from the laser.

The typical divergence of a dye laser is $\sim$ 0.5 mrad. In this case the focal spot size for a lens with a 5 cm focal length would be $\sim$ 0.0025 cm (25 μm).

Another important parameter of the focused laser beam is its confocal length, b, which describes the effective distance (in the propagation direction) over which the focused laser beam exhibits a maximum intensity:

$$b = \pi d^2/(2\lambda) \tag{5.8}$$

As an example, let us estimate the power density delivered by a pulsed laser for typical powers employed in the experiments described in this book. A single 10 nsec laser pulse containing 10 mJ of energy has 1 MW of peak power. The power density at the focal spot of a 10 cm focal length lens for a laser beam having a divergence of 0.5 mrad would be 5×10^{10} W/cm^2 or 5×10^{14} W/m^2. Using S = 5×10^{14} W/m^2 in equation 5.5 gives 1.2×10^{13} V/m which is on the order of the electric field within a molecule.

A detailed description of nonlinear optical processes can be found in a number of classic textbooks.[2–6] For our purposes we consider the simplest interpretation. Intense laser light interacting with matter will induce a dipole moment or polarization of the medium. The polarization per unit volume of a nonlinear medium, P(t), can be related to the driving electric field, E(t), through the linear, $\chi^{(1)}$, and nonlinear, $\chi^{(2)}$, $\chi^{(3)}$, etc. susceptibilities by a Taylor series expansion

$$P(t) = \chi^{(1)}E(t) + \chi^{(2)}E^2(t) + \chi^{(3)}E^3(t) + \cdots \cdots \tag{5.9}$$

It is assumed here that the medium responds instantaneously to the applied field, there are no dispersion or losses, and the incident light is not in resonance with any electronic states of the medium. For simplicity we assume the electric vector of the light at frequency ω is given by

$$E(t) = E_0 \sin \omega t \tag{5.10}$$

Substitution of this into equation 5.6 gives

$$P(t) = \chi^{(1)}E(t) + \chi^{(2)}(E_0)^2 \sin 2\omega t + \chi^{(3)}(E_0)^3 \sin^3 \omega t \tag{5.11}$$

Using the trigonometric identities $\sin^2 x = (1/2)(1- \cos 2x)$ and $\sin^3 x = (3/4) \sin x - (1/4)\sin 3x$ in equation 5.11 results in

$$P(t) = \chi^{(1)}E_0\sin \omega t + \chi^{(2)}E_0^2(1/2)(1- \cos 2\omega t) + \chi^{(3)}(1/4)E_0^3(3 \sin \omega t - \sin 3\omega t) + \cdots \tag{5.12}$$

The first term gives rise to a re-radiated wave at the same frequency as the incident light and traveling with a velocity, v = c/n, where *n* is the index of refraction of the medium and *c* is the velocity of light in a vacuum. The second term results in a polarization oscillation at twice the incident frequency. Thus the re-radiated light has a small component at twice the incident laser frequency. This process is given the name second harmonic generation or SHG. Also, note that an even smaller component of the re-radiated beam emerges at three times the frequency of the incident light. This is called third harmonic generation or THG.

In cases where multiple lasers are co-propagating in a nonlinear dielectric medium, one can also produce beams which possess sum and difference frequencies of the incident beams. For illustration we consider the *mixing* of two different beams of frequency ω_1 and ω_2. The electric vector of the light then becomes

$$E(t) = E_{01} \sin \omega_1 t + E_{02} \sin \omega_2 t \qquad (5.13)$$

Substitution of this into equation 5.9 and considering only the contributions involving second order ($\chi^{(2)}$) terms gives

$$P^2(t) = \chi^{(2)} [E_{01}^2 \sin^2 \omega_1 t + E_{02}^2 \sin^2 \omega_2 t + 2 E_{01} E_{02} \sin \omega_1 t \sin \omega_2 t] \qquad (5.14)$$

Again, using the trigonometric identities described above the first two terms result in second harmonic generation at $2\omega_1$ and $2\omega_2$ as expected from above and using the identity $\sin x \sin y = 1/2 [\cos (x - y) - \cos (x + y)]$ gives

$$P^2(t) = \chi^{(2)} \left[E_{01}^2 \left(\tfrac{1}{2} \right) (1 - \cos\, 2\omega_1 t) + E_{02}^2 \left(\tfrac{1}{2} \right) (1 - \cos\, 2\omega_2 t) \right.$$
$$\left. + 2 E_{01} E_{02} \left(\tfrac{1}{2} \right) (\cos (\omega_1 - \omega_2)t - \cos(\omega_1 + \omega_2)t) \right] \qquad (5.15)$$

The product $(\sin \omega_1 t)(\sin \omega_2 t)$ corresponds to sum $(\omega_1 + \omega_2)$ and difference $(\omega_1 - \omega_2)$ frequencies. Thus the polarization contains frequencies of $2\omega_1$, $2\omega_2$, $\omega_1 + \omega_2$, $\omega_1 - \omega_2$) and a term which is constant, $\omega = 0$. The $\omega = 0$ term corresponds to a DC polarization and was called *optical rectification* by its discoverers.[7] This method of wave mixing to produce sum and difference frequencies is the basis for optical parametric oscillators (OPO; see Chapter 4) which can deliver widely tunable laser light covering the IR to the UV region of the spectrum.[8]

Second harmonic generation, SHG

The first nonlinear optics investigation was the observation of second harmonic generation from a quartz plate by Franken et al. in 1961.[9] Upon passing a ruby laser beam (6943 Å) through crystalline quartz they observed the faint trace of a beam on a photographic plate at 3472 Å which was dispersed from the strong 6943 Å beam. The editing process at *Physical Review Letters*, thinking the faint trace was a smudge, erased the image in the journal publication. Fortunately an arrow was pointed toward the "spot" and, since this experiment could be easily demonstrated by anyone possessing a laser of sufficient intensity, an erratum to the article was not necessary.

For SHG production in crystals the inversion symmetry property of the medium is important. It is easy to see from equation 5.8 that for a material which possesses a center of inversion, i.e. is centrosymmetric, the second-order suscep-tibility must be identically zero in the electric dipole approximation. Considering only $P^2(t)$ from equation 5.9

$$P^2(t) = \chi^{(2)} E^2(t) \qquad (5.16)$$

Reversing the direction of the electric field (i.e., $E \rightarrow -E$) for a medium pos-sessing inversion symmetry also reverses the polarization and therefore $P^2(t) = -P^2(t)$ which requires that $P^2(t) \equiv 0$. Thus $\chi^{(2)}$ must be identically zero for a centrosymmetric crystal in the electric–dipole approximation. For example, sec-ond harmonic generation will *not* occur in an isotropic medium such as pyrex glass. Thus, no even harmonics can be produced in media which possess a center of symmetry (inversion center).[10] The production of efficient SHG in non-centrosymmetric crystals must also take into account the phase matching of the pump ω and second harmonic 2ω waves. If the wavelength of the light is λ_0 then the wave vectors of the pump and SHG waves are $k(\omega) = 2\pi n_\omega/\lambda_0$ and $k(2\omega) = 4\pi n_{2\omega}/\lambda_0$. Because of the phase difference between the two waves they are only in phase over a certain length called the coherence length, L_c. Quickly following the publication by Franken et al. [9] which introduced SHG, Maker et al. [11] demonstrated the effects of dispersion and focusing on SHG. They considered a plane wave of frequency ω traveling in a medium with an index of refraction n_ω giving rise to a second harmonic wave of frequency 2ω with an index of refraction $n_{2\omega}$. The SHG intensity is given by

$$S(2\omega) = 2\pi c P^2 (k(2\omega)/\Delta k)^2 \sin^2 \Delta kx \qquad (5.17)$$

where c is the speed of light and $\Delta k = k(2\omega) - k(\omega) = 4\pi n_{2\omega}/\lambda_0 - 2\pi n_\omega/\lambda_0$.

Maker et al. [11] passed a ruby laser beam through a quartz platelet and the SHG light was recorded photoelectrically in the setup shown in Figure 5.1.

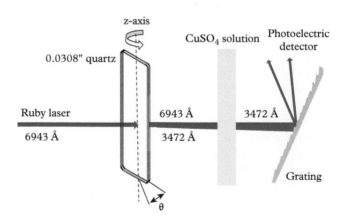

Figure 5.1 *Experimental arrangement described by Maker et al. [11] used to demonstrate phase matching in second harmonic generation. The* $CuSO_4$ *solution was employed to block the xenon lamp radiation from the ruby laser.*

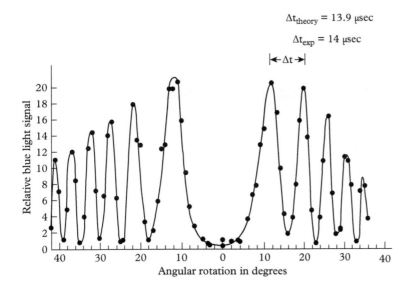

$\Delta t_{\text{theory}} = 13.9 \ \mu\text{sec}$

$\Delta t_{\text{exp}} = 14 \ \mu\text{sec}$

Figure 5.2 *Relative intensity of the second harmonic (2ω, 3472 Å) light from a ruby laser (ω, 6943 Å) passing through a quartz platelet with the z-axis parallel to the face being the axis of rotation through angle θ (Reproduced from [11].)*

The 0.0308″ thick quartz platelet was rotated through an angle θ about the z-axis thereby increasing the thickness of the platelet resulting in the data shown in Figure 5.2.

From equation 5.17 the SHG signal I_{SHG} is given by

$$I_{\text{SHG}} \approx \sin^2\{2\pi(n_\omega - n_{2\omega})t/\lambda_0\}/[n_\omega - n_{2\omega}]^2 \tag{5.18}$$

The peaks in Figure 5.2 result when the second harmonic wavelets interfere constructively. Using the trigonometric identity $\sin^2 x = \frac{1}{2}(1 - \cos 2x)$, the function described by equation (5.17) has a maximum value when

$$x_{\text{max}} \equiv L_c = (\lambda_0/4) \ |\lambda_0/n_\omega - n_{2\omega}| \tag{5.19}$$

The intensity of SHG is at maximum when the thickness $t = \frac{1}{4}(\lambda_0/|n_\omega - n_{2\omega}|)$, the coherence length, L_c. Phase matching is accomplished by rotating the sample; as the sample is rotated the thickness of the crystal, t, increases giving rise to the so-called "Maker fringes" shown in Figure 5.2.

A simple experiment demonstrating second harmonic generation and Maker fringes such as that shown in Figure 5.1 can be carried out in an undergraduate laboratory that has a reasonably intense pulsed laser. A thin quartz crystal can be used as above. However, we have also observed strong SHG and many "Maker fringes" from "home grown" crystals of sodium chlorate. Information on the growth and properties of sodium chlorate crystals can be found in [12]. The use of the fundamental of a Nd:YAG laser at 1064 nm produces an easily observable 532 nm green beam. The use of the 532 nm second harmonic of the Nd:YAG is safer but the wavelength of the SHG would then be 266 nm. The 1064 nm line cannot be seen and therefore its use in the laboratory can be very dangerous to the eye (*use with extreme caution*). Laser safety is discussed in Chapter 6.

Third harmonic generation, THG

Returning to equation 5.12, we see that the third-order polarization per unit volume of a nonlinear medium is

$$P^3(t) = \chi^{(3)} \left(\tfrac{1}{4}\right) E_0^3 (3 \sin \omega t - \sin 3\omega t) \qquad (5.20)$$

The first term $\left(\tfrac{1}{2}\right)\chi^{(3)} E_0^3 (3\sin \omega t)$ gives rise to a nonlinear contribution to the real part of the refractive index, i.e. $n = n_0 + n_2 I$, where I is the intensity of the incident wave $I = (12\pi^2/n_0^2 c)\ \chi^{(3)}$. The intensity dependent, nonlinear refractive index results in a self-focusing of the laser beam (see [2] and [3] for an excellent discussion of self-focusing). If n_2 is positive, a laser beam passing through a medium will focus into the medium or outside of the medium. The index is greater at the center of the beam and it will be diffracted to a point inside or outside the medium depending upon the intensity of the laser and the length of the material in the beam direction.

The second term in equation 5.20 describes the generation of a third harmonic wave at 3ω. This is illustrated in Figure 5.3 for xenon in which three photons give rise to one photon at three times the energy of each incident photon.

It should be noted that $\chi^{(3)}$ is generally 10^7 times smaller than $\chi^{(2)}$ and therefore THG intensities are considerably smaller than SHG. Experiments involving THG to produce tunable VUV light are described in Chapter 26. In the case of gases, if the energies of any three of the photons are near an electronic state of the atom or molecule, resonance enhancement of the THG will occur. However, in certain cases if the three photons generate THG which is out of phase with the three-photon excitation of an electronic state, the THG will cancel the three-photon excitation of the state as discussed below.

Multiphoton excitation and multiphoton ionization spectroscopy

Multiphoton excitation and multiphoton ionization (MPI) processes are special cases of nonlinear optics. Multiphoton excitation involves the collective interaction of a number of photons with an atom or molecule. If the n photons are in resonance with an excited state of the atom or molecule, the rate of absorption is proportional to I^n. If further photon absorption, e.g., m photons, leads to ionization (multiphoton ionization) the ionization signal will be proportional to I^{n+m}. However, when the laser intensity is high, the ionization signal

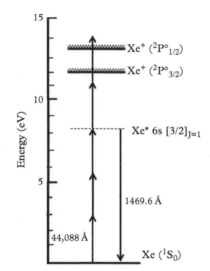

Figure 5.3 *Illustration of third harmonic generation in xenon. The three photons are shown to be in resonance with the $6s[3/2]_{J=1}$ state. However, THG only occurs on the blue side (shorter wavelength) of the $6s[3/2]_{J=1}$ state.*

may be saturated and the power law dependence may be smaller than $n + m$. If the laser is tightly focused the ionization may be determined by the beam waist of the focused laser and the power dependence is closer to $I^{3/2}$. In Chapters 16 and 18, experiments are presented in which multiphoton ionization of alkali atoms and iodine molecules are described using tightly focused laser beams. Multiphoton ionization of xenon can be illustrated by the three-photon resonant five-photon ionization scheme shown in Figure 5.3. In this case three photons are resonant with the $6s(3/2)_{J=1}$ excited state from the $^1S_0$ ground state of Xe. Absorption of two more photons produces ionization into the continuum of both fine-structure levels $Xe^+(^2P^0_{1/2})$ and $Xe^+(^2P^0_{3/2})$. This MPI process is labeled a three-plus-two resonantly enhanced multiphoton ionization or [3+2] REMPI process. MPI of many gases such as xenon have been studied employing the simple experimental setups described in Chapters 16 and 18. Such MPI experiments can be carried out even with low-power nitrogen laser-pumped dye lasers.

The [3+2] REMPI of xenon produces both fine-structure levels of the xenon ion. This results in electrons of two different energies. Many researchers have determined the energies of MPI electrons using energy analyzers or time-of-flight methods to record what is called multiphoton ionization photoelectron spectroscopy or MPI-PES (see [13] for a review of this field). In addition, the ability to rotate the linear polarization of a laser beam allows one to measure MPI angular distributions of energy-resolved photoelectrons. The MPI-PES spectrum for molecules often provides important vibrational and sometimes rotational information as well.

We have used the xenon atom to illustrate MPI; however, xenon is also a special case which can serve to illustrate an important connection between MPI and harmonic generation in a dispersive gas. MPI of xenon at low pressure [14] shows a strong enhancement for the case of three photons tuned to the $6s[3/2]_{J=1}$ resonance, as shown in Figure 5.3. However, for pressures greater than ~ 0.3 Torr it was found that the MPI through the $6s[3/2]_{J=1}$ disappears as the third harmonic light appears.[15] It was clear from this that THG and MPI were in competition. Glownia and Sander [16] provided further convincing evidence for the competition between multiphoton excitation and THG by reflecting circularly polarized light (CPL) into the ionization chamber resulting in restoration of the [3+2] REMPI. THG is not possible with CPL. Reflection was necessary to obey parity in the three-photon excitation. Jackson and Wynne [17] explained these two interesting observations by showing that the disappearance of the MPI is due to an *interference* between the two coherent pathways to the $6s[3/2]_{J=1}$ state: the three-photon excitation driven by the electric field at frequency ω and a one-photon excitation driven by the electric field at the third harmonic frequency 3ω. The three-photon and one-photon excitation have equal amplitudes and *opposite* signs resulting in the cancellation of the excitation and subsequent ionization. Their theoretical analysis was confirmed by Jackson and Wynne [17] who showed that reflecting the linearly polarized laser beam back upon itself with a mirror restores the MPI. Absorption of two photons in one direction and one from the opposite direction (reflected beam) produces excitation of the $6s[3/2]_{J=1}$ state without the accompanying THG and the ionization reappears. Such interference effects have been reported in all of the rare gases,[18] other atoms, and many molecules as well.

Nonlinear optical rotary dispersion

Experiments on optical rotary dispersion (ORD) of chiral liquids are discussed in Chapter 31 (Inversion of Sucrose by Acid Catalyzed Hydrolysis) and Chapter 20 (Optical Rotary Dispersion of a Chiral Liquid (α-pinene)). These experiments were performed under single photon collision conditions. Using relatively intense pulsed lasers it is possible to investigate nonlinear effects on CD and ORD.

Optical rotation is defined as the rotation of the plane of polarization of linearly polarized light as it propagates through a medium and was first discovered in quartz crystals by the French physicist Dominique Arago. Soon after, Biot and others noted that various natural substances in liquid form also exhibit optical rotation. This led to the discovery of molecular dissymmetry or optical activity in molecules by Pasteur. Since that time, much effort has been directed towards the theoretical and experimental investigation of single-photon optical rotary dispersion (variation of the rotation with wavelength), as well as to single-photon circular dichroism (CD), which is a differential absorption of right- and left-circularly polarized light. Optical rotation and CD are related through the Kramers–Kronig relations.[19,20] Multiphoton CD can be easily understood as a difference in absorption of two or more photons for right- and left-circularly polarized light. The Kramers–Kronig relationship can then be used to obtain a multiphoton optical activity from the multiphoton CD. Until recently, very few studies have been reported in the area of multiphoton optical activity. Optical activity refers to both ORD and CD. Theoretical treatments of nonlinear two-photon excitation have appeared [21–23] as have two experiments describing nonlinear circular dichroism in multiphoton ionization of 3-methylcyclopentanone.[24,25] Below we focus on nonlinear ORD effects.

The plane of polarization of a linearly polarized light beam will be rotated to the right (dextrorotary) or left (levorotary) to an observer looking directly towards the light source. The magnitude of the rotation, Φ, in radians can be expressed by one of Fresnel's equations.[26]

$$\Phi = (n_\ell - n_r)l/\lambda \qquad (5.21)$$

where, n_ℓ and n_r are the indices of refraction for left- and right-circularly polarized light, respectively, λ is the wavelength of the incident light, and l is the path length.

Two experiments have been reported that describe two-photon [27] and three-photon [28] optical rotation effects. In the experiment of Cameron and Tabisz,[27] shown in Figure 5.4, optical rotation of a linearly polarized beam of

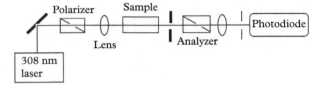

Figure 5.4 *Geometry of the experiment by Cameron and Tabisz [27] used to study the two-photon optical rotation of uridine and sucrose.*

a 308 nm excimer laser beam passing through uridine and sucrose was studied as a function of laser power. In uridine, a specific two-photon rotation of -0.7 ± 0.1 deg was measured for a 50 mW laser beam. For sucrose at the same power level the specific two-photon rotation was found to be -0.16 ± 0.05 deg. This corresponds to a two-photon rotation of -0.013 ± 0.002 deg/mW for uridine and $-3.2 \times 10^{-3} \pm 9 \times 10^{-4}$ deg/mW for sucrose.

Gedanken and Tamir [28] employed a similar setup to measure the optical rotation at specific wavelengths (ORD) for camphorsulfonic acid using a dye laser as shown in Figure 5.5. Rotations were measured with and without the focusing lens for both + and − enantiomers of camphorsulfonic acid. Typical results for rotation were 5.22 deg without the lens and 4.41 deg with the lens, for a difference of −0.8 deg at 650 μJ pulse energy for the (+) enantiomer. The (−) enantiomer gave +0.77 difference. A plot of ln(rotation) versus ln(laser power) gave a slope of 2.9 ± 0.1, confirming the three-photon nature of the nonlinear process.

Optical activity in crystals is somewhat different than optical activity in solutions. In solutions, the rotation is a direct result of dissymmetry in the molecule, i.e., the mirror images of the molecules are non-superimposable. However, the rotation in crystals is not due to molecular dissymmetry, but rather the molecules that make up the crystal are arranged in such a fashion as to exhibit a specific handedness, or helicity. Thus, it is not necessary that the molecules making up the crystal be chiral. Such is the case for $NaClO_3$ and $NaBrO_3$ crystals.

Crystals of $NaClO_3$ and $NaBrO_3$ have good optical quality, i.e., they transmit from 220 nm to the near-infrared, and have a high transmission exponent. The crystals used for this experiment were grown from aqueous solutions of the salts. Interestingly, $NaClO_3$ and $NaBrO_3$ have the same crystal structure but exhibit opposite rotations in the visible region of the spectrum.[29] To record an ORD

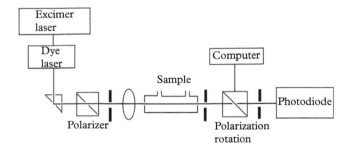

Figure 5.5 *Geometry of the experiment by Gedanken and Tamir [28] used to study the three-photon ORD of camphorsulfonic acid.*

Figure 5.6 *Experiment to measure the ORD of chiral samples such as NaClO₃. The light source and spectrometer can be replaced by a tunable dye laser. The second polarizer is rotated relative to the first to observe the optical rotation.*

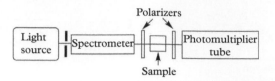

curve for $NaClO_3$ and $NaBrO_3$, a simple optical spectroscopic method can be devised, as shown in Figure 5.6.

A Fibre-Lite model 3100 cw lamp was used for the incident light source. A 2-mm aperture limited the measuring light beam size and improved beam quality. A $\frac{1}{4}$-meter monochromator was employed to select the wavelength of light. A polarizer (Glan prism) was used for polarizing the incident light beam, while a second polarizer served as the analyzer. The sample was placed on a rotatable plate which could be heated to 100 °C. The thicknesses of the samples of l-$NaClO_3$ and d-$NaBrO_3$ were 3.28 and 9.0 mm, respectively.

The first polarizer was fixed while P2 was rotated until the transmitted light intensity reached a minimum value. Figure 5.7 shows a typical plot of intensity versus angle between the polarizers. The difference in angles between the polarizers at the specific wavelengths from 470 to 650 nm were recorded and used to construct an optical rotatory dispersion curve. The ORD curves for d-$NaClO_3$ and l-$NaBrO_3$ in the wavelength range of 470 to 650 nm were fit to the Drude equation

$$\Phi(\lambda) = A/(\lambda^2 - \lambda_0^2) \tag{5.22}$$

where A and λ_0 are constants and λ is the wavelength. Figure 5.8 shows the measured ORD for both crystals. The yellow line represents the fit to the experimental data of Chandrasekhar and Madhava [30] represented by the equation

$$\Phi(\lambda) = \frac{1.2528}{(\lambda^2 - (0.100)^2)} - \frac{0.1546}{(\lambda^2 - (0.210)^2)} \tag{5.23}$$

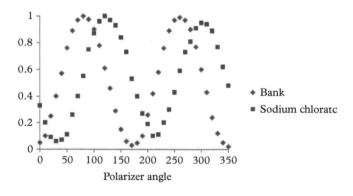

Figure 5.7 *Typical data recorded for blank polarizer and sodium chlorate crystal. Maxima and minima were found by fitting data to a quadratic function at the peaks and minima.*

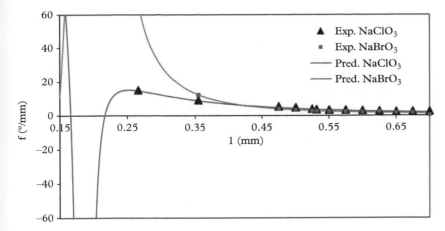

Figure 5.8 *Experimental ORD measurements for sodium chlorate, NaClO₃, and sodium bromate, NaBrO₃, at low laser power. Overlayed on the figure are the experimental fits (equations 5.23 and 5.24) from Chandrasekhar and Madhava [29] and Kizel et al. [30].*

The green curve represents a fit to previous data by Kizel et al. [18] for the same conformation of sodium bromate.

$$\Phi(\lambda) = \frac{1.449}{(\lambda^2 - (0.100)^2)} - \frac{1.887}{(\lambda^2 - (0.224)^2)} \quad (5.24)$$

The Drude equation essentially describes the ORD curve for wavelengths that are outside a single absorption region. To obtain the constants for each crystal, a plot was made of $1/\Phi$ versus λ^2 and a linear least squares fit was performed. For NaClO₃, the constants were found to be $A = 1.11 \times 10^6$ and $\lambda_0 = 86.2$. For NaBrO₃, $A = 5.00 \times 10^5$, while $\lambda_0 = 316.0$. From Figure 5.7, it can be seen that in this particular wavelength region, the optical rotation of the l-NaClO₃ crystal ranged from 2.58 ± 0.89 deg/mm to 4.78 ± 0.89 deg/mm, while the rotation of the d-NaBrO₃ ranged from -1.56 ± 0.05 to -4.13 ± 0.05 deg/mm. It is interesting to note that both optical and anomalous X-ray diffraction studies show that NaClO₃ and NaBrO₃ crystals which have identical handedness rotate the plane of polarized light in the opposite sense.[31] Unlike quartz, these crystals are also isotropic.

Using the 1064 nm Nd:YAG laser and its harmonics, 532 and 355 nm, we have found nonlinear optical rotation effects for both sodium chlorate and sodium bromate crystals. Rather simple experiments using focused and unfocused laser light reveal both positive and negative changes in the optical rotation depending upon the crystal and the wavelength. Notice the divergence of the yellow and green line curves in Figure 5.8 at short wavelength to see why the rotations are expected to change due to multiphoton effects. A plot of ln (change in angle) versus ln (laser power) gives the order on nonlinearity from the slope of this line.

....

REFERENCES

1. M. Goeppert-Mayer, "Über Elementarakte mit zwei Quantensprüngen," *Annals Phys.* **9**, 273 (1931).

2. R. W. Boyd, *Nonlinear Optics*, third edition, Academic Press, San Diego, 2008.

3. Y. R. Shen, *The Principles of Nonlinear Optics*, John Wiley & Sons, New York, 1964.

4. N. Bloembergen, *Nonlinear Optics*, fourth edition, World Scientific, London, 1996.

5. D. C. Hanna, M. A. Yuratich, and D. Cotter, *Nonlinear Optics of Free Atoms and Molecules*, Springer-Verlag, Berlin, 1979.

6. P. N. Butcher and D. Cotter, *The Elements of Nonlinear Optics*, Cambridge University Press, Cambridge, UK,1991.

7. M. Bass, P. A. Franken, J. F. Ward, and G. Weinreich, "Optical rectification," *Phys. Rev. Lett.* **9**, 446 (1962).

8. J. A. Giordaine and R. C. Miller, "Tunable coherent parametric oscillation in $LiNbO_3$ at optical frequencies," *Phys. Rev. Lett.* **14**, 973 (1965).

9. P. A. Franken, A. E. Hill, C. W. Peters, and G. Weinreich, "Generation of optical harmonics," *Phys. Rev. Lett.* **7**, 118 (1961).

10. SHG has been observed in the case of centrosymmetric crystals made of chiral molecules (E. W. Meijer, E. E. Havinga, and G. L. J. A. Rikken, *Phys. Rev. Lett.* **65**, 37 (1990)) which is ascribed to transitions that are *both* magnetic and electric dipole allowed.

11. P. D. Maker, R. W. Terhune, M. Nisenoff, and C. M. Savage, "Effects of dispersion and focusing on the production of optical harmonics," *Phys. Rev. Lett.* **8**, 21 (1962).

12. R. M. Pagni and R. N. Compton, "Asymmetric synthesis of optically active sodium chlorate and bromate crystals," *Cryst. Growth Des.* **2**, 249 (2002).

13. R. N. Compton and J. C. Miller, "Multiphoton ionization photoelectron spectroscopy: MPI-PES," in *Laser Applications in Physical Chemistry*, ed. by D. K. Evans, Marcel Dekker, Inc., New York, 1989.

14. R. N. Compton, J. C. Miller, A. E. Carter, and P. Kruit, "Resonantly enhanced multiphoton ionization of xenon: Photoelectron energy analysis," *Chem. Phys. Lett.* **71**, 87 (1980).

15. J. C. Miller, R. N. Compton, M. G. Payne, and W. R. Garrett, "Resonantly enhanced multiphoton ionization and third harmonic generation in xenon gas," *Phys. Rev. Lett.* **45**, 114 (1980).

16. J. H. Glownia and R. K. Sander, "Experimental evidence for the competition between resonantly enhanced multiphoton ionization and third harmonic generation in xenon," *Phys. Rev. Lett.* **49**, 21 (1982).

17. D. L. Jackson and J. J. Wynne, "Interference effects between different optical harmonics," *Phys. Rev. Lett.* **49**, 543 (1982); D. L. Jackson, J. J. Wynne, and P. H. Kes, "Resonance-enhanced multiphoton ionization: Interference effects due to harmonic generation," *Phys. Rev. A* **28**, 781 (1983).

18. J. C. Miller and R. N. Compton, "Third harmonic generation and multiphoton ionization in rare gases," *Phys. Rev. A* **25**, 2056 (1982).

19. H. A. Kramers, "La diffusion de la lumiere par les atomes," *Atti Cong. Intern. Fisici, (Transactions of Volta Centenary Congress) Como* **2**, 545 (1927).

20. R. de L. Kronig, "On the theory of the dispersion of X-rays," *J. Opt. Soc. Am.* **12**, 547 (1927).

21. I. Tinoco, Jr. "Two-photon circular dichroism," *J. Chem. Phys.* **62**, 1006 (1975).

22. E. A. Power, "Two-photon circular dichroism," *J. Chem. Phys.* **63**, 1348 (1975).

23. A. Rizzo, B. Jansík, T. B. Pedersen, and H. Agren, "Origin invariant approaches to the calculation of two-photon circular dichroism," *J. Chem. Phys.* **125**, 64113 (2006).

24. U. B. von Grafenstein and A. Bornschlegl, "Circular dichroism laser mass spectrometry: Differentiation of 3-methylcyclopentanone enantiomers," *Chem. Phys. Chem.* **7**, 2085 (2006).

25. R. Li, R. Sullivan, W. Al-Basheer, R. M. Pagni, and R. N. Compton, "Linear and non-linear circular dichroism of R-(+)–3-methylcyclopentanone," *J. Chem. Phys.* **125**, 144204 (2006).

26. See E. Hecht, *Optics*, fourth edition, Pearson/Addison Wesley, San Francisco, 2001.

27. R. Cameron and G. C. Tabisz, "Observation of two-photon optical rotation by molecules," *Molec. Phys.* **90**, 159 (1997).

28. A. Gedanken and M. Tamir, "Multiphoton optical rotary dispersion," *Rev. Sci. Instrum.* **58**, 950 (1987).

29. S. Chandrasekhar and M. Madhava, "Optical rotatory dispersion of crystals of sodium chlorate and sodium bromate," *Acta Cryst.* **23**, 911 (1967).

30. V. A. Kizel, Y. I. Krasilov, and V. N. Shamraev, "Study of the optical activity of the crystalline state I.," *Optics and Spectrosc.* **17**, 470 (1964).

31. S. C. Abrahams, A. M. Glass, and K. Nassau, "Crystal chirality and optical rotation sense in isomorphous $NaClO_3$ and $NaBrO_3$," *Solid State Commun.* **24**, 515 (1977).

6

Laser Safety

Introduction

Lasers are used in many areas of modern chemistry and physics, in optical and electronic instrumentation, to initiate chemical reactions and to study the progress of reactions in both laboratory and field environments. Lasers are also used by surgeons for a wide variety of applications from delicate repairs of tears in the retina to removal of plaque from the arteries. The most familiar applications of lasers are as light sources for spectroscopy, but new laser devices now provide more intense light in different regions of the spectrum, from the terahertz, to infrared, to visible and ultraviolet, and even to X-ray regions. Physical chemists use lasers for photodissociation and to follow the time-resolved evolution of fragments produced, but also organic chemists use UV excimer lasers to initiate photochemistry for synthesis. Pulsed lasers were also responsible for the first laboratory production of fullerene molecules which helped spawn the new area of nanomaterials. Excimer lasers are also used for photolithography, in which patterned metal films deposited by inorganic or organometallic complex decomposition are used to construct microelectronic circuits used in computers, cell phones, etc. Ultraviolet nitrogen lasers or excimers are used by biochemists and polymer chemists for neat laser desorption, or matrix assisted laser desorption ionization (MALDI), to obtain mass spectra of biomolecules and polymers. These short wavelength lasers are also commonly employed in LASIK (*in situ* keratomileusis), which is a common surgery used to correct vision in people who are nearsighted, farsighted, or have astigmatism.

Laser fluorescence lies at the heart of many imaging microscopes used in materials science and biology. In physics, lasers provide the basis for atom trapping as well as atom cooling. Atom cooling led to the demonstration of Bose–Einstein condensation. In atmospheric science, laser-based LIDAR instruments interrogate the atmosphere, probing for radicals such as hydroxyl as a function of altitude. Laser diodes and fiber-optic systems are employed for phone communication and digital data storage and retrieval. Finally, the National Ignition Facility is exploring the use of 192 giant pulsed laser beams to produce power from nuclear fusion. These applications of lasers are by no means complete, but serve to illustrate how pervasive these systems are throughout science and technology in the modern world. It is more important than ever to be educated in the capabilities of lasers and the issues of their safe operation.

Laser Experiments for Chemistry and Physics. First Edition. Robert N. Compton and Michael A. Duncan.
© Robert N. Compton and Michael A. Duncan 2016. Published in 2016 by Oxford University Press.

As the applications of lasers become more widespread, their usage is extending outside the core of physical scientists that have used these systems for many years and have experience in their safe operation. New users are by definition less experienced and likely less knowledgeable in the issues that affect laser safety. This is particularly true for students in undergraduate classes who encounter lasers and laser-based experiments for the first time. This chapter is designed to point out some of the most important concepts that affect laser safety in the lab, with the student in mind. For more extensive coverage of laser safety issues, the reader should consult any one of several courses available in on-site training or web-based platforms. Students planning to do laser experiments in the lab, as well as their professors and teaching assistants, should receive an appropriate local or online laser safety training course. The American National Standards Institute (ANSI), in coordination with the Laser Institute of America (LIA), has developed extensive laser safety standards and protocols. This information is available for purchase via their website at https://www.lia.org/store/ANSI+Z136+Standards. The Z136.1 document describes general safety issues, while a series of additional documents are available for more specific applications of lasers. These documents provide information that can be included in a course on laser safety, which is highly recommended for all laser users. Another excellent source of laser safety information is the Wikipedia site at http://en.wikipedia.org/wiki/Laser_safety.

Critical safety issues in laser operation

The most obvious issues regarding laser safety involve burns to the skin and especially the eye. These are serious concerns, but not the only ones. In addition to these, laser operators must be aware of the safety protocols for the chemicals used with lasers, of which there are several. The other critical area involves possible contact with high voltages that are present in almost every electronic circuit used for laser operation. We will touch on these latter areas first, before discussing tissue and eye concerns in more detail. The discussion below is by no means complete, but instead focuses on those areas most likely to be encountered using the lasers discussed in this book.

Chemical exposure

Not all lasers use dangerous chemicals, but those which do require special precautions. Of the gas lasers most commonly encountered in research labs, helium-neon, nitrogen, and CO_2 lasers use non-toxic gases and do not require any special gas handling. However, excimer lasers use mixtures containing hydrogen chloride (HCl) or molecular fluorine (F_2) gases. Both of these are highly corrosive. Fluorine in concentrated form is also an explosive hazard, subject to spontaneous detonation if left in metallic containers for extended periods.

Because of these issues, both are usually used in dilute mixtures (3–5%) in helium. Pure HCl or fluorine should not be used, as the gas mixture recipes used in excimer lasers are based on the dilute pre-made mixtures. In either case, the gas bottles used to fill the tank of an excimer laser should be situated in proper chemical gas cabinets with ventilation and/or pumping. If such a vacuum cabinet is not available, gas bottles may also be mounted in an appropriate rack situated within a standard chemical laboratory fume hood. The gas mixtures within excimer lasers have different useful lifetimes, depending on the rates of halogen loss by adsorption on the walls and other component surfaces in the cavity, and the production of volatile by-products of halogen chemistry, which absorb light at the excimer wavelengths and quench its emission. The lasers have been engineered over time to extend these lifetimes, but after a few weeks of operation the gas mixture must be purged out and replaced. When the gas is dumped, it is "scrubbed" by appropriate chemical filtering on the output of the laser gas tank, and the contents of this filter must be monitored and replaced as needed to maintain their activity. After it passes through such filters, the remaining gas has most of its halogen content removed, but it should be directed via sealed gas lines to an appropriate laboratory hood for final disposal.

A second area of chemical concern with laboratory lasers involves the solutions used in dye lasers. There are dozens of dye molecules used for this purpose, and these go by different names depending on the source from which they are purchased. All of these have the general characteristic of efficient light emission at visible wavelengths, which is accomplished in highly conjugated aromatic ring systems (see Chapter 4 section on dye lasers). Although the full dangers of exposure to each of these dye molecules are not completely established, especially regarding long-term chronic exposure, these systems are generally regarded to be carcinogens, and contact with skin or ingestion should be completely avoided. This is easily accomplished by wearing an appropriate lab coat and gloves during the preparation and handling of dye solutions. The solutions should be manipulated within secondary containment pans so that spillage, if any, can be contained and prevented from contaminating the work area. Laser dyes lose their activity over time because of photochemical degradation, especially when ultraviolet pump lasers are employed. Appropriate chemical waste practices must be used for collecting, storing, and disposing of expended dye solutions.

The main solvent used with laser dyes is methanol, which is of course flammable. An unusual solvent sometimes used in certain dye solutions is p-dioxane, which has special dangers for flammability. This solvent has a particularly low dielectric constant, and its circulation through insulating tubing, such as the Teflon most often used for dye circulation systems, leads to a large build-up of static charge. This charging can be great enough to cause arcing to nearby grounded metal surfaces that can puncture the tubing, causing leakage outside the tubing and ignition of this solvent in air from subsequent arcs. Many newer dye lasers have grounded wires inserted within the circulation tubing to remove the charge when this solvent is used.

High voltages

In every laser system used, energy has to be put into the lasing material to excite the atoms or molecules which emit light. In most laser systems this is accomplished either directly or indirectly using high-voltage discharges. In gas lasers (CO_2, nitrogen, argon ion, excimer), an electrical discharge excites the gas directly, using fast electron collisions to produce atomic and molecular species in their excited states. In solid-state lasers (Nd:YAG, Ti:sapphire, etc.), a discharge excites gases in a flashlamp (usually containing xenon), and the broadband output of this flashlamp excites the dopant ions in the lasing crystal. However, the high-voltage circuit needed for the flashlamp is very similar to that used to excite gas lasers. *The high voltages used in these applications are lethal!* The circuits used to apply high voltages to discharges or to flashlamps generally employ high-voltage capacitors, which store charge even when the power to the laser device is not turned on. Untrained personnel should therefore never open a power supply or laser head where high-voltage wires or circuit elements are exposed, whether or not the laser is turned on. In the event that such power supplies must be tested or adjusted in any way, make certain that the capacitors have been fully discharged, even if the main power has been unplugged.

Eye and skin damage

Because lasers represent a concentrated form of energy, they can burn human skin or eyes. While avoiding both of these is most desirable, the long-term effects of eye damage are far more serious and detrimental to quality of life. Skin burns may be painful and they may leave scars, but they will heal. Eye damage is permanent. We therefore focus our concerns here on issues regarding eye safety. The dangers to human eyes from laser radiation depend on the power/pulse energy of the laser, the rate of energy deposition, which can be quite different for pulsed versus continuous lasers, and on the wavelength of the laser radiation in question. Depending on these conditions, some lasers are relatively safe to operate and others are extremely dangerous. It is therefore essential that laser users develop an understanding of the danger involved with the specific laser system in use and then employ appropriate safety measures.

According to standards developed by ANSI, lasers are classified by their level of danger in terms of the *maximum permissible exposure* (MPE) that is safe for human eyes. Class I and II lasers include those which have such low power levels as to pose insignificant danger. Diode laser pointers were formerly in class II, but higher power levels from these have become available recently, placing some of these in class III or even class IV depending on the power level. Class V lasers are those which are completely enclosed so that their light, even though it may be high power, cannot pose any threat. Class III and IV lasers include many laboratory systems that may be encountered in a teaching or research lab. These Roman numeral classifications have recently been changed to English numbering, but many lasers are still labeled with the Roman numeral system. Class III lasers are generally those for which direct viewing of the beam is regarded as dangerous to human eyes, but diffuse reflected light is not a significant danger. This includes most low-powered diode lasers and helium-neon lasers. Class IV lasers, on the other hand, are highly dangerous, both in direct viewing and in the form of partial beam reflections and diffuse scattering. These include argon ion, CO_2, nitrogen, Nd:YAG, excimer, and dye lasers, as well as optical parametric oscillators. Experience and common sense indicate that invisible beams in the infrared or ultraviolet region of the spectrum are more dangerous than beams in the visible regions that can be seen with the human eye. Table 6.1 provides a more complete list of common laboratory lasers and their operating characteristics.

Class IV lasers are therefore more dangerous than class III lasers because of their higher power or pulse energies. However, it is important to emphasize that the greatest danger from such lasers is not from direct impact of the main beam. Most laboratory workers with basic common sense will not stare directly into a visible laser beam, and this kind of accident is not common.

Table 6.1 *Characteristics of typical laboratory lasers.*

Laser	P/CW	Wavelength(s)	Energy/power
He-Ne	CW	632.8 nm	1–10 mW
argon ion	CW	488 nm	0.1–20 W
Ti:sapphire	CW	far red	<10 W
N_2	P	337 nm	<10 mJ/pulse
excimer	P	308 (XeCl), 248 (KrF), 193 nm (ArF)	10–1000 mJ/p
Nd:YAG	P/CW	1064, 532, 355, 266 nm	10–3000 mJ/p or 0.1–20 W
dye	P/CW	visible, UV	<100 mJ/p or <10 W
CO_2	P/CW	10.6, 9.6 μ	0.1–10 J/p or 0.1–100 W
diode	CW	near IR, visible	mW–10 W
F-center	CW	IR (near 3 μ)	mW
OPO	P	IR, visible, UV	<100 mJ/pulse
FEL	P	IR, visible, UV	<100 mJ/pulse

The most common dangers result from *unanticipated* beams from invisible (IR or UV) lasers, or from accidental reflections of the main beam or its partial reflections. Unanticipated reflections can come from the surfaces of metal tools, jewelry, optical mounts, etc., that are inadvertently inserted into a working laser beam. It is important to note that laser mirrors are usually constructed of smooth shiny metal surfaces. Metal tools (wrenches, screw drivers, etc. used to adjust or tighten mounts), rings, watches, metal optical mounts, etc., therefore have the same kind of reflecting surfaces as mirrors, except that the surfaces are irregular and the direction of light reflection is unpredictable. Another common problem is partial reflectance of laser beams off optical surfaces that the user expects to be transmitting. Referring to Chapter 2 and the section on surface reflections, it can be seen that the surfaces of typical quartz or glass optics, such as prisms, lenses, windows on a sample holder, etc., reflect about 4% of the light at each interface. For a class IV laser, with high power or pulse energy, even reflections that are 4% (or 8% from two surfaces of a glass plate) of the main laser beam intensity are powerful enough to cause serious eye damage. The curved surfaces of lenses or the mounting angles of other optics may bounce such reflections upward off the laser table, or to unexpected sideways angles, potentially hitting users or other lab personnel in the face or eyes.

Pulsed lasers are far more dangerous than those whose light output is CW. For example, a typical argon ion laser might operate at a power of 1 W. The main beam would burn the eyes or skin, but reflections from optics would be much less likely to do so. However, a pulsed Nd:YAG laser might operate at 10 pulses per second, each with a pulse energy of 100 mJ, resulting in the same 1 W of average power. However, this is very misleading, because the laser is usually off. Each pulse lasts only 5 nsec, and the "on" time is then only 50 nsec out of every second. However, when it is on, during the 5 nsec of one pulse, the *peak* power is $(0.1 \text{ J})/(5 \times 10^{-9} \text{ sec})$, or 20 MW! Although the laser is only on for a short period, at this high peak power it can cause significant burn damage. Indeed, pulse energies 10 times less than this are employed to vaporize metal for the production of gas-phase metal atom clusters (see Chapter 12). Lasers used for undergrad lab experiments may have pulse energies of only 10–20 mJ/pulse, but even a 4% reflection of this beam is still enough to cause serious eye damage. Another aspect of pulsed laser usage is the glare that results from small amounts of scattered light from dirty optics, secondary reflections, etc., that tend to light up the whole laser area in pulsed flashing light. Even though there may be no direct hit into a user's eyes that would result in any discernable burns, the pulsed glare is well known to cause headaches to users, especially if this occurs in an otherwise darkened room. Laser labs are often darkened to make it easier to read oscilloscope screens or other electrical instrument displays. However, in such a darkened room, human eyes dilate to improve light collection, but they cannot respond on a fast timescale to reduce this heightened sensitivity when a nanosecond laser flash lights up the room. This can result in repeated brief moments of severe eye saturation. Unfortunately, there is no good documentation of the long-term effects of chronic exposure to such pulsed glare.

A special note of caution applies to pulsed lasers focused in air or on a surface, such as those used in Laser Induced Breakdown Spectroscopy studies (see Chapter 11). Such a pulsed focus laser can produce a bright plasma, containing a broad range of wavelengths, extending into the ultraviolet region and beyond. Such light has been shown to damage the pixels of a camera used to image these plasmas and it is possible that similar damage could occur to human eyes.

The use of short-wavelength lasers, such as the 355 nm third harmonic of a Nd:YAG laser, the 337 nm radiation from a nitrogen laser, or especially the UV output of an excimer laser, could pose other hidden dangers. UV light from the sun or that from a welding arc is well known to produce an increased probability of cataracts of the eye. Even indirect exposure to these UV lasers or their scattered light over the long term is likely to lead to similar problems. There are also debates over whether coherent radiation may be more dangerous than incoherent light in this regard. Also because Rayleigh scattering is proportional to the fourth power of the frequency of light, such scattering from the molecules in air may lead to significant UV exposure over time, increasing the risks of both cataracts and skin cancer.

Invisible lasers are far more dangerous than visible lasers, and the kind of eye damage they can do is also different. As noted above, direct impact accidents with visible lasers are not common because the user can see where the main beam is and can also track down accidental reflections or scattering. However, infrared or ultraviolet lasers, such as the beams from a CO_2 laser, Nd:YAG fundamental beam, or an excimer laser, are all invisible to the human eye. The main beam cannot be seen, nor can stray reflections from tools or optics. The human eye transmits not only visible beams, but also near-IR wavelengths. Laser light from such beams is transmitted through the lens and cornea, and hits the retina or optic nerve, where burns can cause irreversible damage. Ultraviolet lasers, on the other hand, are not transmitted through the lens and cornea, but rather are absorbed there, causing surface damage to the exterior of the eye. Ironically, this injury is somewhat reversible, as surgery can replace the lens of the eye. Infrared beams are therefore more dangerous in this sense than ultraviolet wavelengths, but both are worse than visible lasers.

Recommended laboratory practice

Experience in laser labs has made it possible to recognize several rules and guidelines that make it possible to work safely with lasers without the need for costly and inconvenient safety equipment. These guidelines are not intended to replace a proper course in laser safety, but they provide reasonable, common-sense practices that can be effective in providing a safe working environment.

1) *Limit laser lab access to trained personnel.* In a normal university environment, students and faculty come and go freely throughout most laboratory buildings. It is important to make laser labs distinctly different from other

work areas by limiting the access to qualified trained users. This can be done by locking lab doors, using warning signs, and especially using flashing warning lights when lasers are in operation. Laser signs are available at modest cost from suppliers such as Lab Safety Supply. Warning lights are somewhat more expensive, and require electrical expertise for installation, but again the costs are not prohibitive. Red light bulbs and protective cages are available from ordinary electrical supply stores or from university physical plants. Many lasers have switched power outlets, activated when the laser power is on, that can be used to switch on a warning light. A low-cost blinker inserted into the socket of the light provides a flashing bulb. Figure 6.1 shows an example of a lab door at the University of Georgia with signs and homemade warning lights.

2) *Design the work area to minimize laser exposure.* Lasers should be located in rooms away from other experiments, to the extent possible. They should be mounted on tables, bolted down to avoid accidental movement of

Figure 6.1 *Laser warning signs and lights installed on a laser lab door to limit the access of unauthorized personnel.*

beams, at waist level relative to standing users. Avoid beams at eye level or periscope optics that shoot beams upward off the table whenever possible. Lasers should also be isolated from office areas, where personnel not involved in experiments could be hit by reflected beams. If lasers are mounted at waist level for standing operators, this is at eye level for someone sitting at a desk nearby.

3) *Enclose the laser work area.* All lasers and their beams should be enclosed with curtains around the laser table, or within colored plastic or blackened sheet-metal boxes. Although commercial laser curtains are available and function well, less costly options are just as effective. Heavy grade canvas fabric can be cut into curtain sections and sewn with a loop along one end, and curtain rods can be constructed from electrical conduit and metal posts (see Figure 6.2). Aluminum sheet-metal or dark plastic boxes can be constructed at low cost. Easily mountable black metal sheets that are designed to block reflected beams are available from vendors such as Thorlabs. If laser beams have appropriate enclosures, they can be aligned onto the sample at low laser power, and then covered with curtains, boxes, etc., before turning up the power to higher levels for the experiment.

4) *Wear goggles when appropriate.* Although wearing goggles is the first laser safety practice that comes to mind for many people, this is not always the most appropriate approach. Many laser labs use multiple wavelengths at the same time, and standard laser goggles only block one specific wavelength; wearing multiple pairs of goggles is not a feasible alternative. Full-wavelength goggles, if such a thing did exist, would make it impossible for the user to see anything, and then it would be impossible to actually work in the lab. The problem with multiple wavelengths is common in

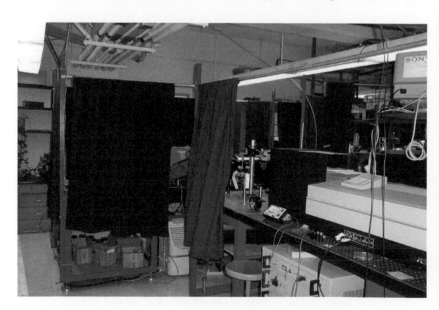

Figure 6.2 *Homemade laser curtains and metal boxes enclosing the laser area.*

experiments using tunable dye lasers or OPOs, which necessarily involve a pump laser and the tunable output of the dye or OPO, perhaps together with other pump or tunable laser harmonics. In this case, a viable alternative is to wear dark glasses such as those used by welders. Ordinary sunglasses can also provide attenuation of visible beam colors, and the plastic of these will often also block UV beams that might be present. Laser-specific goggles should definitely be worn when high-power invisible beams are in use, such as those from the Nd:YAG fundamental (1064 nm); general UV goggles are sufficient for excimer lasers. In all of these various cases, it is recommended that laser and optic alignment for the experiment be done at the lowest laser power for which the beam can be seen. Then, it should be possible in most cases to confine or enclose the laser system as discussed above, before turning up the power for conduction of the experiment. In this way, goggles would not be needed except during the initial alignment/setup phase of the work.

5) *Do not insert hands, tools, optics, or metal objects into a working laser.* As discussed above, metal objects or optics inserted into a laser beam can cause unanticipated reflections that can be just as dangerous as the main beam. It is therefore highly recommended that users remove all jewelry (watches, rings, etc.) from the hands and arms before working around a laser. Likewise, all insertion or alignment of optics in the laser path should be done either before the laser is activated or when it is operating at extremely low power, when the user and all people in the vicinity are wearing goggles. Never attempt to insert anything into or near the beam of a high-power (class IV) laser when it is in operation! In some cases, it is useful to see the spatial profile and/or alignment of a beam when it is not convenient to turn the laser power down, such as in the case of a frequency-doubled dye laser. If a viewing card is inserted into a UV beam, the user should be wearing UV goggles. If a visible beam is to be inspected this way, it should be done with a darkly colored card, rather than a white one.

The guidelines for laser safety presented here are by no means complete. It is indeed impossible to anticipate here every situation that might occur in a working laser lab. However, perhaps the most important tools available are *awareness of* and *attention to* the problem, together with the creative ideas of lab personnel. If all users recognize the potential dangers, it is usually true that acceptable safety practices and equipment can be worked out.

....

FURTHER READING

1. K. Barat, *Laser Safety Management*, CRC Press, Boca Raton, FL, 2006.
2. R. Henderson and K. Schulmeister, *Laser Safety*, second edition, Taylor & Francis, New York, 2012.
3. https://web.stanford.edu/dept/EHS/prod/researchlab/radlaser/laser/program/program.pdf.

Part II

Laser Experiments for Thermodynamics

The Speed of Light

7

Introduction

As described in Chapters 1 and 2, the speed of light is a fundamental constant with far-reaching connections throughout chemistry and physics. The speed of light in different materials determines their refractive index, which is critically important in optics and therefore in all instruments and measurements involving lasers. We therefore present here experiments designed to measure this critical parameter.

Capacitance measurement of the speed of light

In physics and chemistry the International System of Units (Le Système International d'Unités) or SI units are used to calculate the electric force between two charged particles, Q_1 and Q_2, separated by a distance r using Coulomb's Law:

$$F = \frac{Q_1 Q_2}{4\pi\varepsilon_0 r^2} \tag{7.1}$$

Likewise, in order to calculate the magnetic force between two wires of length L separated by a distance r and carrying currents I_1 and I_2 Ampere's Law is employed

$$F = \frac{\mu_0 I_1 I_2}{2\pi} \frac{L}{r} \tag{7.2}$$

The proportionality constants ε_0 and μ_0 are related to another fundamental constant of nature, the speed of light. In Chapter 1 we saw from Maxwell's Equations that the speed of light *in vacuo*, c, the permittivity of free space, ε_0, and the permeability of free space, μ_0, are all constants and related through

$$c = \frac{1}{\sqrt{\mu_0 \varepsilon_0}} \tag{7.3}$$

Since μ_0 is defined in SI units as $4\pi \times 10^{-7}$ newtons/ampere^2, a measurement of the electrical permittivity of free space ε_0 provides a direct measurement of c.

The dielectric constant ε_0 can be determined by accurately measuring the capacitance of a parallel-plate capacitor *in vacuo*. In 1906 Rosa and Dorsey [1] of the National Bureau of Standards obtained $\varepsilon_0 = 8.84 \times 10^{-12}$ c²/Nm². Combining this with the defined value for μ_0 completely verified the relationship $c = \frac{1}{\sqrt{\mu_0 \varepsilon_0}}$ with $c = 3.0002954 \times 10^8$ m/sec, very close to the value known accurately today. This experiment can be repeated to determine the permittivity of air, which allows a student to estimate the velocity of light in air from a simple two-plate capacitor. The relative permittivity, ε_r, is defined as $\varepsilon_r = \varepsilon/\varepsilon_0$ where $\varepsilon_0 = 8.854187817620\ldots \times 10^{-12}$ F/m⁻¹. Since the relative permittivity of air is very close to 1 (ε_r(air) = 1.00058986 @ STP and 0.9 MHz) a measurement of ε can be used to determine the speed of light in a vacuum to 0.05%. The capacitance between two conducting plates each having an area A and separation D is given by

$$C = \varepsilon A/D \qquad (7.4)$$

where ε is the permittivity of air, A is the area of either disc, and D is the separation between the two discs. In this experiment three small holes are placed symmetrically around the outside of two thin round metal discs. The two metal discs are separated by placing three small (dia.= 0.0125 in.) sapphire balls into the three holes, as shown in the right side of Figure 7.1. The inner distance between the two plates in this example was determined by measuring

Figure 7.1 *Experimental setup used to determine the permittivity of air (or any liquid). On the right the plates are shown in a small glass container with a Teflon cap machined to fit snugly into the glass dish. On the left the meter shows the capacitance between the two plates in air.*

the distance between their outside edges and subtracting their thickness to yield D = 0.0415 in. (1.05×10^{-3} m). The radius of the capacitor plate in this example was R = 2.86 cm, giving A = 2.57×10^{-3} m^2. The capacitance between the two plates was then determined as shown in the left side of Figure 7.1.

After correcting for the permittivity of the sapphire balls and the capacitance of the leads, etc., a value of C = 22 picoFarad was obtained. Taking ε from equation 7.4 and ε = CD/A, one obtains $\varepsilon = 8.98 \times 10^{-12}$ F/m. Using the defined value for μ_0 gives a speed of light as 2.98×10^8 m/sec. This is very close to the accepted value of 299,792,458 m/sec.

The simple setup in Figure 7.1 can be used to measure the permittivity of a liquid as well. The Teflon lid shown in Figure 7.1 is used when measuring the permittivity of a volatile liquid. For example, if one is to determine the permittivity of benzene, it is necessary to prevent evaporation which results in cooling of the liquid. The value of ε depends strongly on the temperature.

Pulsed laser determination of the speed of light

We now describe a simple but accurate method for measuring the velocity of light in air and in various media for different wavelengths in order to test some of the properties of light discussed in Chapter 1. A nitrogen laser (Laser Photonics LN1000) with a pulse duration of ~500 psec (5×10^{-10} sec) and $\lambda = 337.1$ nm, or a dye laser pumped by this nitrogen laser, is used as the light source. The time of travel of the *centroid* of this laser pulse is determined in air and in a fiber-optic cable, as well as various liquids. For air and most weakly dispersive media the group and phase velocities, v_g and v_p, are essentially the same. Strongly dispersive media may distort the pulse shape as a result of the finite frequency spread of the light pulse and the variation of speed with frequency. These experiments are very easy to perform and need no further explanation. A schematic of the experimental setup is shown in Figure 7.2.

Although Figure 7.2 shows only two passes between the mirrors, the laser pulse can make multiple passes between the two parallel mirrors before arriving at the photodiode. It is not difficult to obtain as many as 10 of these reflections. Figure 7.3 shows a sample of the time-of-flight data for light pulses traveling in air recorded with a Hewlett-Packard Infinium sampling oscilloscope. For this run the distance was ~10.5 m corresponding to a time difference of 35.3 nsec. By accurately measuring the distance between the two mirrors and using multiple passes, students are able to determine the velocity of light in air to 1% or less (in this case $c = 10.5/35.3 \times 10^{-9}$ m/sec = 2.97×10^8 m/sec). The main error is the uncertainty in the last path from the mirror to the detector, thus using multiple passes reduces this error.

Figure 7.2 *Experimental velocity of light measurement setup. The beam splitter, a clear window of glass or quartz, provides a weak trigger pulse to the photodiode. The light makes a path between two parallel mirrors back to the photodiode. The angle of reflection of the mirror making the first reflection can be adjusted to allow many reflections between the two parallel mirrors.*

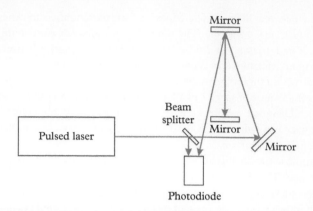

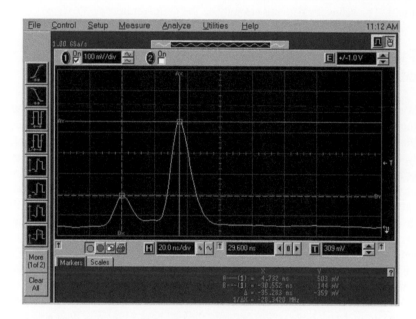

Figure 7.3 *Typical oscilloscope trace of the time-of-flight of a laser pulse from the experimental setup shown in Figure 7.2.*

Speed of light in iodine vapor

The index of refraction of air is 1.00029 and does not play a role in the measurements above. In order to measure the index of refraction of air, one would need to measure the velocity of light to a very high precision (i.e., with a longer path length). However, there is one interesting variation of this experiment that can be used to explore the index of refraction of a gas. As shown in later experiments, the iodine molecule has many rovibrational transitions between the X and B states in the region of the laser operating in the rhodamine 6G dye laser region. If one places iodine crystals into a 2-meter evacuated tube, and the tube is heated with heating tape, the iodine vapor can reach reasonably high densities (see also

Chapter 17). Tuning the light near one of the rovibrational lines of the $X \rightarrow B$ transition such that the light arriving at a photodiode detector is faint but measurable, it is possible to detect changes in the time of arrival of laser pulses down the tube (again using multiple passes) as the laser wavelength is tuned through these transitions. The time of flight increases as one nears a resonance due to the increasing index of refraction ($v = c/n$). The iodine is removed by simply freezing the vapor onto a cold finger. With some attention, this method could be developed into a new kind of spectroscopy, which could be called multipass index-of-refraction spectroscopy.

Many students who have performed this experiment with iodine vapor notice that the intensity of the multipass laser beam is greatly attenuated when the laser frequency is tuned near an absorption resonance. This observation forms the basis of what is commonly called cavity ring-down spectroscopy (CRLAS).[2–4] In this spectroscopy, a single laser pulse makes many (hundreds) passes through the sample tube. The resulting intensity of each pass is recorded and the slope of this intensity versus time of passage is used to determine the absorption cross-section. CRLAS is much more sensitive than direct absorption measurements, and can be employed to determine sample (pollutant) concentrations to the level of parts per trillion.[2–4] This method could be used to determine weakly absorbing features in the iodine molecule. Further development of this idea could lead to an excellent senior research project for an interested student.

Speed of light in a fiber-optic cable

The index of refraction of a fiber-optic cable (or of a liquid in a long tube) can be readily determined by sending a laser pulse down one end of the cable and returning the other end of the cable near the entrance, as shown in Figure 7.4. By properly positioning the photodetector to see the transmitted pulse as well as a small amount of reflected light from the entrance to the cable, one can determine the time of transit of the laser pulse as shown in Figure 7.5. Using this method, the dependence of the index of refraction upon wavelength can be determined as shown in Figure 7.6.

In Figure 7.6 there is a modest decrease in the index of refraction, n, with wavelength over the visible region of the spectrum. A perfect fiber cable would have a constant index of refraction with wavelength. The value determined here is consistent with those in the literature.[5] A variation of this experiment could also provide accurate determination of the index of refraction for liquids. This would be done by first measuring the speed of light in an evacuated $\frac{1}{4}$- or $\frac{1}{8}$-in. diameter tube of glass in order to get a speed in air, and then adding a liquid to the tube to perform the same experiment. The difference in flight time of the laser pulse with and without the liquid could be employed to get the index of refraction of the liquid (see also Chapter 10). This is similar to the method described above for the measurement of the index of refraction of iodine vapor.

Figure 7.4 *Experimental setup to measure the transit time of a dye laser pulse through a fiber-optic cable. The pulse from the dye laser enters the cable where some light is also scattered to the detector at the right. This is the start pulse for the time-of-flight through the cable. The output of the cable is positioned near the detector and provides the exit pulse.*

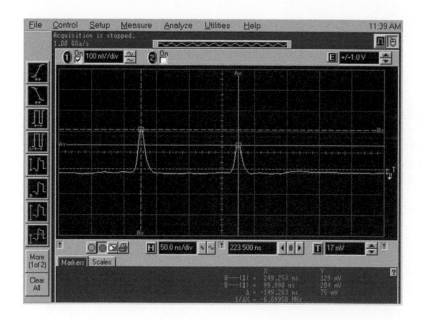

Figure 7.5 *Oscilloscope trace of the passage of a dye laser pulse through a 29.5-m fiber-optic cable. The first pulse is the reflection of light off the entrance to the cable and the second pulse is the light pulse exiting the cable.*

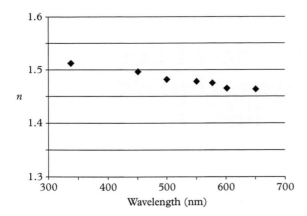

Figure 7.6 *Wavelength dependence of the index of refraction, n, of a fiber-optic cable determined by speed-of-light measurements as a function of the laser wavelength. The combined errors (length and time) for these measurements are approximately the size of the diamond symbols.*

REFERENCES

1. E. B. Rosa and N. E. Dorsey, "A new determination of the ratio of the electromagnetic to the electrostatic unit of electricity," *Bull. Bur. Stand.* **3**, 433 (1907); see also "A comparison of the various methods of determining the ratio of the electromagnetic to the electrostatic unit of electricity," *Bull. Bur. Stand.* **3**, 605 (1907).
2. A. O'Keefe and D. A. G. Deacon, "Cavity ring-down optical spectrometer for absorption measurements using pulsed laser sources," *Rev. Sci. Instr.* **59**, 2544 (1988).
3. P. Zalicki and R. N. Zare, "Cavity ring-down spectroscopy for quantitative absorption measurements," *J. Chem. Phys.* **102**, 2708 (1995).
4. G. Berden, R. Peeters, and G. Meijer, "Cavity ring-down spectroscopy: Experimental schemes and applications," *Int. Rev. Phys. Chem.* **19**, 565 (2000).
5. J. Crisp and B. Elliott, *Introduction to Fiber Optics,* third edition, Newnes, Oxford, 2005.

8

The Speed of Sound in Gases, Liquids, and Solids

Introduction

Sound waves in gases and liquids propagate as longitudinal mechanical waves, i.e., the wave oscillates in the direction of its propagation. There is no physical mechanism (restoring force) in gases and liquids to support a transverse wave. Waves that stimulate the ear and the brain to cause the sensation of hearing have frequencies in the audible range (20 to 20,000 cycles/sec). Frequencies above and below this are called ultrasonic and infrasonic, respectively. Sound waves in solids can propagate both as longitudinal and transverse waves. Many previous sound velocity measurements in undergraduate laboratories involved a determination of the wavelength, λ, for a fixed frequency, ν, source of sound. The sound velocity is then the product of these two numbers ($U = \lambda\nu$). In the present experiments, however, we measure the time it takes for the sound wave to travel a known distance through a material. We use pulsed lasers to generate a sound at a precise position and sharp moment in time and to synchronize this event electronically with its detection at a point some well-defined distance away. This method is convenient to measure the speed of sound in gases, liquids, and solids.

The speed of sound in gases

The finite speed of sound propagating through air has been recognized for many years, and is in fact easily seen in everyday experiences. When lightning strikes, the flash of light reaches us before we hear the sound of thunder. The time delay can be used to estimate how far away the lightning struck. Likewise, soldiers on a battlefield or hunters in a large open field are familiar with seeing the puff of smoke from a far-away gun well before hearing the sound. These observations prompted scientists to investigate the speed of sound in air many years ago. It was then a logical extension of this to consider the same phenomenon in other gases.

Laser Experiments for Chemistry and Physics. First Edition. Robert N. Compton and Michael A. Duncan.
© Robert N. Compton and Michael A. Duncan 2016. Published in 2016 by Oxford University Press.

It can be shown that the speed of motion of a longitudinal wave, U, in a gas is related to the variation of pressure, P, with a density ρ, according to the following equation:

$$U^2 = \frac{\partial P}{\partial \rho} \tag{8.1}$$

Newton was one of the first to derive this equation in 1729 (see *Principia*) and used the expression to obtain a relation between the sound velocity and the molecular mass and gas temperature. Newton assumed that the compressions and rarefactions causing the sound waves were isothermal in nature (i.e., the temperature does not change):

$$U^2 = \left(\frac{\partial P}{\partial \rho}\right)_T \tag{8.2}$$

We can rewrite equation (8.2) as:

$$\left(\frac{\partial P}{\partial \rho}\right)_T = \left(\frac{\partial P}{\partial V}\right)_T \left(\frac{\partial V}{\partial \rho}\right)_T \tag{8.3}$$

and since $\rho = M/V$ where M is molecular weight and V is molar volume, it follows that $V = M/\rho$ and: $(\partial V/\partial \rho)_T = - V/\rho$. Also, since PV is constant for isothermal compressions of an ideal gas, $(\partial P/\partial \rho)_T = -P/V$. According to Newton, this leads to:

$$U^2 = \left(-\frac{P}{V}\right)\left(-\frac{V}{\rho}\right) = \frac{P}{\rho} \tag{8.4}$$

and using

$$P = \frac{RT}{V}, \quad \rho = \frac{M}{V}$$

Newton concluded that:

$$U^2 = \frac{RT}{M} \text{ or } U = \sqrt{\frac{RT}{M}} \tag{8.5}$$

Agreement between the speed of sound predicted by Newton and that determined by direct measurements, e.g. cannon blasts observed at great distances, was poor. Newton's value for U was about 16% too low. The reason for this discrepancy is that the assumption that the process is isothermal is incorrect.

One hundred years later, Laplace (1822) explained the source of this discrepancy. He noted that the compressions and rarefactions occur so rapidly that heat cannot flow between the compressed hot spots and rarefied cool spots, i.e., the compressions and rarefactions occur adiabatically. As a consequence, equation 8.2 should read:

$$U^2 = \left(\frac{\partial P}{\partial \rho}\right)_q = \left(\frac{\partial P}{\partial \rho}\right)_s \tag{8.6}$$

Now

$$\left(\frac{\partial P}{\partial \rho}\right)_q = \left(\frac{\partial P}{\partial V}\right)_q \left(\frac{\partial V}{\partial \rho}\right)_q \tag{8.7}$$

and as before:

$$\left(\frac{\partial V}{\partial \rho}\right)_q = \left(\frac{\partial V}{\partial \rho}\right)_T = -\frac{V}{\rho} \tag{8.8}$$

However, for an adiabatic expansion

$$PV^\gamma = \text{constant, so that } \left(\frac{\partial V}{\partial \rho}\right)_q = -\frac{\gamma P}{V}$$

where γ is the ratio of heat capacity at constant pressure to that at constant volume. Thus,

$$U^2 = \left(\frac{-\gamma P}{V}\right)\left(\frac{-V}{\rho}\right) \text{ where } \gamma = \frac{C_p}{C_v} \tag{8.9}$$

But, as before $P/\rho = RT/M$, so that:

$$U^2 = \frac{\gamma RT}{M} \text{ or } U = \sqrt{\frac{\gamma RT}{M}} \tag{8.10}$$

M is the molecular weight of the gas, which has a mean value of about 29 amu for air. γ is the ratio of the specific heat of a gas at constant pressure to that at constant volume. A number of these values are given in Table 8.1.[1]

If U is the velocity of sound at 0 °C, then the velocity at any temperature T (°C) is

$$U_T = U_a \sqrt{\frac{273 + T}{273}} = U_a\sqrt{1 + 0.00366T} \tag{8.11}$$

Note that if T is not too large we can approximate

$$U_T = U_a(1 + 0.00183T) \tag{8.12}$$

and for air

$$U_T = 331.5 + (0.607)\,T(°C)\,\text{m/sec} = 1088 + (1.1)T(°F)\,\text{ft/sec} \tag{8.13}$$

Table 8.1 *Heat capacities and heat capacity ratios for different gases.*

No. of atoms	Substance	C_p (cal/mol·C)	C_v (cal/mol·C)	γ
Monatomic	He	4.97	2.98	1.666
	Ar	4.97	2.98	1.666
Diatomic	N_2	6.95	4.955	1.402
	O_2	7.03	5.03	1.396
	H_2	6.865	4.88	1.408
Polyatomics	H_2O	8.20	6.20	1.32
	CO_2	8.83	6.80	1.299

Table 8.2 *The speed of sound in different gases.[1]*

Gas	Speed (m/sec)	Speed (ft/sec)
Air	331.5 (0 °C)	1088
	346.3 (25 °C)	1136
N_2	317.2 (0 °C)	1041
CO_2	259 (0 °C)	850
He	965 (0 °C)	3166
Ar	323 (27 °C)	1060

As shown in Table 8.2, the speed of sound varies significantly with different gases. Because of this, it is logical that it will also vary when two different gases are mixed. Since the velocity of sound differs for gases in proportion to their heat capacity ratios, the velocity differences can be correlated directly with the gas concentrations in a mixture. From equation (8.10),

$$U^2 = \frac{\gamma\,RT}{M} \equiv \frac{\left(\frac{C_p}{C_v}\right)}{M}RT \tag{8.14}$$

For a mixture of two gases, 1 and 2, $M = X_1M_1 + X_2M_2$, $C_p = X_1C_{p1} + X_2C_{p2}$, and $C_v = X_1C_{v1} + X_2C_{v2}$, where X_1 and X_2 are mole fractions of the two gases and M_1 and M_2 are the molar masses. This gives

$$U = \sqrt{\frac{\dfrac{X_1C_{p1} + X_2C_{p2}}{X_1C_{v1} + X_2C_{v2}}RT}{X_1M_1 + X_2M_2}} \tag{8.15}$$

Since $X_1 + X_2 = 1$ we can also write:

$$U = \sqrt{\frac{\dfrac{X_1 C_{p1} + (1 - X_1)C_{p2}}{X_1 C_{v1} + (1 - X_1)C_{v2}} RT}{X_1 M_1 + (1 - X_1)M_2}} \qquad (8.16)$$

Thus, if the heat capacities of the two gases are known, the mole fractions can be determined from the sound velocity measurement.

The speed of sound changes greatly with temperature and is independent of the pressure. Many researchers have considered the decrease in sound velocity with relative humidity (see e.g., [2] and [3]). Using the heat capacities for air and water one can calculate that the speed in dry air at 20 °C is 343.36 m/sec and it increases slightly to 344.37 at 80% relative humidity.

Because the speed of sound is related to basic thermodynamic properties, its measurement has been a popular component of undergraduate laboratories for many years.[4] In some of the earliest configurations, the velocity of sound in a gas was determined by measuring the wavelength of the sound wave for a particular frequency. Every physics student before 1970 employed the Kundt's Tube method in which soft powder (often cork filings) was placed in a glass tube.[5] Sound from a tuning fork or a speaker with a known frequency was introduced into the tube. If the sound was intense enough the power would redistribute itself inside the tube showing the nodes and anti-nodes of the waves. The speed of sound was then calculated from the product of frequency times the wavelength. An alternate method often used in the early physics laboratories used an adjustable water level in a glass tube together with a tuning fork of known frequency or a variable sound source and speaker. The sound was introduced into the tube and the water level was raised and lowered and distinct antinodes could be heard (from the increased sound intensity). A better way to vary the air column is to mount a cork on a long wooden dowel. The diameter of the cork is chosen to fit snugly into the horizontal glass tube. Using a variable frequency sound generator or tuning fork at the open end of the tube, one can hear the resonances as the cork is moved down the tube. The first antinode is heard when the air column is $1/4$ of the wavelength of the sound, thus $U = 4\upsilon L_1$ (L_1 = length of the antinode). The next resonance occurs at a distance L_2, three times L_1, or $u = 4\upsilon L_2/3$. Many antinodes can be determined with this method and velocities of sound in air can be determined to better than 10% if the experiment is carefully performed. However, these two methods do not easily allow for the study of different gases.

A somewhat more recent variation of this allows the speed of sound to be determined for many different gases, again using a pipe of fixed length, with a speaker generating the sound at one end and a microphone detecting it at the other. Using an audio oscillator, sounds with different frequencies can be generated and detected. If the source wave is detected on one oscilloscope channel (i.e., in the *y* direction) and the transmitted wave is detected on the other axis (i.e., *x*), then the resultant of the two waves generates a so-called Lissajous pattern on the oscilloscope determined by the phase difference between the two waves. Adjustment of the frequency to find diagonal patterns, when the waves are exactly in phase, allows for a determination of the sound velocity.

These traditional methods focus on the frequencies and wavelengths of sound and their relationship with the length of the pipe in use. A much newer method focuses specifically on the generation of a sound at a precise moment at a precise position and the measurement of its transmission in *time* down the pipe. This method was first described for measurements in gases by Baum, Compton, and Feigerle.[6] The apparatus used for this experiment is shown in Figure 8.1, and a photo of the setup at the University of Georgia is shown in Figure 8.2. A sound that is sharp in time is generated with the fast pulse of a laser focused onto the surface of a metal or graphite rod. Pulsed nitrogen or CO_2 lasers can be used, but a small pulsed Nd:YAG laser is highly recommended for this, using the second harmonic (green) light at 532 nm. (Caution: the 1064 nm Nd:YAG laser cannot be seen by the human eye but damage to the eye can easily occur. For this reason use of the second harmonic light is recommended.) The pulse from this laser lasts only about 5 nsec, so it is essentially instantaneous in time. The rod

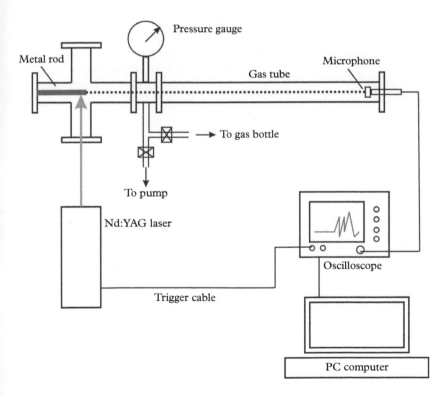

Figure 8.1 *The speed of sound experimental configuration.*

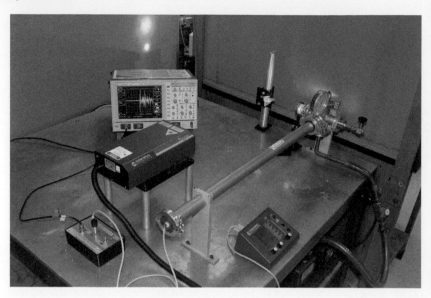

Figure 8.2 *A photograph of the speed of sound experiment at the University of Georgia.*

surface absorbs the green light from the laser and this generates a sharp snap inside the gas tube. The tube is filled with the gas of interest, and the sound from the laser snap goes down the tube to the opposite end where it is detected with a microphone (an inexpensive transducer can be purchased from Radio Shack). The time between the snap and its detection on the microphone is measured with a digital oscilloscope. The moment of the laser firing is provided electronically as the so-called "trigger out" pulse that comes from the laser. This is used to trigger the oscilloscope to start a time trace. The sound detected at the microphone produces an electrical signal that is detected on the signal line of the oscilloscope, and the time t is read directly from the oscilloscope screen. If the distance between the source of the sound (the point where the laser hits the rod) and the microphone is L, then a simple calculation provides the velocity of sound in that gas, $U = L/t$. If desired, the scope trace can be digitized and transferred to a computer for storage and/or plotting.

Safety notes

1. Be sure to read the laser safety section of this book and receive appropriate laser safety training before beginning this experiment! Figure 8.2 demonstrates the main laser safety issue with this experiment. The green Nd:YAG laser beam is quite bright and there is much reflection (see back wall) as it passes through windows and general glare from the optics. It is strongly recommended that the entire laser area be enclosed under a black box to block the glare and reflections.

2. This experiment uses pressurized gases. If the gas tube is over-pressurized, windows, gas fittings, or microphone pieces can be launched at high velocity from the experiment. Because of this, laboratory safety glasses must be worn at all times.

Procedure

Assemble the SOS tube and use an ordinary mechanical vacuum pump to evacuate it. Make sure that a "good" vacuum is obtained (pressure less than 0.1 Torr) so that leaks will not introduce impurities to the gases to be studied. Introduce gases one at a time using standard gas cylinders and regulators. Adjust the gas pressure for each to be 1.0 atm and record the room temperature.

Make measurements of the time it takes for the signal to travel down the tube as well as the length of the tube, and then calculate the speed of sound. Collect data traces in the form of graphs of signal on the microphone versus time on the digital oscilloscope. Transfer these to the computer and save them on the hard drive and/or a flash drive for future use. These will usually be in the form of a spreadsheet, such as those produced with Excel, Origin, SigmaPlot, etc. Make measurements on different gases as time allows (e.g., He, N_2, CO_2, Ar, air). For air, correct the measured velocity for temperature and humidity. For one selected gas, study the speed of sound at different pressures of gas in the tube. Sound velocities should be independent of pressure (see equation 8.10) unless a gas is used that may form dimers (e.g., NO_2, NO). Compare the measured values for U with those in the literature. Determine values of γ and compare these to predictions of the equipartition of energy theorem.

Representative data

Figure 8.3 shows an example of representative data for this experiment, using nitrogen gas with a gas tube length of 0.883 m. The oscilloscope trace begins at the moment of the laser firing, triggered by the synch-out pulse from a small Nd:YAG laser (New Wave Polaris II). The positive-going signal detected by the microphone first appears at 2.492 msec (see arrow marking onset). This represents the sound waves which followed the shortest path (i.e., a direct line) from the laser impact point to the detector. The unwanted signal coming after this is unavoidable. It represents other sound waves following longer paths to the detector, involving reflections off the walls and interferences. The resulting speed of sound in nitrogen is 354.3 m/sec. This is greater than the value indicated for nitrogen in Table 8.2 because the room temperature was about 25 °C, and the gas temperature in the tube could have been warmer than this. (A heating tape

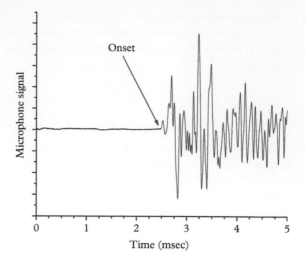

Figure 8.3 *Representative data from a measurement of the speed of sound in nitrogen. The distance from the impact point to the detector was 88.3 cm, and the onset occurs at 2.492 msec, resulting in a velocity of 354.3 m/sec.*

could be wrapped around the metal tube in order to examine the temperature dependence of the speed of sound (U ~ $\sqrt{T}$).)

If the laser pulse is employed to start the timing for the sound pulse one might ask "is there a time delay between the laser pulse and the sound pulse?" There probably is a small difference but the time required to generate the sound wave in gases is small compared to the other measurement uncertainties. This is not the case for liquids. We will see that measurements of the speed of sound in liquids using this method require measurement of the time between the laser firing and the generation of sound.

Effects of the tube diameter

The speed of sound in gases using hollow tubes can be affected by the tube diameter because the sound velocity depends on the frequency of the waves traveling in a tube of finite diameter. This phenomena was investigated over 100 years ago by Helmholz [7] and by Kirchoff.[8] Kirchoff found that the velocity U depends on the tube radius and sound frequency through a complicated relationship,

$$U = U_0\left(1 - \frac{\mu^{1/2} + n^{1/2}\gamma^{-1/2}\,(\gamma - 1)}{2a\sqrt{\pi\nu}}\right) \tag{8.17}$$

Where U_0 is the true velocity $\left(U_0 = \sqrt{\frac{\gamma RT}{M}}\right)$, μ is the kinematic viscosity, γ is the heat capacity ratio C_p/C_v, n is the thermal diffusivity of the gas, a is the radius of the tube, and ν is the sound frequency. These corrections are small if the tube radius is large. A number of experimental tests of Kirchoff's theory have been published over the years. For example, the interested reader should refer to the recent paper of Yazaki, Tashiro, and Biwa.[9] This study reports measurements of sound velocities in air at seven points along copper tubes as long as 40 m. Copper was employed to reduce the temperature gradients. Using such long tubes minimized the effects of reflections back down the tube since the sound was attenuated over such a long distance. They observed the effects of the parameters in equation 8.17 for tube radii of 0.6, 1, and 2 mm and for pressures from 2×10^3 to 10^5 Pa at various frequencies. These experiments verified Kirchoff's theory over a wide range of thermodynamic conditions (isothermal to isentropic). The use of large diameter tubes in the experiments described in this chapter simplifies the treatment of sound velocities. The interested student could examine the sound velocity in tubes of much smaller diameters using the laser sound method described here. The beauty of the laser sound method is that very small tubes can be employed. For example, experiments at the University of Tennessee have used very long coiled tubes of PVC for this experiment.

Questions surrounding the diameter of the tube can be addressed by performing the sound velocity measurements in an open room. To eliminate wall effects, a student at UT, Ben Graves, was able to focus a pulsed laser onto a glass rod in order to produce a loud sound wave in open air in the lab. Using a 2-m stick to allow accurate measurements of the distance to a Radio Shack transducer, he was able to determine the speed of sound in air by carefully measuring the time it took for the sound pulse to travel accurately measured distances ($\sim \pm 0.5$ mm) along the meter stick. The speed of sound in air was then determined from the slope of the distance versus the time, as shown in Figure 8.4. Determining the slope reduces many of the errors in this experiment.

Unfortunately, the temperature (or humidity) was not determined in this experiment. The temperature dependence of U for air is approximately $U = 331.3 + 0.606 \ T$ m/sec. Using the measured speed in air of 345.7 m/sec

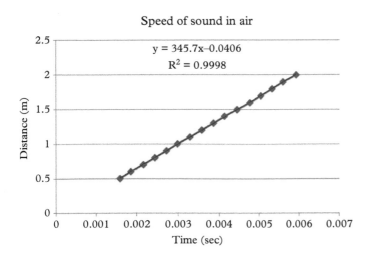

Figure 8.4 *The measured time of arrival of sound waves versus the distance from the pulsed sound source in open air. The slope of distance versus time gives* $U = 345.7 \ m/sec.$

predicts a temperature of 23.7 °C, a realistic value for the summertime when the measurements were performed.

The speed of sound in dry air can be calculated from the known composition of air using only N_2 (78.08%), O_2 (20.95%), Ar (0.93%), and CO_2 (0.033% and rising) at 300 K using the relationship shown in equation 8.18

$$V = \sqrt{\frac{\left(\dfrac{X_{N_2}C_{pN_2} + X_{O_2}C_{pO_2} + X_{Ar}C_{pAr} + X_{CO_2}C_{pCO_2}}{X_{N_2}C_{vN_2} + X_{O_2}C_{vO_2} + X_{Ar}C_{vAr} + X_{CO_2}C_{vCO_2}}\right)RT}{X_{N_2}M_{N_2} + X_{O_2}M_{O_2} + X_{Ar}M_{Ar} + X_{CO_2}M_{CO_2}}} \tag{8.18}$$

where X, M, C_p, and C_v, represent the mole fraction, molar mass in kg, constant pressure heat capacity (J mol^{-1} K^{-1}), and constant volume heat capacity (J mol^{-1} K^{-1}) of each gas. T is the ambient temperature (300 K) while R is the gas constant expressed in J mol^{-1} K^{-1}. The speed of sound is calculated under these conditions to be 347.3 m/sec compared with the measurement of 345.7 m/sec from Figure 8.4.

Measurements of the speed of sound in two component mixtures can be used in a number of applications. For example, the speed of sound in a gas above a liquid can be used to determine the vapor pressure of the liquid. This was illustrated in [6] where the vapor pressure of carbon tetrachloride (CCl_4) in nitrogen gas was accurately determined. The measured speed of 254 m/sec (compared to 352 for pure N_2) gave a vapor pressure of 126 Torr which was within experimental error of the accepted value. Another application of this method is the determination of CO_2 in mixed gases as shown below.

Measuring CO_2 content in human breath

Human breath contains substantial carbon dioxide resulting from respiration. Expelling breath into the speed of sound tube results in a marked decrease in the speed of sound. The amount of CO_2 exhaled into the sound tube can be calculated from the familiar equation below:

$$V = \sqrt{\frac{\left(\dfrac{(1-X_{CO_2})C_{pair} + X_{CO_2}C_{pCO_2}}{(1-X_{CO_2})C_{vair} + X_{CO_2}C_{vCO_2}}\right)RT}{(1-X_{CO_2})M_{air} + X_{CO_2}M_{CO_2}}} \tag{8.19}$$

This experiment has been performed by many undergraduate students as well as summer high school students over the past 15 years at UT. In one set of laboratory experiments by undergraduate student Brad O'Dell, an average of ten measurements yielded a speed of 342.6 ± 0.2 m/sec. If human breath is treated solely as carbon dioxide enriched air, the mole fraction of carbon dioxide can be determined by solving equation 8.19, yielding a CO_2 mole fraction of approximately 0.11.[10,11]

This measurement can be used to estimate the contribution of all humans to the relative concentration of CO_2 in the atmosphere. Considering that the average adult human exhales 6 l of gas each minute,[12] each person generates 1.6 kg of carbon dioxide per day. Multiplying this amount by the adult human population, 4.7932 billion people,[13] yields an annual human carbon dioxide production of 2.8×10^{12} kg. The US Environmental Protection Agency reports that global fossil fuel combustion released 2.704×10^{13} kg of carbon dioxide into the atmosphere in 2004.[14] Annual human carbon dioxide release is only one tenth of that resulting from burning fossil fuels. Thus one might conclude that the CO_2 released by humans contributes to global warming. However, it is important to note that the CO_2 produced by humans is a part of the planet's natural carbon cycle, thus making no net contribution to carbon dioxide buildup in the atmosphere.[15] Nevertheless, humans do release other greenhouse gases (e.g., methane). (For this reason beans are not served on a submarine!)

The speed of sound in liquids

Sound also travels as a longitudinal wave in liquids. Formally, the speed of sound, U, is related to the isentropic (adiabatic) compressibility K_s of the fluid given by

$$K_s = -\frac{1}{V}\left(\frac{\partial V}{\partial P}\right)_s = \frac{1}{\rho U^2} \tag{8.20}$$

where V, P, and ρ represent the volume, pressure, and density of the liquid. The isentropic compressibility is related to the isothermal compressibility using

$$K_T = K_s + \alpha^2 T \frac{V_m}{C_{p,m}} \tag{8.21}$$

where α is the isobaric expansivity, V_m is the molar volume, and $C_{p,m}$ is the molar heat capacity at constant pressure. The velocity of sound, U, in a liquid having density, ρ, and compressibility, K_s, is given by

$$U = \sqrt{\frac{1}{\rho K_s}} \tag{8.22}$$

The density of a fluid can be measured separately; therefore the measurement of the sound velocity allows the determination of the compressibility. Measurements of sound velocities have been used to study the physicochemical properties and especially molecular interactions in liquids. A number of empirical, semi-empirical, and statistical methods are employed to predict sound velocities using various approximations to relevant molecular interactions: Flory–Patterson theory (FPT),[16] the Free Length method of Jacobson,[17] collision factor theory

(CFT) of Schaaffs,[18] the Nomoto theory,[19] and the method of Avsec and Marcic,[20] among others.

Many methods have been employed to measure the speed of sound in liquids. The method described here was devised by a former undergraduate student at the University of Tennessee (now Dr. Andy Fischer).[21] Although many methods have been used to determine the speed of sound *relative* to that of another liquid (most often water), this method allows a determination of the *absolute* speed. Because it is a non-intrusive method using light, it can even be used to determine the speed of sound for highly corrosive liquids under extreme conditions of temperature, pressure, or chemical environment. This basic method can be applied to small samples (e.g., in a capillary) or at high pressure. The one requirement is that the medium be transparent to laser light. However, even this can be relaxed if the laser is focused onto a metal foil holding the liquid and a second laser is used to determine the arrival of the sound at a second metal foil a distance away. Bouncing a collimated laser off the second foil into a pinhole through a detector provides a time of arrival of the sound pulse.

The basic principle of this method involves the focusing of a pulsed laser beam into a liquid to produce a spark. The spark produces a heated gas followed by a collapse of the liquid giving a spherical outgoing sound wave. A CW laser at a known distance away is deflected by the light scattering from the distortion of the index of refraction of the liquid as the sound wave passes. This deflection of the laser beam is a result of Brillouin scattering due to the change in the index of refraction with time. The experimental setup is shown schematically in Figure 8.5.

The second harmonic (532 nm) of a Quanta-Ray DCR Nd:YAG laser (10 Hz repetition rate, pulse width ~7 nsec) is reflected down into a liquid solution through the air–liquid interface via a turning prism. The laser is focused into the liquid using a quartz cylindrical lens with a focal length of 4 cm. A pyrex cell ($4 \times 2 \times 8$ cm) containing the liquid is mounted rigidly together with the turning prism on a translation stage. The micrometer has incremental readings of 0.08 mm. Measurements are made with an uncertainty of half this value. The translation stage is capable of extending to 50 mm, but for this study, the maximum distance used is 15 mm. A Metrologic He-Ne laser (1 mW), which is well collimated by four apertures (two before and two after), passes through the 2 cm length of the cell. The He-Ne laser is fixed to a laser table. The turning prism, focal lens, and glass cell are all attached to a precise translation stage in order to accurately vary the distance between the He-Ne focal point and the Nd:YAG laser focal spot.

Two photodiodes (Thorlabs Model DET1-SI) are used to measure the time between the production of the sound wave from the Nd:YAG laser and the detection of the sound wave. One photodiode detects the initial light pulse and is positioned next to the turning prism. The second photodiode records the time behavior of the light from the He-Ne laser. As shown in Figure 8.5, the He-Ne beam must pass through four apertures before reaching the photodiode.

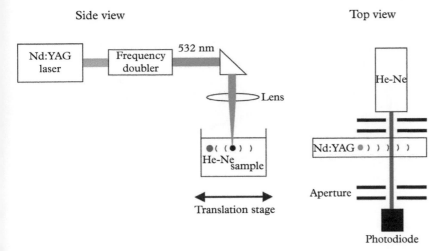

Side view   Top view

Figure 8.5 *Experimental arrangement used to determine the speed of sound in liquids. The glass cell (4 × 2 × 8 cm) and prism are mounted on a precisely controlled translation stage. The 532 nm pulse from the Nd:YAG and the CW He-Ne beam are at right angles to each other. The turning prism and focusing lens are translated relative to the fixed He-Ne laser and detector and liquid.*

The positioning of the apertures dictates that any disturbance within less than 0.06 mm of the He-Ne beam would not pass into the photodiode. Timing corrections due to the finite speed of light are negligible.

The initial laser pulse serves as a start pulse for the sound velocity measurement and is recorded on a 500 MHz Infinium digital oscilloscope. The time of arrival of the sound pulse at one position of the He-Ne laser beam is recorded as a dip in the He-Ne light intensity. The time of arrival of the wave (either the onset or the minimum in the He-Ne intensity) is recorded for various distances, x, from both sides of the laser focus. Since the speed of sound is determined from the slope of Δx versus Δt, the results are independent of which point is measured on the waveform of the sound pulse, as long as the location is consistent (i.e., onset or minimum). Typically, the width of the dip at the minimum was < 9.0 nsec, which is estimated as the uncertainty of the time. The minimum is determined by observing the intensity of the signal as a function of the time steps on the cursor readout of the oscilloscope, which can be determined to 1 nsec. The minimum, rather than onset, is always much easier to determine precisely and was used in all of the measurements reported herein. Figure 8.6 shows a typical recording of the Nd:YAG laser pulse and the time of transit of the sound pulse.

When the He-Ne laser travels directly though the laser focus (x = 0) the time profile dramatically changes into a broad structured waveform with ~100 μsec width and a long tail (~300 μsec). This broad, structured, and complicated time profile is a result of an initial thermal spike introduced into the solution as a result of the formation of a bubble, its subsequent collapse, and local heating of the liquid. This feature appears and disappears (i.e., its linear dimension or diameter) over a distance of 0.16 mm. A second minimum in the He-Ne intensity can be seen at ~130 μsec and may represent an oscillation of the bubble.

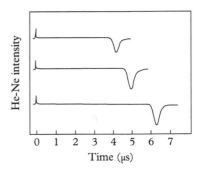

Figure 8.6 *Plot of the minimum He-Ne light intensity (see Figure 8.5) as a function of distance from the laser focal spot. The time intercept at x = 0 is interpreted as the time required to produce an initial sound wave following the laser–water interaction. The slope of x versus t gives the speed of sound.*

Figure 8.7 *A representative plot of the time corresponding to maximum deflection (i.e., minimum in the He-Ne intensity) in water relative to the laser firing versus the micrometer reading on both sides of the focus* (x = 0). x = 0 *is defined as the midpoint of the position in which the focal spot is over the He-Ne laser beam.*

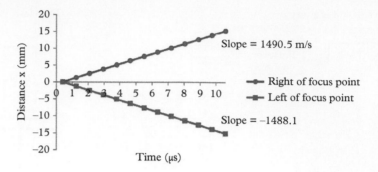

Figure 8.7 shows a plot of the time of flight as defined by the minimum in Figure 8.6 versus the distance as measured relative to the focal spot of the laser. Data are recorded on both sides of the laser focus.

The speed of sound in pure water at 23 °C is determined from the slope of either (or both) of these lines in Figure 8.7. The average of the two slopes for the data give U = 1489 m/sec. These measurements represent an absolute value. Agreement with well-accepted values for H_2O is within 0.06%. Based upon the estimated error in the measurement of distance ($\pm$ 0.04 mm) and time ($\pm$ 9.0 nsec), we derive an error of 0.5% in the determination of the speed of sound. Based upon the agreement (~ 0.06%) with accepted values for pure water, the actual error is expected to be less than 0.5%. Measurements of the velocity of sound in pure water at various temperatures and laser intensities were reproducible over a period of one year's study. These studies also show that the laser does not measurably alter the temperature or chemical composition of the liquid. The intercept in the time at zero distance (~ 0.5 µsec) represents the time between laser absorption and sound formation, and this time varies depending upon several factors (laser power, concentration, etc.). Initial laser absorption at the focal spot produces local heating, dissociation, and even ionization depending upon the laser intensity. A micro-bubble is formed followed by its collapse and the creation of the sound wave. At high laser power, light emission can be observed coming from the plasma produced from the focal volume.

The speed of sound in solids

As described in the introduction, sound waves in a solid can be both longitudinal and transverse. In fact, sound can propagate through a solid in four possible modes depending upon the manner in which the particles making up the solid oscillate. Longitudinal waves are compression waves in the direction of the wave similar to the case for gases or liquids. Transverse or shear waves are also possible in which the particle oscillations are at right angle (transverse) to the direction of propagation. At surfaces and interfaces other types of oscillations (elliptical or

Rayleigh) and transverse surface modes are possible. Sound waves in thin solids are described by plate waves. In what follows we describe measurements of longitudinal and shear waves. The general equation for the speed of sound in any elastic media is given by

$$U = \sqrt{\frac{C}{\rho}} \tag{8.23}$$

where C is a coefficient of "stiffness." For a gas, C is the modulus of bulk elasticity. Applying this formula to water at 0 °C gives $U = \sqrt{\{(2.06 \times 10^9 \text{ N/m}^2)/(999.8 \text{ kg/m}^3)\}} = 1435.4$ m/sec where C and ρ are 2.06×10^9 N/m² and 999.8 kg/m³, respectively. The accepted value at 0 °C is 1403 m/sec.

Sound waves generate volumetric compressions and shear deformations which are called longitudinal waves and shear waves, respectively. Solids possess a stiffness for both longitudinal and shear deformations. Thus, for a solid the stiffness C corresponds to both the bulk (longitudinal) and shear modulus, K and G, respectively. The bulk modulus is defined as negative of the volume times the rate of change of pressure with volume or the equation

$$K = -V \frac{dP}{dV} \tag{8.24}$$

The speed of sound can be defined for a longitudinal wave through the bulk often defined as C_1 and a shear wave defined as C_s

$$C_1 = \sqrt{\frac{K + \frac{4}{3}G}{\rho}} = \sqrt{\frac{Y(1-\upsilon)}{\rho(1+\upsilon)(1-2\upsilon)}} \tag{8.25}$$

and

$$C_s = \sqrt{\frac{G}{\rho}} \tag{8.26}$$

where ρ is the density of the medium and υ is Poisson's ratio (negative of the ratio of transverse to axial strain).

A simple method for determining the speed of sound in a glass rod consists of focusing a laser onto the surface of a long rod obtained from a glass shop. The sound exiting the end of the rod can be detected with a microphone identical to that used in the speed of sound in gases experiment. Placing the microphone close to the end of the rod allows the detection of many reflection signals within the glass rod. The distance between the peaks attributed to the sound waves corresponds to twice the length of the glass rod. A typical experimental setup is shown in Figure 8.8 and a photo of the experiment is shown in Figure 8.9.

Figure 8.10 shows a representative dataset for the speed of sound measurement in a glass rod. Seven sound reflections through the glass rod are shown

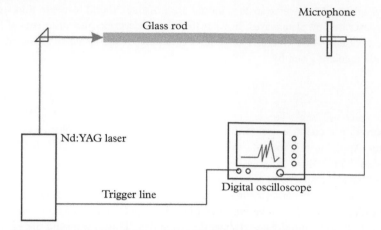

Figure 8.8 *The configuration for the speed of sound measurement in a solid rod.*

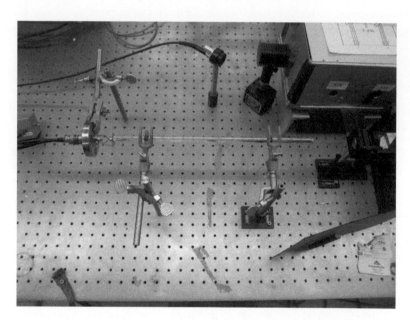

Figure 8.9 *Photo showing the experiment to measure the speed of sound in a solid rod at the University of Tennessee.*

corresponding to 12 passes along the length of the rod. Sound waves in the air traveling around the rod are also seen along with reflections from objects on the laser table and room. It was possible to reduce these signals significantly with proper baffles.

The length of the glass rod was 0.448 m corresponding to a path length of 0.896 m and the average time between the 6 passes was 0.164 msec, giving a longitudinal speed of sound of 5463 m/sec, which is comparable to the results from previous measurements of sound velocity in pyrex. The speed of sound for pyrex can be calculated from the equation for C_1 above. Poisson's ratio for pyrex

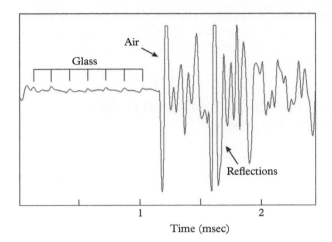

Figure 8.10 *Waveform of the sound wave passing through a pyrex rod. The large intensities due to the sound traveling around the rod through the air (and reflections from objects in the room) can be greatly reduced by placing sound absorbing baffles. Blowing up the signal due to the multiple reflections in the glass may allow for attenuation coefficient measurements.*

is 0.2, therefore the factor $(1-\upsilon)/[(1+\upsilon)(1-2\upsilon)]$ above becomes 1.11. A typical value for the Young's Modulus for pyrex is 65 GPa (65×10^9 N/m^2), the density of pyrex is 2.23 g/cm^3 (2230 kg/m^3), and the longitudinal speed C_1 is calculated to be 5690 m/sec. This value is greater than that of our measurement in a glass rod. However, if the wavelength of the sound is large compared to the diameter of the rod the longitudinal speed reduces to

$$C_1 = \sqrt{\frac{Y}{\rho}} \qquad (8.27)$$

Using this equation for a long rod one calculates C_1 = 5399 m/sec, which is 4% lower than the measured value. Of course a measurement of the speed of sound in a rod and independent knowledge of the density of the rod would allow a determination of the Young's Modulus for the material.

This method could be improved by the employment of transducers placed on the exit of the glass rod. Although we have not done this, it should be possible to use this method to determine the speed of sound in solids other than glass. By measuring the attenuation of the peaks upon reflections up and down the rod, attenuation coefficients could also be determined. The attenuation of the wave follows a simple exponential relationship given by

$$A(z) = A_0 e^{-\alpha z} \qquad (8.28)$$

Where A_0 is the sound intensity at $z = 0$, $A(z)$ is the intensity at z, and α is the attenuation coefficient (decibels per centimeter, db/cm).

This study was carried out by undergraduate student Ben Graves at the University of Tennessee as part of a summer internship. Although this has not been performed, we suggest that bouncing a He-Ne laser off the end or sides of the rod in order to detect the time of flight of the sound wave similar to the detection of the sound wave in a liquid (see Figure 8.5) might allow for the determination of the longitudinal and shear velocities, respectively. This simple method could be used to detect the position of flaws in metals, complementing other standard techniques for this.

Speed of sound in plasmas, BECs, and neutron stars

The experiments above describe measurements of the speed of sound in three familiar states of matter: solid, liquid, and gas. The speed of sound in the fourth state of matter, a plasma, is somewhat more complicated and depends on many physical conditions. In the simplest case of ionic sound velocity in the absence of magnetic fields and under conditions in which the electron and ion densities are the same and only singly charged ions are present, a rather simple relationship for the ionic sound velocity can be derived (see [22–25] and other references therein). It is interesting to note that one neglects collisions between ions. In this case, the longitudinal wave produces density fluctuations resulting in electric fields which provide the necessary restoring force to sustain the wave. Thus if one approximates the plasma as an isotropic, adiabatic fluid a dispersion relation can be derived which looks very similar to the equations derived above for sound velocities in gases

$$U = \sqrt{\frac{(\gamma_i k T_i + \gamma_e k T_e)}{m + M}} \tag{8.29}$$

Where γ_i and γ_e represent the heat capacity ratios for the ions and electrons, k is Boltzmann's constant, T_i and T_e are the ion and electron temperatures, and m and M are the electron and ion masses, respectively. In most plasmas the electron temperature is much higher than that of the ions. The heat capacity ratio, γ_e, sometimes called the adiabatic index for the electrons, is given by

$$\gamma_e = 1 + 1/n \tag{8.30}$$

where n is the number of degrees of freedom. Since n is large for fast-moving electrons, we can set $\gamma_e = 1$. Also neglecting the mass of the electron relative to that of the ion and employing $T_e \gg T_i$. Equation 8.29 reduces to

$$U = \sqrt{\frac{k T_e}{M}} \tag{8.31}$$

Thus the wave velocity is determined by the electron temperature and the ionic mass. These waves are often referred to as ion acoustic waves or IAWs. As one example, a time-of-flight method was employed to determine the IAW for an argon plasma and found $U = 2.96 \times 10^5$ cm/sec, which is 9.1 times faster than the sound velocity in argon.[25] This corresponds to an electron temperature of 3.65 eV or 4.24×10^4 K.

In 1995 Eric Cornell and Carl Wieman discovered a new state of matter called a Bose–Einstein condensate (BEC). First predicted in a collaboration between Bose [26] and Einstein,[27,28] the BEC state of matter is a consequence of the interaction between particles having zero or integer intrinsic spin (Bose particles). Bose particles tend to coalesce whereas Fermi particles (half-integer intrinsic spin) tend to avoid each other (Pauli exclusion principle). Einstein noted that at low temperatures particles obeying Bose–Einstein statistics would settle into the lowest energy state, i.e., coalesce into what is now known as a BEC. Using laser-trapping techniques Cornell and Wieman cooled rubidium atoms down to 170 nanokelvin to produce the first BEC.[29] Together with Wolfgang Ketterle, who demonstrated a number of unique properties of the BEC, they shared the 2001 Nobel Prize in Physics. One of the many interesting properties of a BEC is the speed of sound. Density fluctuations propagate according to a wave equation from which Bogoliubov [30] and later Lee et al. [31] extract a sound velocity as

$$U(r) = \sqrt{\frac{n(r)V}{m}} \qquad (8.32)$$

where $V = \frac{4\pi\hbar^2 a}{m}$ corresponds to the repulsive interactions of bosons with a mass m and scattering length a. Andrews et al. [32] were able to devise a clever method to measure the sound velocity for various densities in a trap containing sodium atoms. The temperatures in the trap were 1.6 and 0.4 μK depending upon the trap conditions. The density n_0 of the BEC was determined from

$$n_0 = \frac{\pi m^2 d^2 \nu^2}{8\hbar^2 a} \qquad (8.33)$$

where d is the axial length of the BEC and ν is the resonant frequency of the trap. Figure 8.11 shows the experimental data versus the theoretical prediction in equation 8.32.

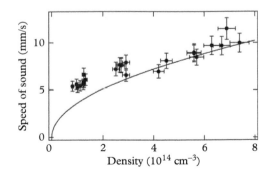

Figure 8.11 *Speed of sound in a BEC of sodium atoms versus the condensate peak density. Error bars indicate the statistical errors only. Reprinted with permission from M. R. Andrews, D. M. Kurn, H.-J. Miesner, D. S. Durfee, C. G. Townsend, S. Inouye, and W. Ketterle, Phys. Rev. Lett. 80, 2967 (1998). Copyright 1998 by the American Physical Society.*

In the density range from 10^{14} to 6×10^{14} cm^{-3} the sound velocity ranges from ~4 to 12 mm/sec. The very slow speed is due to the low temperature and the properties of the BEC.

Finally, we mention the sound velocities in matter having enormous densities such as that in a neutron star. The property of matter above the nuclear saturation density, $\sim 0.16/\text{fm}^3$, is of great interest in cosmology and quantum chromodynamics or QCD. At very large densities, QCD predicts that matter is composed of free quarks, with a speed of sound equal to the speed of light divided by $\sqrt{3}$. Previously, there was no evidence for matter with a speed of sound larger than this value. However, it has recently been pointed out that the existence of neutron stars with masses equal to two solar masses implies that the speed of sound in neutron stars must exceed this value.[33] Neutron star matter presently holds the record for the fastest sound in the universe. However this may change when we know more about Dark Matter.

....

REFERENCES

1. *CRC Handbook of Chemistry and Physics*, D. R. Lide, editor, 79th edition, CRC Press, Boca Raton, FL, 1998, pp 14–38 to 14–41.
2. G. S. K Wong and T. F. W. Embleton, "Variation of the speed of sound in air with humidity and temperature," *J. Acoust. Soc. Am.* **77**, 1710 (1985).
3. O. Cramer, "The variation of the specific heat ratio and the speed of sound in air with temperature, pressure, humidity, and CO_2 concentration," *J. Acoust. Soc. Am.* **93**, 2510 (1993).
4. C. W. Garland, J. W. Nibler, and D. P. Shoemaker, *Experiments in Physical Chemistry* eighth edition, McGraw-Hill, Boston, 2009.
5. M. Ference, Jr., H. B. Lemon, and R. J. Stephenson, *Analytical Experimental Physics*, University of Chicago Press,1956.
6. J. C. Baum, R. N. Compton, and C. S. Feigerle, "Laser measurement of the speed of sound in gases: A novel approach to determining heat capacity ratios and gas composition," *J. Chem. Ed.* **85**, 1565 (2008).
7. H. Helmholtz, *Crelles Journal* **57**, 1 (1859).
8. C. Kirchhoff, "Über den Einçuss der Warmeleitung in einem Gase auf die Schallbewegung," *Pogg. Ann.* **134**, 177 (1868).
9. T. Yazaki, Y. Tashiro, and T. Biwa, "Measurements of sound propagation in narrow tubes," *Proc. Royal Soc. A* **463**, 2855 (2007).
10. M. W. Chase, Jr., *J. Phys. Chem. Ref. Data, Monograph 9*, **27**, 1 (1998).
11. D. R. Williams, *Earth Fact Sheet*, National Aeronautics and Space Administration, 2007.
12. V. C. Scanlon and T. Sanders, *Essentials of Anatomy and Physiology*, fifth edition, Davis, Philadelphia, 2007.
13. *CIA World Factbook*, US Central Intelligence Agency, Washington, DC, 2008.

14. "U.S. Greenhouse Gas Inventory Reports," U.S. Environmental Protection Agency, 2007.

15. "Frequently Asked Global Change Questions," Carbon Dioxide Information Analysis Center, Oak Ridge National Laboratory, 2007.

16. D. Patterson and A. K. Rasogi, "The surface tension of polyatomic liquids and the principle of corresponding states," *J. Phys. Chem.* **74**, 1067 (1970).

17. B. Jacobson, "Intermolecular free lengths in the liquid state. I. Adiabatic and isothermal compressibilities," *Acta Chem. Scand.* **6**, 1485 (1952).

18. W. Schaaffs, "Zur Bestimmung von Molekülradien organischer Flüssigkeiten aus Schallgeschwindigkeit und Dichte," *Z. Phys.* **114**, 110 (1939).

19. O. Nomoto, "Deviation from linearity of the concentration dependence of the molecular sound velocity in liquid mixtures," *J. Phys. Soc. Jap.* **13**, 1524 (1958).

20. J. Avsec and M. Marcic, "Calculation of the speed of sound and other thermophysical properties," *J. Thermophys. Heat Transfer* **14**, 39 (2000).

21. A. T. Fischer and R. N. Compton, "Laser-based measurements in water and aqueous D,L and DL alanine solutions," *Rev. Sci. Instrum.* **74**, 3730 (2003).

22. A. Fridman and L. A. Kennedy, *Plasma Physics and Engineering*, Taylor and Francis, New York, 2004.

23. N. A. Krall and A. W. Trivelpiece, *Principles of Plasma Physics*, McGraw Hill, New York, 1973.

24. J. L. Shohet, *The Plasma State*, Academic Press, New York, 1971.

25. N. S. Suryanarayana, J. Kaur, and V. Dubey, "Study of propagation of ion acoustic waves in argon plasma," *J. Mod. Phys.* **1**, 281 (2010).

26. S. Bose, "Plancks Gesetz und Lichtquantenhypothese," *Z. Phys.* **26**, 178 (1924).

27. A. Einstein, "Quantentheorie des einatomigen idealen Gases," *Sitzungsber. Kgl. Preuss. Akad. Wiss.* **1924**, 261 (1924).

28. A. Einstein, "Quantentheorie des einatomigen idealen Gases," *Sitzungsber. Kgl. Preuss. Akad. Wiss.* **1925**, 3 (1925).

29. M. H. Anderson, J. R. Ensher, M. R. Matthews, C. E. Wieman, and E. A. Cornell, "Observation of Bose-Einstein condensation in a dilute atomic vapor," *Science* **269**, 198 (1995).

30. N. N. Bogoliubov, "On the theory of superfluidity," *J. Phys.* **11**, 23 (1947).

31. T. D. Lee, K. Huang, and C. N. Yang, "Eigenvalues and eigenfunctions of a Bose system of hard spheres and its low-temperature properties," *Phys. Rev.* **106**, 1135 (1957).

32. M. R. Andrews, D. M. Kurn, H.-J. Miesner, D. S. Durfee, C. G. Townsend, S. Inouye, and W. Ketterle, "Propagation of sound in a Bose-Einstein condensate," *Phys. Rev. Lett.* **79**, 553 (1997). Erratum: *Phys. Rev. Lett.* **80**, 2967 (1998).

33. P. Bedaque and A. W. Steiner, "Sound velocity bound and neutron stars," *Phys. Rev. Lett.* **114**, 031103 (2015).

9

Thermal Lens Calorimetry

Introduction

The thermal lens effect results when a laser beam passing through a sample is absorbed, causing heating of the sample along the beam path. As shown below, the sample's heating may in turn modify the transmittance of the same laser which caused the heating. Specifically, sample heating may cause expansion of the transmitted laser spot size in an effect known as "thermal blooming" or "thermal lensing." This relatively simple effect has been described in the literature as the basis for ultrasensitive analytical techniques.[1,2] At least one prior experiment has been described for an undergraduate instrumental laboratory.[3] The utility of thermal lens calorimetry is derived from the brightness of laser light sources and the sensitivity with which laser beams may be detected. In the experiment described here, a simple experimental apparatus using inexpensive components is employed to observe the thermal lens effect and to derive useful experimental data from its magnitude and time dependence. Specifically, the heat capacity of different solvents and the extinction coefficients of blue dye solutions are determined.

The origin of the thermal lens effect is easy to understand. Virtually any kind of laser may exhibit this effect in liquid, solid, or gaseous samples. The first requirement is a sample which absorbs the laser radiation to be used. In the present experiment, we use a red He-Ne laser at 632.8 nm with various blue solutions. Blue solutions absorb a red laser beam passing through them, resulting in the desired heating along the laser path. The total amount of heating depends on how strongly the solution absorbs the laser and on the power of the laser. However, since the spatial intensity profile of the laser is non-uniform, the heating will not be evenly distributed throughout the sample. It is greatest at the center of the laser beam and less toward the outer edges. For the common case of a Gaussian beam profile, which applies to many continuous lasers including the He-Ne, the spatial variation in intensity is a well-characterized function. The heating caused by absorption of a Gaussian beam has radial symmetry along the laser path which creates a corresponding radial temperature gradient in the sample. The refractive index of most materials, including the solvents used here, decreases with increasing temperature. Therefore, corresponding to the temperature gradient in the sample there is a gradient in the refractive index, with the refractive index lowest at the center of the beam path. Effectively, then, the laser path is shorter at the beam center. This effect, with the radial symmetry, makes a diverging lens

Laser Experiments for Chemistry and Physics. First Edition. Robert N. Compton and Michael A. Duncan.
© Robert N. Compton and Michael A. Duncan 2016. Published in 2016 by Oxford University Press.

out of the sample. If the concentration of the solution is low enough so that the laser is not absorbed completely, part of its light will be transmitted through the sample, experiencing the lens. The transmitted part of the same laser that heats the sample to cause lens formation can be observed to diverge, or "bloom," as it passes through the sample. Blooming can be detected visually or instrumentally with a photodiode positioned on the center of the laser beam axis. When blooming occurs, the diode detects a loss in laser intensity which is proportional to the absorption strength.

Another interesting aspect of thermal lensing is that the heating and subsequent lens formation is not instantaneous. It takes a finite time to develop depending on the laser power and the thermal properties of the sample. In solution, the heat capacity and thermal conductivity determine the time for blooming to occur (typically milliseconds). To best observe the thermal lens effect, laser light is focused with a lens at a precise position on the sample in the laser path. The light from the CW He-Ne laser is turned off and on with a rotating chopper. By correct choice of components, lens formation will occur during the "on" cycle of the laser through the chopper, and it will dissipate by cooling during the "off" cycle, so that the effect can be observed repetitively. Detailed equations predicting the magnitude and time behavior of thermal lensing are given below, which make it possible to select components and optimize the performance of the experimental system.

Theory

To derive useful data from the thermal lens waveform, it is important to consider briefly some more detailed concepts about the focusing properties of laser beams.[4] As shown in Figure 9.1 (see also discussion in Chapter 2), a laser focused with a lens does not achieve an arbitrarily small spot size. In Gaussian laser beams, the actual spot size is limited by diffraction to a minimum beam waist, ω_0, defined as half the beam diameter. This minimum beam waist is related to the waist before focusing, ω_i, by

$$\omega_0 = (\lambda f) / (\pi \omega_i) \tag{9.1}$$

λ is the wavelength of the laser used and f is the focal length of the lens. Therefore, a shorter focal length lens results in a smaller spot size at the focus. While this equation demonstrates the effects of the diffraction limit, it cannot be applied quantitatively unless the incoming laser beam diameter and lens thickness fall within the limits of the so-called "thin lens approximation." The idealized thin-lens equation is presented here to show how the focal length varies with wavelength, etc. For a real laser beam with finite lens thickness, the minimum spot size should be determined experimentally. Simple observation of the laser spot is difficult because of its brightness. A good way to determine the spot size in practice is

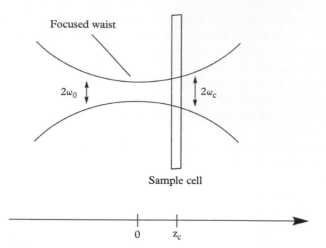

Figure 9.1 *The focusing of a Gaussian laser beam.*

to place an adjustable aperture on the beam at its focus and close this until the transmitted beam begins to lose its brightness. The laser can then be blocked, and the aperture size measured with a micrometer.

As shown in Figure 9.1, the focused beam achieves a minimum diameter, maintains nearly this same diameter for some distance, and then diverges again. A convenient measurement of the focusing is the "confocal length," z_c, which is the distance from the center of the minimum waist region to a point downstream where the beam size is $\sqrt{2}$ larger, i.e., $\omega_c = \sqrt{2}(\omega_0)$. It can be shown that the maximum thermal lens signal occurs when the sample is placed at the point z_c.[1]

When the sample is positioned at z_c, a simple expression can be used to obtain the absorbance, A, of the solution,

$$\Delta I_{bc}/I_{bc} = -2.303P(dn/dT)A/(\lambda\kappa)$$
$$= 2.303 \text{ EA} \tag{9.2}$$

Here, I_{bc} is the initial intensity measured at the center of the laser beam and ΔI_{bc} is the loss in intensity after the blooming has reached its steady state value. P is the laser power in watts, dn/dT is the variation of the refractive index with temperature, λ is the wavelength, and κ is the thermal conductivity of the solvent. E represents an enhancement factor reflecting the improved sensitivity of the thermal lens effect over simple absorption experiments. It depends on the wavelength and the laser power used. E values for selected solvents are given in Table 9.1. As shown, the enhancement is greatest for nonpolar solvents which usually have high dn/dT and low thermal conductivity. The equation above applies strictly to the case of a "weak" thermal lens. When the relative change in I_{bc} is greater than 0.1, a higher order quadratic term should be included in the analysis, i.e.,

$$\Delta I_{bc}/I_{bc} = 2.303 \text{ E A} + (2.303 \text{ E A})^2/2 \tag{9.3}$$

Table 9.1 *Thermo-optical data for solvents used in thermal lens experiments.*

Solvent	C_p (J/mol·K)	κ (mW/cm·K)	dn/dT 10^{-4}/K	E/mW	ρ (g/cm^3)
acetone	126.4	1.60	−5.0	2.16	0.7899
benzene	135.6	1.44	−6.4	3.05	0.8765
CCl$_4$	132.6	1.02	−5.8	3.88	1.5867
diethyl ether	172.0	1.30	−5.0		0.7138
ethanol	113.0	1.67	−3.9		0.7893
methanol	81.6	2.01	−3.9	1.33	0.7914
toluene	156.1	1.33	−5.6		0.8669
water	75.3	6.11	−0.8	0.09	1.000
cyclohexane	152.3	1.24	−5.4	3.6	0.7785

As mentioned above, the time variation in the thermal lens effect can also be used to determine the heat capacity, C_p, of the solvent. The 1/e lifetime of the exponential decay, t_c, can be determined from

$$t_c = \omega_0{}^2 \rho C_p/(4\kappa) \qquad (9.4)$$

where ω_0 and κ are defined above and ρ is the density of the solvent.[1]

Methodology

A diagram of the thermal lens experiment is shown in Figure 9.2. The laser used is a He-Ne (Uniphase model 1202) operating at 632.8 nm. This laser provides a nominal output of 10 mW with a divergence of about 1.0 mrad. The laser is passed through a 60 Hz chopper (homemade), bounced with a mirror mounted on an x,y tilting mount, and then focused with a positive lens (focal length 80–100 mm). Although alignment without the mirror mount can in principle be achieved by moving the He-Ne laser body, utilization of the mirror is highly recommended. At a critical distance beyond the focus of the laser beam, which must be adjusted to optimize the lens effect, the diverging laser passes through a 2-mm thick cuvette containing the sample at a carefully measured concentration. Adjustment of the sample-to-lens separation is accomplished with a basic translation stage. At a point about 1 m beyond the sample, an adjustable aperture is positioned on the laser beam. The light transmitted through this aperture is detected by a photodiode. The output of the photodiode is displayed on an oscilloscope, from which the experimental data of light intensity versus time is acquired and transferred to a PC for plotting and analysis.

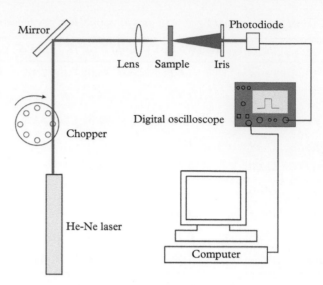

Figure 9.2 *The schematic diagram of the thermal lens experiment.*

Table 9.2 *Compounds which exhibit the thermal lens effect with a He-Ne laser.*

Compound	Mol. Wt.	Solvent	Conc. (M/l)	ε(l/mol·cm)
acid blue 25	416.4	MeOH	2.6×10^{-4}	4540
indophenol blue	276.3	MeOH	2.3×10^{-4}	3260
bromophenol blue	691.9	acetone	2.0×10^{-4}	2475
azulene	128.16	acetone, MeOH	2.4×10^{-3}	254

The thermal lens effect may be observed in a variety of blue colored solutions using various solvents. [5,6] Table 9.2 lists suggested compounds and typical concentrations for this experiment. Table 9.1 provides a list of potential solvents for these experiments and their thermo-optical properties. Several solute/solvent combinations from these lists are possible. If alternate solutions are used, the concentrations may have to be adjusted.

It is easiest to align the optical system initially without the sample or the focusing lens in place. Bounce the unchopped laser with a mirror mounted in the tilting holder through the aperture onto the photodiode. The output of the diode should be connected to one of the oscilloscope channels. With the scope triggering set on "auto" or "line," it should be possible to observe a DC voltage offset resulting from light detected by the photodiode. This level will be reduced if the aperture limits the laser spot size, and decreased to zero if the laser is blocked. These simple tests ensure that the photodiode and the oscilloscope are working properly. A 60-Hz level superimposed on the detected laser light level may be observed in rooms illuminated with fluorescent discharge lighting. This unwanted noise can be eliminated by turning the room lights out.

When the chopper is activated, and the scope triggering is set to "internal," the DC level should be replaced by a series of equally spaced square waves. The number of waves per time period and their separation can be used to verify the frequency and duty cycle of the chopper. A 60-Hz rate is convenient for the experimental components described here. However, it may be useful to vary this rate slightly to optimize performance. When the sample is inserted into the optical path, the square wave pattern on the scope will be diminished in intensity due to absorption of the laser beam by the sample. Adjustment of the voltage scale on the scope will restore the waveform to full scale on the viewscreen. If the sample is too concentrated, the transmitted light will not be bright enough to detect easily with the diode and the scope may not trigger properly. Erratic or irregular scope traces and/or low square wave intensity can be corrected with a more dilute sample. However, if the sample is too dilute it will be difficult to locate the thermal lens signal. Some experimentation may be required to find an acceptable compromise in concentration.

The focusing lens should be the final optical element inserted into the beam path. If the lens is not positioned with its center on the laser beam, the laser will be deflected off the diode detector. The lens position should be adjusted to find the center of the optical axis already defined by the other components. Either the lens or the sample should be mounted on the translation stage to adjust the sample-to-lens spacing. Again, if the lens is not centered on the laser beam, or if the translation stage is not straight, adjustment of the lens-sample separation will cause the laser beam to "walk" off the detector. The desired thermal lens effect is extremely position sensitive. It is best to begin with the sample located near the focal length of the lens. By small displacements to either side of this position, the desired waveform shown in Figure 9.3 can be obtained on the oscilloscope.

As shown, the detected voltage versus time waveform rises initially as it did with no sample, but it falls exponentially toward later time, eventually converging to some final value. As shown below, the time dependence and the final value of this waveform are the desired quantities from which the final results will be calculated. At other lens positions nearby, an inverse lens effect may be observed which gives rise to a similar curved waveform, but one which rises in intensity with time. This is not the desired shape. With stable optical mounts and a little patience the correct waveform can be achieved.

Once the thermal lens waveform has been optimized to obtain the correct shape and maximum depletion, two significant parameters are derived. The first is the ratio of the loss in intensity at the beam center, ΔI_{bc}, to its initial value, I_{bc}, i.e. $\Delta I_{bc}/I_{bc}$. This ratio will be used to determine the extinction coefficient of the solute, as described below. The second parameter of interest is the time constant for the decay of intensity, t_c. Since the time decay is exponential (i.e., a first-order decay process), the time constant is defined as the point where the I_{bc} value has fallen to a value $1/e$ times the total intensity change. In other words, at t_c the

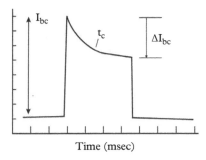

Time (msec)

Figure 9.3 *The oscilloscope waveform showing the time-dependent thermal lens effect.*

intensity will be I_c, where $I_c = I_{bc} - [\Delta I_{bc}(1-1/e)]$. This value can be estimated visually from the oscilloscope. If the data is collected with a storage scope, the waveform can be transferred to a computer and fit to an exponential decay via computer data analysis software, e.g., in an Excel worksheet.[7] The final value of t_c will be used to calculate the heat capacity, as described below. $\Delta I_{bc}/I_{bc}$ and t_c values for several solutions in different solvents can be accumulated in a single lab period.

Data analysis

The first step of this experiment is to determine the laser spot size at the sample. To do this, measure the focused laser spot size. This is ω_0. Then determine the spot size at the sample. The maximum thermal lens effect is obtained at the point one confocal length beyond the minimum waist, where the beam has expanded to about $\sqrt{2}$ times the minimum waist. Then, using the exponential lifetime and ΔI_{bc} values, determine the heat capacity for the solvent and the extinction coefficient for the dye solution. Compare the data obtained to the appropriate literature values.

Safety

The toxicological effects of the organic dye compounds used in this experiment are not fully documented and they should therefore be handled with due caution. In particular, contact with skin or eyes should be avoided. Proper safety precautions are especially important when operating any laser equipment. The main concern from a He-Ne laser such as that recommended here is retinal damage which may result from direct on-beam viewing of the laser. Side-on viewing of the light scattered from dust particles in the air, viewing of the spot of the laser striking a darkened card, or laser impact on skin are not generally considered hazardous with low-powered He-Ne systems. To prevent accidental ocular impact, protective eyewear, or goggles, are highly recommended. *Laser-Gard* model LGS-HN, which are designed for use with He-Ne lasers, provide enough transmission (20%) to view and align the beam safely. See Chapter 6 for details about laser safety.

....

REFERENCES

1. J. M. Harris and N. J. Dovichi, "Thermal lens calorimetry," *Anal. Chem.* **52**, 695A (1980).
2. H. L. Fang and R. L. Swofford, in *Ultrasensitive Laser Spectroscopy*, D. S. Kliger, ed., Academic Press, New York, 1983, p.175.

3. S. R. Erskine and D. R. Bobbitt, "Detecting trace amounts of a colored compound by He-Ne laser-based thermal lens spectroscopy," *J. Chem. Educ.* **66**, 354 (1989).

4. D. O'Shea, W. R. Callen, and W. T. Rhodes, *Introduction to Lasers and their Applications*, Addison-Wesley, Reading, MA, 1977.

5. J. E. Salcido, J. S. Pilgrim, and M. A. Duncan, "Time-resolved thermal lens calorimetry with a helium-neon laser," *Physical Chemistry: Developing a Dynamic Curriculum*, American Chemical Society, Washington, DC, 1993, p. 232.

6. K. Seidman and A. Payne, "The determination of the heat capacities of liquids with time resolved thermal lens calorimetry: A more accurate procedure," *J. Chem. Educ.* **75**, 897 (1998).

7. The exponential decay in this data converges to its long-time limit at a vertical intensity whose numerical value is non-zero. However, an exponential fit usually requires convergence to zero at long time. To account for this, the data can be off-set via subtraction of a vertical constant before exponential fitting.

10

Laser Refractometry

Introduction

One of the most basic optical properties of matter is the refractive index. Different materials have individual values for the refractive index because light travels at different speeds through them. The speed of light in air is only slightly slower than the vacuum value, but the values for solids or liquids can be significantly lower than the vacuum value. The refractive index, n, is the ratio of the speed of light in vacuum to that in a particular material, $n = c/v_i$. In Chapter 7, we reported pulsed laser flight-time methods to measure the speed of light in different materials (air, iodine vapor, fiber-optic cable). These measurements are equivalent to determining the refractive index of these materials. Similar experiments could be done for a variety of liquid, solid, or gas samples. In the present experiment, we use a different method to determine the refractive index involving the refractive bending of a laser beam. Similar experiments have been described previously in the literature.[1–4]

Because light travels at different speeds in different materials, light rays refract, or bend, as they travel through an interface between two materials. The degree of refraction is described by Snell's Law, which relates the angle of incident and refracted rays (measured from the normal of the interface) to the refractive indices of the two materials: $n_1 \sin \alpha = n_2 \sin \beta$. The angles for refraction are defined in Figure 10.1.

There are many designs for refractometers to measure the refractive indices of different materials. All of these devices have the common feature of a collimated beam of light passing through the material of interest, with measurement of the angle of refraction. The most commonly used refractometer is the *Abbe refractometer*. In the present experiment, however, we illustrate the design and use of a simple homemade refractometer which employs a He-Ne laser. Similar devices have been described previously for the measurement of liquid samples and laser based *Michelson interferometer* devices have been described for gaseous samples.

The design for the laser refractometer to be used here is shown in Figure 10.2. It consists of a He-Ne laser which provides collimated light at 632.8 nm, a semicircular dish to contain the liquid sample (Ward's Science—formerly CENCO), a pinhole aperture to center the laser, and a display board on which to measure the deflection angle of the laser. The sample dish is mounted on a sliding block which fits into a track on the mounting board, with motion orthogonal to the general direction of the laser light. All of these components are mounted firmly in

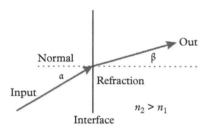

Figure 10.1 *Refraction at an interface between two materials with different refractive indices.*

Laser Experiments for Chemistry and Physics. First Edition. Robert N. Compton and Michael A. Duncan.
© Robert N. Compton and Michael A. Duncan 2016. Published in 2016 by Oxford University Press.

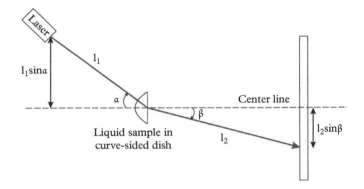

Figure 10.2 *The schematic of the He-Ne laser refractometer for liquid sample measurements.*

place together on a rigid mounting board. Special care must be taken in how the sample is situated in the laser beam for a refraction measurement. First of all, it is apparent from Snell's law that there is no refraction at normal incidence. Thus, for example, if a liquid sample in a cuvette with rectangular cross-section is placed in the laser path with the light normal to the cuvette face, the laser will simply pass directly through the sample without deflection. If the incident laser light approaches such a rectangular cuvette at an angle away from normal, refraction will occur but it takes place at both the incoming and outgoing interfaces by complementary amounts and there will again be no change in the laser angle at a point beyond the sample. The special curved-dish sample holder makes it possible to observe a net deflection after passing through a liquid sample. The sample dish is movable to position the incoming laser beam normal to the outside curved wall of the dish. At this point the normal incidence results in no refraction and the laser passes straight forward into the sample. At the flat exit surface, however, the laser beam will be away from normal incidence and there will be net refraction. The angle of the outgoing beam can then be used to determine the refractive index of the liquid sample via Snell's Law.

Procedure

First position the laser aperture at the center of the mounting board flush with the surface of the sample dish exit face in a position on the center line of the mounting board. With no sample dish in place, turn on the laser and adjust it so that it passes through this aperture and hits the downstream display board. Fix the aperture and clamp the laser firmly in place to make sure that the laser light continues to pass through the aperture as the sample dish position is adjusted. Add the appropriate liquid sample to the curved dish and place the dish on its rectangular mounting block within the slide tracks on the mounting board. Slide the sample across the board into the path of the laser. As the laser hits the curved exterior of this dish, there will be a small amount of reflected light which bounces back into the general direction of the laser. Slide the sample to a point so that this reflected light spot goes exactly back into the exit aperture hole on the front of the laser. At this position, when the reflected light follows the exact reverse path as the incoming light, the laser beam is normal to the exterior surface of the curved dish and there is no refraction at that surface. The laser beam should now be in a position to pass through the liquid sample with refraction at the outgoing surface, and it should still be passing through the centering aperture and hitting the downstream display wall.

It is now possible to determine the two angles from which one can calculate the angle sines necessary for the application of Snell's Law by measuring distances on the mounting board with a ruler. The incident angle (α) is determined via the dimensions of the imaginary triangle defined by the laser beam path from the laser exit hole to the center aperture, a line perpendicular to the center line which intersects the laser output hole, and the segment of the center line from the center aperture to the perpendicular. α is then the angle between sides of the triangle, and $\sin \alpha$ is given by the ratio of side lengths. In a similar way, an imaginary triangle on the exit side of the sample can be defined by the center line from the aperture to the downstream wall, the laser path distance from the aperture to the display wall and the displacement of the spot on the display wall from the center line. β is then defined by the side's length. The n_1 value in this configuration is that for the sample, and the n_2 value is that for air. Substitution of the measured and known quantities allows determination of the unknown value n_1. In this way, simple length measurements with a ruler allow the determination of the refractive index for the liquid sample (at 632.8 nm).

Make refractive index determinations for several liquid samples such as water, methanol, benzene, etc. Include some heavier oils such as household olive oil. Compare the values you obtain to those from the literature (e.g., [5]). Be sure to notice that the literature values are generally cited for the standard wavelength of 589 nm (the wavelength of the sodium "D" line), whereas the measurements here are with a He-Ne laser at 632.8 nm. The refractive index varies with wavelength, and this can be studied by substituting the red He-Ne laser wavelength with the green or blue light from modern laser pointers, which have high enough power (10–50 mW) for many such applications. Study the dependence of the refractive index on solutions, such as salt water. If possible, study the dependence of refractive index on temperature, by warming or cooling the sample, and derive dn/dT. Discuss the systematic and random errors associated with this kind of refractometer and describe how the device could be modified to improve its performance.

The apparatus described here is attractive because of its simplicity. However, the temperature dependence of the refractive index is a particularly interesting aspect of this experiment. In a more advanced version of the experiment, the present curved-dish liquid sample holder could be replaced with a somewhat more elaborate container allowing better insulation and precise temperature control of the sample. In particular, it is interesting to investigate temperatures near the critical point of a material, where the index of refraction of the liquid approaches that of the vapor.

....

REFERENCES

1. B. Spencer and R. N. Zare, "Laser based measurement of refractive index changes: Kinetics of 2,3-epoxy-1-propanol hydrolysis," *J. Chem. Ed*. **65**, 835 (1988).
2. E. D. Noll, "Measuring the index of refraction of liquids with a laser. Experiment I," *Phys. Teach*. **11**, 307 (1973).
3. E. D. Noll, "Measuring the index of refraction of liquids with a laser. Experiment II," *Phys. Teach*. **11**, 309 (1973).
4. G. R. Van Hecke, K. K. Karukstis, and J. M. Underhill, "Using lasers to demonstrate the concept of polarizability: Variation in the refractive indices of the *o*-halobenzenes," *The Chem. Educ*. **2**, S1430-4171(97)05147-X (1997).
5. W. M. Haynes, ed., *CRC Handbook of Chemistry and Physics*, ninety-sixth edition, CRC Press, Boca Raton, FL, 2015–2016.

Part III

Laser Experiments
for Chemical Analysis

Laser-Induced Breakdown Spectroscopy

Introduction

This chapter introduces a very simple but highly useful method of laser spectroscopy called Laser-Induced Breakdown Spectroscopy (LIBS). LIBS has become a standard tool in many areas of analytical spectroscopy. Two recent publications fully describe the science underlying LIBS and highlight many of the applications in this established and growing field.[1,2]

In this technique a sufficiently intense pulsed laser beam is focused into a gas or onto a liquid or solid in order to ablate the sample into atoms, excited atoms, ions, excited ions, and electrons. If the electric field of the laser exceeds the dielectric strength of the material, dielectric breakdown occurs. The electrons are rapidly heated by inverse bremsstrahlung (photon absorption by an electron in proximity to another atom, ion, or electron in order to conserve momentum). The temperature of this plasma can approach 10^6 K. Over a short period of time after the initial plasma formation the electrons and ions recombine to form ground- and excited-state atoms. Aided by the presence of a buffer gas, the excited atoms can later recombine into ground- and excited-state molecules. An approximate timescale for these processes is illustrated in Figure 11.1.

In the initial electron–ion phase in the developing plasma, energy is dissipated through bremsstrahlung continuum emission, radiation from excited ions,

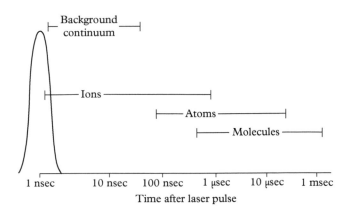

Background
continuum

Ions

Atoms

Molecules

1 nsec   10 nsec   100 nsec   1 μsec   10 μsec   1 msec

Time after laser pulse

Figure 11.1 *Approximate representation of the time sequence of events following laser-induced breakdown of a solid sample. The laser pulse is ~1 nsec in duration as shown. The background continuum is black-body radiation and bremsstrahlung radiation generated in the initial electron–ion plume. Ions, atoms, and molecules follow after the initial laser matter interaction.*

and diffuse interstellar clouds.[8] The C_2 molecule has six singlet and seven triplet states which have been well characterized. These states produce nine band systems extending from the UV to the IR region: Swan band $(d^3\Pi_g \rightarrow a^3\Pi_u)$, Phillips band $(A^3\Pi_u \rightarrow X^1\Sigma_g^+)$, Mulliken band $(D^1\Sigma_u^+ \rightarrow X^1\Sigma_g^+)$, and the Deslandres–d'Azambuja band $(C^1\Pi_g \rightarrow A^1\Pi_u)$. The Swan system lies between 420 and 770 nm and is the strongest and most easily observed of these bands. Figure 11.3 shows the energy levels giving rise to the vibrational bands of the Swan system.

Transitions between the v' vibrational levels of the upper $d^3\Pi_g$ state to the vibrational levels v" of the lower $a^3\Pi_u$ state are depicted approximately in Figure 11.4. The experimental arrangement used to perform LIBS spectroscopy of graphite is shown in Figure 11.5.

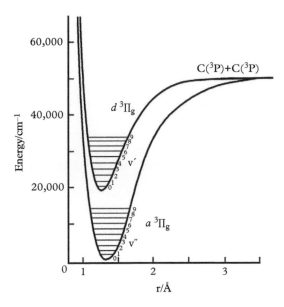

Figure 11.3 *Potential energy curves showing the $d^3\Pi_g \rightarrow a^3\Pi_u$ states of the C_2 molecule which gives rise to the Swan bands.*

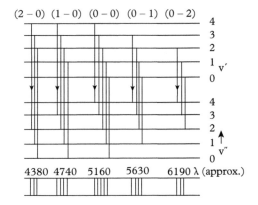

Figure 11.4 *Approximate energy level diagram and wavelength scale showing the v' to -v" transitions in the Swan band system $d^3\Pi_g \rightarrow a^3\Pi_u$.*

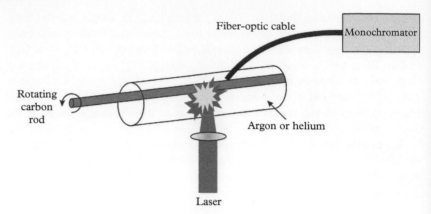

Figure 11.5 *Experimental layout of the geometry for recording the LIBS spectroscopy of laser ablated graphite.*

The graphite rod can be slowly rotated by hand or a stepping motor arrangement to expose new surface for each laser pulse during the experiment. Without rotation the focused laser will dig holes into the rod which reduces the volume of the plume and therefore the light generated. The intense background light in the early stage of the plume can be reduced in two ways. The detector can be gated to record spectra at various times after the laser firing in order to reduce the continuum background, or a slit can be employed to image only the later part of the plasma in order to enhance the observation of atomic and molecular features. Figure 11.6 shows the Swan bands for $\Delta v = 0$ using two different buffer gases.

A spectrum showing the Swan band corresponding to $\Delta v = -1$ is shown in Figure 11.7.

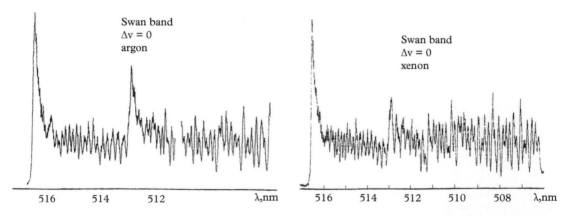

Figure 11.6 *LIBS spectroscopy of graphite resulting in $\Delta v = 0$ for the Swan band. The two different buffer gases argon (left) and xenon (right) were used to cool the expanding plasma. A XeCl excimer laser (308 nm) having 15 nsec pulse duration at 30 mJ energy was focused to produce 5 J/cm^2.*

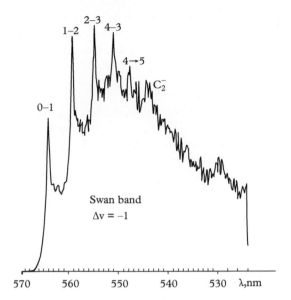

Figure 11.7 *Swan band corresponding to* $\Delta v = -1$. *The feature labeled* C_2^- *may correspond to transitions between vibrational levels of the* C_2^- $B \rightarrow X$ *system [B* $^2\Sigma_u^+ \rightarrow X\,^2\Sigma_g^+$*].[10] A XeCl excimer laser (308 nm) having 15 nsec duration at 30 mJ energy was focused to produce 5 J/cm².*

Table 11.1 *Spectroscopic constants of* C_2 *in its a and d states.*

	T_e (cm^{-1})	ω_e (cm^{-1})	$\omega_e x_e$ (cm^{-1})	r_e (Å)
$a\,^3\Pi_u$	716.24	1641.35	11.67	1.3119
$d\,^3\Pi_g$	20022.50	1788.22	16.440	1.2661

In these experiments, the light was detected ~3 µsec after the initial laser pulse in order to reduce the background continuum emission. This is not necessary but provides much cleaner spectra. The vibrational features in Figures 11.6 and 11.7 can be used to extract the molecular constants for the $d\,^3\Pi_g$ and $a\,^3\Pi_u$ states of $^{12}C_2$. Also the rotationally resolved spectra in Figure 11.6 can be used to obtain the internuclear bond distance for the states as well. For comparison, the molecular constants for these states are given in Table 11.1 (taken from [9]). Refer to Chapter 14 on the Morse potential in order to extract these constants from the data. More details of LIBS spectroscopy of graphite can be found in the paper by Puretzky et al.[11]

A simple LIBS setup in air

Below we describe a relatively simple LIBS experiment which consists of a Nd:YAG laser operating at 1064 nm and an inexpensive position sensitive compact CCD spectrometer. A Quanta-Ray Nd:YAG laser is focused onto the surface of graphite in air. Light from the laser breakdown is detected by a Thorlabs Compact CCD Czerny-Turner Spectrometer Design Model CCS100 (< $2000). A TTL pulse (+5 V) from a pulse generator can provide a trigger for external synchronization (time delay experiments). The system can accommodate up to 200 scans per second. A picture of the experimental arrangement is shown in Figure 11.8.

A sample spectrum for laser-induced breakdown of graphite in air is shown in Figure 11.9. Easily identified are the 2–0, 1–0, 0–0 and 0–1 (see above) Swan bands. In many cases the spectrum observed for LIBS of a solid sample can depend on the gas above the substrate.

The feature beginning at 388 nm is attributed to emission from CN excited states in the plasma created by reactions between C_2^* molecules in the $d\ {}^3\Pi_u$ excited state (i.e., that responsible for the Swan bands) and N_2 molecules in air to produce CN^* as shown:

$$C_2^*(d\ {}^3\Pi_u) + N_2 + \text{kinetic energy} \rightarrow 2\,CN^*(B\,{}^2\Sigma^+)$$

The emission due to $CN^*\,(B\,{}^2\Sigma^+) \rightarrow CN\,(X\,{}^2\Sigma^+)$ is expanded in Figure 11.10.

The spectroscopic constants for the $B\,{}^2\Sigma^+$ state can be estimated from the spectrum shown in Figure 11.10 and compared with the values listed in Huber and Hertzberg [9]:

$$
\begin{aligned}
&X\,{}^2\Sigma^+\,T_e \equiv 0 && \omega_e = 2068.59\,\text{cm}^{-1} && \omega_e x_e = 13.087\,\text{cm}^{-1} \\
&B\,{}^2\Sigma^+\,T_e = 25,752.0 && \omega_e = 2163.9\,\text{cm}^{-1} && \omega_e x_e = 20.2\,\text{cm}^{-1}
\end{aligned}
$$

Figure 11.8 *Experimental setup of a simple LIBS apparatus consisting of a laser beam (depicted here in red for visual purposes) and a focusing lens before the graphite sample. Light is collected through a fiber-optic cable into the CCD spectrometer. The laptop computer in the background records and displays the data.*

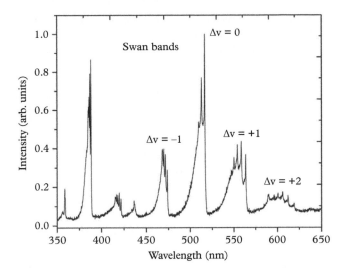

Figure 11.9 *LIBS spectrum of graphite in air recorded with the apparatus shown in Figure 11.8. Note the Swan bands as discussed above and the new feature beginning at ~388 nm which is due to* CN* *(B $^2\Sigma^+$) as discussed in the text.*

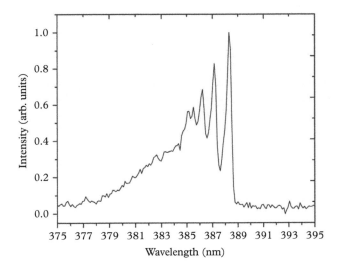

Figure 11.10 *LIBS emission from* CN* *(B $^2\Sigma^+$) → CN (X $^2\Sigma^+$).*

Experiments using LIBS of metals and other solid samples can be performed readily using this simple setup. Also students could try LIBS under different gases to obtain spectra for the molecules made of the substrate and the gas above the substrate. Any sufficiently intense nitrogen laser, excimer laser, or Nd:YAG laser can be employed. Special eye safety care should be taken using the 1064 nm Nd:YAG or excimer UV laser light. In many cases it is best to employ a long-wavelength or short-wavelength laser in order to reduce the amount of light interference in the region of interest. With proper eye safety procedures the observation of the laser plume using a hand-held spectroscope can be used to observe atomic lines from metals such as magnesium or alkali atoms, etc.

....

REFERENCES

1. A. W. Miziolek, V. Palleschi, and I. Schechter, eds., *Laser Induced Breakdown Spectroscopy*, Cambridge University Press, NY, 2006.
2. D. E. Cremers and L. J. Radziemski, *Handbook of Laser-Induced Breakdown Spectroscopy*, John Wiley & Sons, New York, 2006.
3. L. J. Radziemski, T. R. Loree, and D. A. Cremers, "Laser induced breakdown spectroscopy (LIBS): A new spectrochemical technique," *Springer Ser. Opt. Sci.* **39** (Opt. Laser Remote Sens.), 303 (1983).
4. J. Belliveau, L. Cadwell, K. Coleman, L. Hüwel, and H. Griffin, "Laser-induced breakdown spectroscopy of steels at atmospheric pressure and in air," *Appl. Spectros.* **39**, 727 (1985).
5. J. A. Millard, R. H. Dalling, and L. J. Radziemski, "Time resolved laser-induced break down spectroscopy for the rapid determination of beryllium in beryllium-copper alloys," *Appl. Spectros.* **40**, 491 (1986).
6. F. R. Doucet, G. Lithgow, R. Bouchard, P. Kosierb, and M. Sabsabi, "Determination of isotope ratios using laser-induced breakdown spectroscopy in ambient air at atmospheric pressure for nuclear forensics," *J. Anal. At. Spectrom.* **26**, 536 (2011).
7. B. W. Smith, A. Quentmeier, M. Bolshov, and K. Niemax, "Measurement of uranium isotope ratios in solid samples using laser ablation and diode-laser excited atomic fluorescence spectrometry," *Spectrochim. Acta Pt. B: At. Spect.* **54**, 943 (1999).
8. W. Weltner, Jr. and R. J. van Zee, "Carbon molecules, ions and clusters," *Chem. Rev.* **89**, 1713 (1989).
9. K. P. Huber and G. Herzberg, *Molecular Spectra and Molecular Structure: Constants of Diatomic Molecules*, Van Nostrand Reinhold Company, New York, 1979.
10. G. Herzberg and A. Lagerqvist, "A new spectrum associated with diatomic carbon," *Can. J. Phys.* **46**, 2363 (1968). See also P. L. Jones, R. D. Mead, B. E. Kohler, S. D. Rosner, and W. C. Lineberger, "Photodetachment spectroscopy of C_2^- autodetaching resonances," *J. Chem. Phys.* **73**, 4419 (1980).
11. A. A. Puretzky, D. B. Geohegan, R. E. Haufler, R. L. Hettich, X.-Y. Zheng, and R. N. Compton, "Laser ablation of graphite in various buffer gases," *AIP Conf. Proc.* **288**, 365 (1993).

Laser Desorption Time-of-Flight Mass Spectrometry

12

Introduction

Mass spectrometers are one of the most common pieces of laboratory instrumentation used in research and chemical analysis. A basic understanding of mass spectrometry is therefore desirable for all chemists, as well as for those in related areas of science and technology. The tutorial here and the related experiments provide such a basic understanding of mass spectrometry, using a laser desorption time-of-flight instrument (LD-TOF).[1]

There are many kinds of mass spectrometers used throughout chemistry, physics, and biology. The operational details of these different instruments have been discussed often in different textbooks devoted to mass spectrometry.[2,3] The recent text by Watson and Sparkman [3] is particularly good for experimental details. Examples of modern research instruments found in many industrial and university laboratories include magnetic sector, quadrupole, Fourier Transform (also known as "ICR," ion cyclotron resonance), ion trap, and time-of-flight. Each of these kinds of instruments has its own advantages and disadvantages. For example, Fourier Transform (FT-MS) instruments provide the highest possible mass resolution (better than one part in one million!), but they are very expensive and have somewhat limited mass range ($\leq$ 10,000 amu). Magnetic sector instruments and FT instruments require heavy magnets, and are nearly impossible to move once they are placed in a lab. Most FT instruments require liquid helium to function. Ion trap and quadrupole instruments have somewhat limited mass range and moderate resolution, but they can be physically small for bench-top or mobile experiments. We focus here on time-of-flight (TOF) instruments. These have an almost unlimited mass range, but they have lower resolution. However, their simplicity of design and ease of construction and operation make them an attractive option for undergraduate laboratories.

Laser Experiments for Chemistry and Physics. First Edition. Robert N. Compton and Michael A. Duncan.
© Robert N. Compton and Michael A. Duncan 2016. Published in 2016 by Oxford University Press.

Ionization processes

Most mass spectrometers, such as those in a typical chemistry department instrumentation lab, use samples that have sufficient vapor pressures such that they can be studied in the gas phase. Gases can be injected directly into the instrument, or the vapor in equilibrium over a liquid sample can be used. In electrospray ionization (ESI),[4] samples can be injected as liquids into the gas phase. However, samples that are solids at room temperature often have very low vapor pressure and require special handling to achieve sufficient vapor pressure for mass spectrometry. In some cases, simple heating of the sample is sufficient, but in other cases special methods such as laser desorption (LD) are employed. Laser desorption usually is accomplished with a short-pulsed laser such as a nitrogen, Nd:YAG or excimer. In the LD method, the pulsed laser heats the surface and the molecules so rapidly that molecules are vaporized virtually instantaneously. This often avoids decomposition processes that occur when certain samples (e.g., biomolecules) are heated. Because the LD method is pulsed, it is convenient to incorporate it with TOF mass spectrometers, which also operate in a pulsed mode. A variation of the LD method that imbeds biomolecules in an absorbing film is matrix-assisted laser desorption-ionization, known as MALDI.[4] Molecules (and ions) are carried along with the matrix off the surface into the mass spectrometer with minimal dissociation.

After vapor is injected into or produced in the instrument, the next step in any kind of mass spectroscopy is ionization of the sample in the vapor to make positive or negative ions. For normal gas phase molecules, ions are usually produced by electron impact ionization. In this process, electrons produced from a heated wire filament or a metal oxide cathode material are accelerated with an electric field into the source region of the instrument where the gaseous sample is injected. Electron ionization TOF mass spectrometry has been reviewed in detail by Compton and coworkers.[5] The energy of the electron beam must exceed the ionization potential of the atom or molecule to be ionized. Ionization potentials are typically 9–12 eV for most molecules,[2,3] and so the electron beam must have at least this energy. However, the cross-section for electron ionization maximizes at an energy roughly three times the ionization potential.[5] Hence, typical electron beam energies are about 70 eV. The fast electron collision with a molecule then causes ionization. Typical chemical bond energies (see Chapter 1) are 10–200 kcal/mol (about 4–8 eV), thus energetic electrons can also excite antibonding states resulting in fragmentation of the molecule or ion. The analysis of these fragmentation patterns has traditionally been a means for molecular structural analysis.[6] As an example, the ionization process of molecular nitrogen is shown below:

$$N_2 \, (IP = 15.576 \, eV) + e^- (KE = 70 \, eV) \rightarrow N_2^+ + 2e^- \rightarrow N^+ + N + 2e^-$$

Another possible ionization process is photoionization, in which the energy deposited into the molecule by light absorption causes ionization. This is possible if the photon energy exceeds the ionization potential. However, ionization at 9–12 eV requires light in the vacuum ultraviolet (VUV) region of the spectrum (e.g., 9 eV corresponds to 138 nm). Certain kinds of lamps produce VUV light in this region, but the output is weak and the light is absorbed by air, and beam paths must be evacuated, making the experiment more difficult to set up. Lasers are the brightest light sources, however only excimer lasers can deliver VUV photons. Fortunately it is possible to shift visible lasers into the VUV using third harmonic generation (THG) techniques as described in Chapter 26 as well as with Raman up-shifting as described in Chapter 27.

Photoionization is often accomplished with high intensity pulsed lasers using lower energy visible or near ultraviolet photons through the process of "multiphoton ionization" (MPI).[7–9] Experiments on the multiphoton ionization of atoms and molecules are described in Chapters 13, 16, and 18, respectively. The high intensity of pulsed lasers makes it possible for a single molecule or atom to simultaneously absorb more than one photon, and the total energy available for ionization is the sum of these photon energies. Photoionization is more efficient if the wavelength of light used corresponds to the absorption spectrum of the molecule. This leads to the term "resonance enhanced multiphoton ionization," or REMPI. As an example of REMPI, consider the ionization of benzene, shown below:

$$C_6H_6 \text{ (IP = 9.24 eV)} + 2h\nu \text{ (260 nm)} \rightarrow C_6H_6^+ + e^-$$

The wavelength of 260 nm corresponds to the well-known $\pi \rightarrow \pi^*$ resonance in the UV-vis spectrum of benzene. Light at this wavelength has a photon energy of 4.77 eV, and so two photons are sufficient to ionize the molecule. Thus, the ionization process corresponds to one photon absorption producing the excited π^* state, followed by absorption of a second photon resulting in ionization (a so-called 1+1 REMPI process, also known as resonant two-photon photoionization, R2PI), all occurring within the few nanoseconds of the flash of a pulsed laser. The mass spectra of many MPI or REMPI experiments show fragmentation as well. The parent ions are often further dissociated by the absorption of subsequent photons.

These electron impact and photoionization processes work well with normal gas-phase samples, but they are not easily applied to solid samples that have no vapor pressure. Solid samples can include metals, metal compounds (e.g., oxides), salts, polymers, or high molecular weight organics such as fullerenes or polycyclic aromatic hydrocarbons (PAHs). For these species, recent research has shown that it is possible to use a thin film of the sample mounted on the end of a suitable "solids probe" (usually a metal rod with sample on its end) inserted into a mass spectrometer, and then to use pulsed laser radiation to suddenly super-heat the sample and generate vapor. The same laser pulse that causes desorption or vaporization of the sample also causes some fraction of the sample to become ionized. However, the detailed mechanism of how this all happens is still very much in question, and it may vary significantly with different samples.

The mechanism of ionization following pulsed laser desorption of solid samples is believed to involve a complex combination of electron ionization (both positive and negative) and photoionization processes. Multiphoton absorption is definitely possible when a pulsed laser hits a solid, and then the energy of many photons can be deposited instantaneously (within a few nanoseconds) into the molecules in the sample. The material from the super-heated sample then essentially explodes and expands into a plume of material ejected outward into the acceleration region of the mass spectrometer. This plume expansion takes place on the order of 100s of nanoseconds up to a microsecond. Super-heated material may be hot enough to break bonds, resulting in some fragmentation of desorbed molecules. Additionally, the density of the plume can be great enough so that ions collide with neutrals, causing ion–molecule reactions. Finally, most materials in common samples are not perfectly pure and may contain small amounts of salt from glassware. If salt, e.g. NaCl, is present, then its rapid desorption may produce ions directly and detachment of electrons from halide anions may produce electrons in the plume. (These can be seen directly in the negative ion mode of detection which is available on most instruments.) Ions and electrons produced from the solid surface immediately experience acceleration from the fields present in the mass spectrometer source region. This acceleration may induce more energetic ion–molecule collisions in the expanding plume and it may cause electron–molecule collisions. If electron–molecule collisions are energetic enough, electron impact ionization can occur in the plume. In any given experiment, it is actually quite difficult to determine which ionization processes are at work. Different color vaporization lasers can be employed, samples can be cleaned to remove impurities, and it is also possible to desorb material with no acceleration fields turned on and then to pulse-activate the fields after the plume has dissipated (known as "delayed-pulse" or "time-lag" extraction).[5,10] However, these kinds of experiments require additional instrumentation and are not commonly performed. In general practice, the ionization mechanism at work for a particular sample is often not well known.

TOF mass analysis

The basic principles of time-of-flight mass spectrometry (TOF-MS) are simple and were described in the classic paper by Wiley and McLaren in 1955.[10] [5] also covers some of the modern aspects of TOFMS. Ions in a vacuum are accelerated by an electric field. Electric fields are established in a vacuum by situating parallel metal plates, screen wire grids, or some combination of the two in position with electrical feedthroughs providing connections to high-voltage power supplies located outside the vacuum system. The field resulting from two parallel plates is given by the difference in their voltages, V_1 and V_2, and their separation, L, in cm:

$$E = (V_1 - V_2)/L \qquad (12.1)$$

where the field is in V/cm. For example, if plate 1 is at +1000 V and plate 2 is at +500 V and their separation is 2.0 cm, the field is 250 V/cm in the direction of the +500 V plate. The direction of the electric field is defined as the direction a positive charge would move. A charged particle moving through such a field experiences an acceleration in the same way that a ball rolls down a hill (in the absence of any friction). The total energy obtained depends on the initial position in the field and the distance (corresponding to voltage) through which the particle is accelerated. In the case of two plates spaced by 2 cm, a particle (single positive charge) starting at plate 1 would be accelerated toward plate 2. After moving in this field for the full 2 cm, its energy would be 500 eV. If that same ion were produced at the center of the field, its energy after acceleration would be only 250 eV. Various configurations of acceleration plates are used in mass spectrometers, but this simple concept always applies. One additional detail arises when multiply charged ions are produced. It is actually the charge that determines the energy gained by an acceleration field. Therefore, a doubly charged ion moving through the full field above would receive twice the energy, or 1000 eV.

After acceleration through a specific field the final velocity of a particle depends on its mass through the relation,

$$KE = \tfrac{1}{2}\left(mv^2\right) \qquad (12.2)$$

The kinetic energy is given by the acceleration field, as described above, and calculation of the velocity involves a simple units conversion. For example, consider the molecular ion of nitrogen (28 amu) mentioned above, accelerated in an electric field to an energy of 1000 eV. The velocity of this ion is $v = (2KE/m)^{1/2}$. With the unit conversions, this is,

$$v = \left[2(1000\,\text{eV})(1.6 \times 10^{-19}\,\text{J/eV})/(28\,\text{amu})(1.67 \times 10^{-27}\,\text{kg/amu})\right]^{1/2}$$

$$= 8.27 \times 10^4\,\text{m/sec}$$

or 8.27×10^6 cm/sec. Since $v = d/t$, the ion velocity allows the calculation of the time this ion requires to move through a known distance. In TOF-MS, the ions are initially accelerated up to a terminal velocity and then travel through a field-free "drift tube" with a length of about 1 meter. For a 1-meter flight distance, the ion above would take $(1\,\text{m})/(8.27 \times 10^4\,\text{m/sec}) = 1.21 \times 10^{-5}$ seconds, or 12.1 μsec. The benzene ion mentioned above would take 20.2 μsec. Notice that the ratio of the masses of benzene/nitrogen is $78/28 = 2.79$, while the flight time ratio is $t_b/t_n = 1.67$. This demonstrates that the mass is proportional to the square

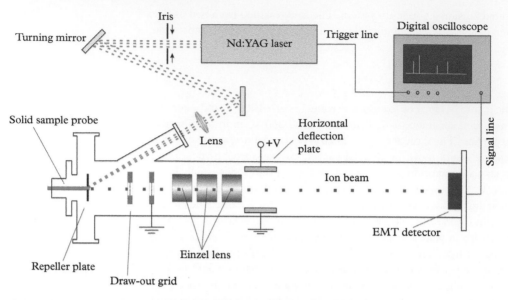

Figure 12.1 *A laser desorption time-of-flight mass spectrometer. (Reprinted with permission from J. Chem. Ed. 91, 291 (2014). Copyright 2014 American Chemical Society.)*

root of the flight time, as indicated in the basic equation above. Thus, the raw data recorded from a TOF spectrometer must be corrected to be presented in a linear mass scale. This is usually done by manipulating the dataset on a computer.

Figure 12.1 shows a schematic of a typical TOF instrument used for laser desorption experiments. The laser hits the sample in the ionization region, where the sample is mounted flush with the rear acceleration plate. This plate has the highest applied voltage. This and other acceleration plates form the two acceleration fields, and beyond this there are deflection plates, an einzel lens, and a charged particle detector at the end of the flight tube. The detector is an electron multiplier tube (Hamamatsu R-595). The two-acceleration field design was first developed by Wiley and McLaren in 1955 to achieve higher mass resolution than the single-stage design.[10] The rear acceleration plate is called the "repeller plate" and the second plate forming the first field zone is the "draw-out grid" (DOG). The repeller plate is usually solid, while the DOG has a screen-covered hole to transmit the ions toward the flight tube. The second field is defined by the DOG on one side, and a second grid on the other, defining the entrance to the field-free flight tube. This flight tube grid also has a screen-covered hole. Deflection plates are used to aim the left–right and up–down direction of the ion beam, while the "einzel lens" (a set of three cylindrical rings with appropriate voltages [11]) acts in the same way that an optical lens does to reduce the radial spread of the ion beam. For TOF mass spectrum measurements, the ionization event (e.g., the laser) must be pulsed to form ions at a precise starting time. A signal pulse at this same time triggers a trace on a digital oscilloscope to measure ion intensity arriving at the detector as a function of time after the ionization pulse, in the microsecond domain. This process is repeated for signal averaging. The signal appears as a series of voltage spikes coming at times corresponding to different masses. To preserve the fast time response (i.e., the sharpness of these spikes), the signal must be terminated into the oscilloscope in 50 Ω (see Appendix II). The raw TOF data is usually transferred to a PC, where the flight-time data can be converted to a mass scale and saved on disk. Our LeCroy digital scope is connected to a PC via an IEEE-GPIB interface. Files can be saved in ASCII format or exported to spreadsheet software such as Excel, Origin, SigmaPlot, etc.

The TOF spectrometer shown here is a "linear" instrument, which is distinguished from the "reflectron" type of instrument now in common usage in many labs.[12,13] A reflectron instrument has two flight tubes connected at a sharp angle by a reflecting field to turn the ion beam down the second tube. This design provides additional mass resolution, but its operation is beyond the scope of our discussion here. TOF mass spectrometers are available from many commercial companies and are a standard instrument found in many chemistry departments. Modular instruments are also available at lower cost for those with experience in constructing such devices.[14,15]

Experimental details

Hazards and safety considerations

The main hazards associated with this lab are exposure to the laser light and to high voltages associated with the mass spectrometer. All personnel should complete an appropriate laser safety training course before operating a Nd:YAG laser (see Chapter 6). Such courses are provided by university safety personnel or are available online. A Nd:YAG is a class IV laser system and severe eye damage is possible from exposure to the main beam or partial reflections from shiny surfaces or optics. If laser safety training is not feasible, exposure can be minimized if the laser is isolated within an appropriate light-tight enclosure. High voltages are typically enclosed in commercial instruments, but may not be in homemade systems. Students are warned against touching any cables or connectors associated with the power supplies. Finally, students should access MSDS forms related to this laboratory and identify hazards and proper handling of the chemicals used.

Instrument operation

The TOF mass spectrometer used for this experiment is shown in Figure 12.1. (The details of instrument operation described here will be different if a commercial instrument is employed.) The laser used for these experiments is a small pulsed Nd:YAG laser (e.g., Continuum Corp., "Minilite" model or New Wave

Research, Polaris II). It includes crystals for the production of light at 532 and 355 nm, with dielectric mirrors to separate the laser harmonics. This laser provides 25 mJ/pulse of light at 532 nm and 4–5 mJ/pulse at 355 nm. The pulse duration of the laser is 5–7 nsec, and it operates at 10 pulses per sec (10 Hz).

Set up the laser for operation at either 532 or 355 nm, as desired for the planned experiment. With the laser adjusted for low-power operation, align the beam so that it passes through the center of the laser window to hit the sample. Even under low-power operation, it is important to wear the appropriate goggles while adjusting the laser alignment. Wear clear plastic UV-blocking goggles for operation at 355 nm or the orange-colored Nd:YAG-blocking goggles for operation at 532 nm. Once the laser is aligned, check the electrical connections. The "trigger out" coaxial cable should be connected from the laser "sync out" or "trigger out" port to the "trigger in" connector on the digital oscilloscope, and the oscilloscope should be configured to accept the "external trigger." This ensures that the oscilloscope begins a data measurement trace at exactly the firing time of the laser pulse and that this process repeats for each of multiple laser shots for signal averaging. After alignment, the laser must be turned up into the higher power mode of operation to measure a mass spectrum. Do this with curtains tightly closed around the laser table or with a box covering the laser area. The signal line from the mass spectrometer detector output must be connected to one of the signal channels of the digital oscilloscope to measure the mass signal as a function of time. Samples are loaded through a vacuum interlock to maintain vacuum in the main mass spectrometer region while the sample is introduced. This is the most critical aspect of the experiment, and one where an accident can crash the entire mass spectrometer. Students should generally not try to load samples without the TA present unless they understand the operation of the interlock system completely. In some instruments, a video camera is available to see the sample when it is in place and to see that the laser is hitting it. If this feature is available, use it to verify sample position and laser alignment. To turn on the mass spectrometer, slowly turn on the high voltages on the different power supplies connected to the repeller plate, DOG, deflection plate, and einzel lens. With the laser turned on, slowly turn up the detector voltage (typically to a level of about 2000 volts). When turning this on, watch the oscilloscope level to avoid saturation of the detector. The TA will indicate the correct voltage levels to use for the different spectrometer grids. In homemade instruments, it is important to avoid contact with the source end of the mass spectrometer when high voltages are on. The voltages used in this experiment are lethal!

Mass spectra are collected and averaged for a selected number of laser shots and displayed with a digital oscilloscope. When a suitable spectrum is obtained, it can be captured and transferred to the computer via the interface cable by running the data collection program on the PC. This saves the data as an x, y array, where the x values represent the TOF and the y values represent the ion intensity. The spectra can be saved and analyzed later. In the analysis, the raw data (signal versus flight time) is converted to a mass spectrum (intensity versus mass).

The manipulations required may be done in the data program provided on the instrument PC, or the data files can be stored on a flash drive and manipulated later using a spreadsheet program such as Excel.

Assignment of TOF mass spectra

Mass calculation

The mass-to-charge ratio of the ions studied determines the kinetic energy they receive upon acceleration in an electric field. In turn, this determines the velocity which the ions have in traversing the flight tube and the TOF, which is the experimentally measured quantity. The analysis presented here assumes singly charged ions.

As described earlier, for an ion with mass m, acceleration in an electric field of value E (in volts per cm) produces an ion at the end of the acceleration region with a kinetic energy, KE, and velocity, v, given by $E = KE = mv^2/2$. In principle, if the electric field is known exactly and the position of ion formation is also known exactly, the transit time to the detector in a TOF experiment could be calculated exactly for a known mass. Alternatively, if the flight time, t, down a flight tube with length, d, is measured, the mass can be determined from:

$$m = 2KE\,(t/d)^2 \qquad\qquad (12.3)$$

To do this exactly would also require that the acceleration time in the source be handled properly. In practice, the exact electric field experienced by the ion is not known because this depends on the position in space where it is formed and first begins to move. The acceleration time in the source region is also inconvenient to calculate. Therefore, the most convenient way to assign a TOF mass spectrum is by comparison to the TOF of a known reference mass. In this situation, since all ions experience the same acceleration field, their kinetic energies are all the same. Therefore, $KE_{ref} = KE_1 = KE_2 = \ldots$ and $m_{ref}v^2_{ref} = m_1 v_1{}^2 = m_2 v_2{}^2 = \ldots$ Since the reference mass is known, an unknown mass can be calculated by the simple ratio:

$$m_1 = m_{ref}(t_1/t_{ref})^2 \qquad\qquad (12.4)$$

The mass analysis program [16] assumes that a reference mass is known, with a corresponding TOF under the same conditions as the data set to be analyzed. The reference mass channel which is input into the program may be one of the peaks in the present mass spectrum, or it may be a channel number determined from a previous spectrum. For example, if two experiments are performed under the same conditions, a known mass channel in one spectrum can be used as the reference for a subsequent spectrum. More commonly, however, the reference mass is a peak in the spectrum under analysis. In practice, it is common to "guess" one channel as the reference channel. A consistent assignment of all the remaining mass peaks confirms that the guess was correct. Such guesses, of course, must be consistent with the chemical species known or believed to be present in the sample.

Accuracy in mass calculations

Two factors limit the accuracy in masses calculated with a simple algorithm like that described above. First of all, an accurate mass determination depends on an accurate TOF measurement. When data is accumulated with typical electronic instruments, errors in the TOF may arise if the zero in the time trace does not correspond to the actual zero time of ion formation. This occurs quite commonly because it is difficult to trigger the oscilloscope at the precise moment of ionization. A trigger delay of a few nanoseconds is common because of the signal transit time of logic pulses through wires of different lengths, or because of the finite speed of the laser light (~1 nsec per foot) through the room to the mass spectrometer. In practice then, equation (12.4) needs to be modified to:

$$m_1 = m_{ref}[(t_1 + t_0)/t_{ref} + t_0)]^2 \qquad (12.5)$$

where t_0 is the offset in the measured time zero. The mass analysis program should therefore allow the user to input an offset in the timescale to correct for an imprecise trigger in the data collection. The exact value of the offset can be determined by a trial and error procedure to get consistent assignments for all the masses, or it can be calculated in the software if two reference masses and their arrival times are entered.

The second limitation on calculated mass accuracy arises from the time channel width used for digital data collection. When data is collected with any digital device, the time channel data used for the mass calculation has propagated errors because each point is no more accurate than the time channel width. For example, if data is collected at a setting of 100 nsec per point, the uncertainty in all data points is ±100 nsec. If a low reference mass (such as helium) is chosen, and its total TOF is 2μsec, the uncertainty is then 5%. When this reference mass is used to calculate an unknown higher mass, the uncertainty is propagated and becomes greater because the reference TOF appears as a squared quantity in the mass calculation. It is therefore likely that the use of a low reference mass will result in a poor accuracy for larger masses, and this error will become progressively worse through the spectrum. If possible, the reference mass should always be chosen near the unknown mass of interest, and the data should be collected with the minimum time channel width to achieve the most accurate TOF measurement and therefore the best mass calculations. If this is done, sub-unit mass assignments can be made. If suitable high-reference masses are not available, errors of several mass units may be obtained in the assignments. The highest mass accuracy is obtained when two knowns are employed whose masses fall to either side of the unknown mass. Analysis of a variety of known mass spectra with different electronics settings and choices for reference masses is recommended to get a feel for the typical accuracy in this software method of mass assignment.

....

REFERENCES

1. R. J. Cotter, *Time-of-Flight Mass Spectrometry: Instrumentation and Applications in Biological Research*, American Chemical Society, Washington, DC, 1997.
2. J. H. Gross, *Mass Spectrometry*, Springer-Verlag, Berlin, 2004.
3. J. T. Watson and O. D. Sparkman, *Introduction to Mass Spectrometry*, fourth edition, John Wiley and Sons, Chichester, UK, 2007.
4. R. B. Cole, ed. *Electrospray and MALDI Mass Spectrometry*, John Wiley, Hoboken, NJ, 2010.
5. N. Mirsaleh-Kohan, W. D. Robertson, and R. N. Compton, "Electron ionization time-of-flight mass spectrometry: Historical review and current applications," *Mass Spectrom. Rev.* **27**, 237 (2008).
6. F. W. McLafferty and F. Turecek, *Interpretation of Mass Spectra*, fourth edition, University Science Books, Mill Valley, CA, 1993.
7. D. H. Parker, J. O. Berg, and M. A. El-Sayed, "Multiphoton ionization spectroscopy of polyatomic molecules," in *Advances in Laser Chemistry*, A. H. Zewail, ed., Springer, Berlin, 1978.
8. L. Zandee and R. B. Bernstein, "Resonance-enhanced multiphoton ionization of molecular beams: NO, I_2, benzene, butadiene," *J. Chem. Phys.* **71**, 1359 (1979).
9. P. M. Johnson, "Molecular multiphoton ionization," *Acc. Chem. Res.* **13**, 20 (1980).
10. W. C. Wiley and L. H. McLaren, "Time-of-flight mass spectrometer with improved resolution," *Rev. Sci. Instrum.* **26**, 1150 (1955).
11. For the operation of an einzel lens see *Building Scientific Apparatus: A Practical Guide to Design and Construction*, second edition, by J. H. Moore, C. C. Davis, and M. A. Coplan, Addison-Wesley, Redwood City, CA, 1989.
12. U. Boesl, H. J. Neusser, R. Weinkauf, and E. W. Schlag, "Multiphoton mass spectrometry of metastables: Direct observation of decay in a high-resolution time-of-flight instrument," *J. Phys. Chem.* **86**, 4857 (1982).
13. D. M. Lubman, W. E. Bell, and M. N. Kronick, "Linear mass reflectron with a laser photoionization source for time-of-flight mass spectrometry," *Anal. Chem.* **55**, 1437 (1983).
14. Comstock, Inc., http://www.comstockinc.com/.
15. R. M. Jordan Company, Inc., http://www.rmjordan.com/.
16. A mass analysis program is usually included as a standard feature on commercial instruments. For those designing and constructing homemade instruments like the one described here, the Duncan lab at the University of Georgia can provide software suitable for running on a PC.

Laser desorption mass spectrometry of fullerenes and PAHs

Introduction

Some of the most unusual organic molecules ever imagined were discovered in the early 1980s when molecular beam experiments used a laser to vaporize solid carbon.[1] A variety of molecules containing only carbon atoms were produced. In the same way that molecules containing only metal atoms are called *metal clusters*, these species are called *carbon clusters*. Analysis of these carbon molecules with a mass spectrometer showed that the cluster containing 60 carbon atoms, C_{60}, is formed with far greater abundance than other sizes and that it is particularly stable. These first experiments that identified C_{60} in molecular beams were done by Professor Richard Smalley and his research team at Rice University, including Professor Robert Curl (another physical chemist at Rice), Professor Harry Kroto (a visiting professor from the University of Sussex), and Rice graduate students Jim Heath and Sean O'Brien. At the time, they could only speculate why C_{60} was so stable and so different from the other clusters. However, Smalley argued that the only structure that could be so stable would be a sphere of carbon formed of interconnected five- and six-membered rings. He and his team then named the molecule "Buckminsterfullerene" after the architect Buckminster Fuller, who specialized in geodesic dome designs.[1] The now-famous structure that he proposed has exactly the same shape as a soccer ball. Because of this, the nickname "Buckyball" soon became associated with C_{60}. Unfortunately, because of the small amount of material produced in molecular beam experiments, the Smalley group could not actually measure this structure to prove that it was correct.

C_{60} might have remained a laboratory curiosity, except for an exciting discovery by Donald Huffman (University of Arizona) and Wolfgang Krätschmer (Max Planck Institute for Nuclear Physics, Heidelberg) in 1990.[2] They found that an electrical discharge made with graphite electrodes in a helium atmosphere generated carbon soot with unusual properties. Part of that soot dissolved in ordinary organic solvents such as benzene and toluene, in which normal soot is completely insoluble. After filtering away the insoluble material and evaporating the solvent, they isolated a yellowish brown powder. With gram quantities available, traditional analysis using an array of specialized instrumental techniques (nuclear magnetic resonance and infrared spectroscopy, X-ray crystallography) became possible. To everyone's delight (especially Smalley's) the powder contained mostly C_{60}, and it had the structure that he had proposed five years earlier! C_{60} thus became the first cluster of a pure element produced initially in the gas phase to be isolated and collected in quantities great enough for traditional chemical experiments. These same experiments also produced several larger, less symmetrical, carbon cage molecules (e.g., C_{70}, C_{84}). Taken together, C_{60} and its analogs are now referred to as the *fullerenes*. A more complete history of the discovery of C_{60}

and the fullerenes has been given in two recent accounts for non-specialists.[3,4] Smalley, Curl, and Kroto won the Nobel Prize in Chemistry in 1996 for this discovery.

The highly symmetrical structure of C_{60} helps to explain many of its unusual properties.[5–8] It is an almost perfect sphere, with 60 atoms arranged in 20 hexagons and 12 pentagons (32 faces in all). Thus, C_{60} has a *truncated icosahedral* structure. Each of the 60 carbon atoms is sp^2 hybridized and occupies an identical site in the cluster. Each has two single bonds and one double bond to its neighbors and is located at the juncture of one five-membered ring and two six-membered rings. This structure is very different from those of other forms of carbon. For example, diamond has tetrahedral bonding to four nearest neighbors. Graphite has an array of planar six-membered rings. However, an array of six-membered rings cannot be bent into a closed cage because of the strain on the bonds. Substitution of precisely the correct number of five-membered rings, as occurs in C_{60}, relieves this strain and makes it possible to form curved structures which can close into cages.

The regular (all faces alike) and irregular (different faces) polyhedral geometries were known mathematically to the Greeks. Euler's Theorem states that the sum of the vertices, V, and faces, F, of any closed polyhedron is equal to the number of edges, E, plus two or V + F = E + 2. For fullerenes which involve five- and six-member rings (faces) it is left to the student to show that the number of five-member rings is always 12 and the number of six-member rings is equal to V/2 − 10. The smallest fullerene is therefore the dodecahedron, C_{20}. It is also found that the most stable fullerenes are those in which the pentagons are isolated, which is known as the *isolated pentagon rule*.

There are numerous equivalent ways to draw the chemical bonds in C_{60}. Equivalent bonding arrangements, known as *resonance structures*, are also found for molecules such as benzene, which gives them their aromatic character. C_{60} is therefore believed by many to be an aromatic molecule, although this is the subject of some debate. Empirical rules have been developed from chemical bonding theory to predict which other molecules similar to C_{60} might be found. In addition to C_{60}, carbon molecules with a multiple of 60 atoms, such as the clusters C_{120}, C_{240}, C_{540}, and C_{960}, are also predicted to be stable aromatic molecules with highly symmetrical open-cage structures.[7]

The spherical cage structure makes C_{60} and the other fullerenes fascinating candidates for all kinds of applications in chemistry and in the preparation of new materials.[5–8] For example, crystals of C_{60} have been prepared in which the molecules arrange themselves in a hexagonal close-packed structure. When alkali metal atoms such as cesium or rubidium are added into the gaps between the balls, the resulting compound is a superconductor when the stoichiometry is M_3C_{60}. The open interior of C_{60} is a cavity about 5 Å wide, and this cavity is large enough to contain other atoms, in particular metals. Numerous research groups are trying to find the conditions necessary for encapsulating atoms with C_{60} or with other fullerenes with larger cavities. The unusual shape of this molecule is of special interest in chemical situations in which molecular shapes determine chemical activity, as in biological molecules, pharmaceutical drugs, or polymers. To investigate these kinds of applications, other functional groups or reactive organic systems have already been attached to C_{60}. Long chains of the form (C_{60})-R-(C_{60})-R ... have also been constructed. This peculiar new molecule, C_{60}, and other members of its family have now emerged into the mainstream of practical chemistry.

In this experiment, we will use a sample of C_{60} prepared by the carbon arc method and extracted with toluene. The carbon arc process produces a mixture of fullerenes and soot, while extraction with toluene isolates a subset of fullerenes which are soluble in toluene. Such mixed fullerene soot is available from many chemical suppliers at low cost. Laser desorption mass spectrometry is now the standard technique for analysis of fullerene materials, and we apply this to the soot extract to analyze what fullerenes it contains. Our instrument and its operation for these experiments has been described in the literature.[9]

Experimental

The soot extract sample is soluble in toluene. Dissolve a small amount of this sample, and apply a drop of this solution onto the sample probe surface. Allow the solvent to evaporate, leaving behind a brown film. Insert the sample probe into the mass spectrometer with the help of the TA. Align the Nd:YAG laser, operating at the 355 nm wavelength, so that it strikes the sample probe, as shown in Figure 12.1. Run the mass spectrometer as described above.

A typical mass spectrum obtain in positive ion mode for a mixed fullerene sample is shown in the upper trace of Figure 12.2. The most abundant ions detected are C_{60}^{+} and C_{70}^{+}. Intense peaks at lower mass are from Na and K

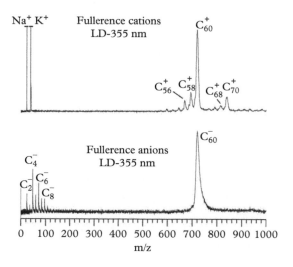

Figure 12.2 *The cation mass spectrum measured for LD of a sample of mixed fullerene powder (top) compared to the anion spectrum measured for a pure C_{60} sample. (Reprinted from Int. J. Mass Spectrom. 354–355, 159 (2013), Copyright 2013, with permission from Elsevier. [10])*

impurities in the sample, which are almost always present. The lower trace of Figure 12.2 shows an anion mass spectrum for a sample of pure C_{60}. In addition to the expected C_{60}^- peak, there are fragment ions at lower masses. These anions survive in the plasma because the small carbon clusters have high electron affinities. Similar small cation clusters are not detected because the neutrals have relatively high ionization energies and any cations formed would transfer charge in the plasma to other species present with lower ionization energies (e.g., C_{60} and C_{70}).[7]

The mass resolution ($m/\Delta m$) seen in Figure 12.2 is approximately 200, which is insufficient to resolve the isotope distribution of the C_{60} mass peak. The poor resolution is a result of the fact that the ions leaving the surface possess a distribution of kinetic energies and corresponding velocities, causing a smear in the arrival time distribution of the ions at the detector. Much higher resolution can be attained by employing a clever technique called delayed-pulse extraction, or DPE.[11–13] To understand DPE, assume that there is initially no draw-out voltage. At the moment of the desorption laser impact, all ions leave the surface. However, after a fraction of a microsecond, those with different velocities will travel different distances out into the vacuum. Specifically, those with higher velocities will travel farther into the vacuum than the slower ones. When the acceleration pulse is applied, forming a field across the source region, the faster ions that are further along the flight path will receive a smaller acceleration energy because they move a shorter distance in this field. The slower ions which did not move as far into the draw-out region will receive a greater acceleration energy as they pass through a longer distance in the field. The resolution is optimized by adjusting the time delay between the laser pulse and the ion draw-out pulse. Figure 12.3 shows the C_{60}^+ mass distribution measured using DPE in the PerSeptive BioSystems departmental mass spectrometer at UT, where the individual

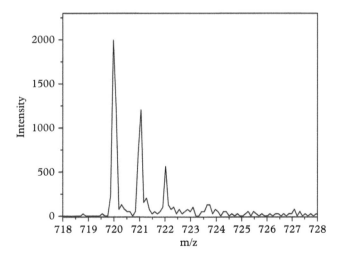

Figure 12.3 *The isotopically resolved mass spectrum of C_{60} measured with a laser-desorption TOF mass spectrometer using delayed-pulse extraction.*

isotopic mass peaks are resolved. Using DPE, mass resolving powers of 1000 are routine and, with special care, 10,000 is possible. DPE (also called time-lag focusing) has made a huge impact in the area of laser desorption and MALDI TOF-MS. The use of DPE is somewhat more expensive because it requires a pulse/delay generator to control the application of the high-voltage pulse.

The intensities of isotopic mass peaks like those seen in Figure 12.3 depend on the statistical probability of forming a molecule with a specific isotopic composition. Such intensities can be modeled, as shown below:

$$F(k, n, p) = [n!/(k!(n-k)!)]p^k (1-p)^{n-k} \qquad (12.6)$$

Where n represents the number of trials, k the number of successes, and p the probability of a success. If one reaches into a box containing Avogadro's number of naturally occurring carbon atoms and pulls out a C_{60} cluster, what is the probability that this cluster is isotopically pure in $^{12}C_{60}$? From equation (12.6) the probability of getting an isotopically pure $^{12}C_{60}$ molecule is:

$$\{60!/60!(60-60)!\}(0.989)^{60}(0.011)^{(60-60)} = 0.51$$

Likewise using ^{13}C as 1.1% abundant in nature, we can calculate that the probability of obtaining a $^{13}C^{12}C_{59}$ molecule is:

$$\{60!/(59!(60-59)!\}(0.989)^{59}(0.011)^{(60-59)} = 0.34$$

Going further, the probability of getting a $^{13}C_2{}^{12}C_{58}$ cluster is

$$\{60!/(58!(60-58)!\}(0.989)^{58}(0.011)^{(60-58)} = 0.11$$

Thus the ratio of the $^{13}C^{12}C_{60}$ to $^{12}C_{60}$ would be 0.66. These isotope ratios are precisely those shown in Figure 12.3. Interestingly, the probability of obtaining $^{13}C_{60}$ is $\sim 3 \times 10^{-118}$, an incredibly small number! The distribution of masses due to atomic isotopes in a molecule is easily seen in mass spectrometry. However, it is very difficult to resolve the isotope effects in vibrations of C_{60}. There are many programs online that calculate the probability distributions for molecules with many different atoms. The IsoPro program is free online (see: https://sites.google.com/site/isoproms/) and not only gives the isotope distributions for any molecule made of atoms from the periodic chart but will also calculate a simulated spectrum given the resolution (mass or spectroscopic) of the instrument used to observe the isotopes. If one uses IsoPro, which uses the ^{13}C abundance as 1.11%, to calculate the isotopic distribution for C_{60}, the distribution is: $^{12}C_{60} = 0.5149647$, $^{13}C^{12}C_{60} = 0.3436570$, $^{13}C_2{}^{12}C_{58} = 0.1127570$.

In addition to the experiment described here, laser desorption and ionization with a pulsed laser can also be used to measure the mass spectra of a variety of polycyclic aromatic hydrocarbons (PAHs) or their complexes with metals.[14] PAHs have multiple aromatic rings like that of benzene connected in a graphitic-like network. These species are found in the soot from flames, in automobile exhaust, and they are believed to be present in interstellar gas clouds. The smallest members of the family, such as naphthalene and anthracene, have enough vapor pressure to be detected with ordinary mass spectrometry. However, larger members of the family, such as pentacene, pyrene, and coronene, have extremely low vapor pressure, just like the fullerenes. These species and their mixtures can be detected easily with LD-TOF measurements.

Questions for consideration

1. How is C_{60} ionized in this experiment?

2. How do fullerenes grow?

3. What experiments can be used to prove that the soccer ball structure of C_{60} is correct?

4. Can you show that the number of pentagons is always 12 and the number of hexagons is always $V/2 - 10$ for any fullerene?

5. Calculate the size, N, of an empty fullerene, C_N which will float in air.

REFERENCES

1. H. W. Kroto, J. R. Heath, S. C. O'Brien, R. F. Curl, and R. E. Smalley, "C_{60}: Buckminsterfullerene," *Nature* **318**, 162 (1985).

2. W. Krätschmer, L. D. Lamb, K. Fostiropoulos, and D. R. Huffman, "Solid C_{60}: A new form of carbon," *Nature* **347**, 354 (1990).

3. J. Baggott, *Perfect Symmetry: The Accidental Discovery of Buckminsterfullerene*, Oxford University Press, Oxford, UK, 1994.

4. H. Aldersey-Williams, *The Most Beautiful Molecule: The Discovery of the Buckyball*, John Wiley & Sons, New York, 1995.

5. R. E. Smalley, "Self-assembly of the fullerenes," *Acc. Chem. Res.* **25**, 98 (1992).

6. W. E. Billups and M. A. Ciufolini, eds., *Buckminsterfullerenes*, VCH, New York, 1993.

7. M. S. Dresselhaus, G. Dresselhaus, and P. C. Eklund, *Science of Fullerenes and Carbon Nanotubes*, Academic Press, San Diego, 1995.

8. K. M. Kadish and R. S. Ruoff, *Fullerenes: Chemistry, Physics and Technology*, Wiley Interscience, New York, 2000.

9. D. S. Cornett, I. J. Amster, M. A. Duncan, A. M. Rao, and P. C. Eklund, "Laser desorption mass spectrometry of photopolymerized C_{60} films," *J. Phys. Chem.* **97**, 5036 (1993).

10. T. C. Cheng, S. T. Akin, C. J. Dibble, S. Ard, and M. A. Duncan, "Tunable infrared laser desorption and ionization of fullerene films," *Int. J. Mass. Spectrom.* **354–355**, 159 (2013).

11. W. C. Wiley and L. H. McLaren, "Time-of-flight mass spectrometer with improved resolution," *Rev. Sci. Instrum.* **26**, 1150 (1955).

12. R. S. Brown and J. J. Lenon, "Mass resolution improvement by incorporation of pulsed ion extraction in a matrix-assisted laser desorption/ionization linear time-of-flight mass spectrometer," *Anal. Chem.* **67**, 1998 (1995).

13. M. L. Vestal, P. Juhasz, and S. A. Martin, "Delayed extraction matrix-assisted laser desorption time-of-flight mass spectroscopy," *Rapid Commun. Mass Spectrom.* **9**, 1044 (1995).

14. T. M. Ayers, B. C. Westlake, and M. A. Duncan, "Laser plasma production of metal and metal-compound complexes with PAH's," *J. Phys. Chem. A* **108**, 9805 (2004).

Laser vaporization to produce silver atom clusters

Introduction

One of the frontier research areas in chemistry and physics today focuses on microscopic particles of metals. Particles so small that they contain only a few atoms are referred to as metal atom clusters, or simply *metal clusters*.[1–5] Metal clusters can be indicated with chemical formulas just like conventional molecules. For example, the clusters Ni_3, Fe_5, and Ag_{17} contain 3, 5, and 17 atoms of nickel, iron, and silver, respectively, but no other elements. It is not yet clear whether clusters of metal atoms have properties like small pieces of solid metal, or whether they are so different that they have a whole family of unusual properties all their own. Until recently these kinds of questions were pointless to ask, because no one could make metal particles this small. Now, however, laboratories around the world can make clusters out of any metal (or other element) in the periodic table. Systematic studies are underway to measure their physical and chemical properties and to determine what applications there might be for these unusual species.

Clusters are most often produced by a technique known as "laser vaporization."[6] In this method, a solid piece of the metal to be studied is mounted in a special holder inside a vacuum chamber. A high-powered pulse of laser light, usually from a Nd:YAG laser, is focused with a lens onto the surface of the metal in much the same way that a magnifying glass can be used to ignite a piece of paper with concentrated sunlight. However, the laser light is so powerful that it generates a temperature of about 10,000 K, which is enough to vaporize a small amount of the metal. The result is an extremely hot "plasma" composed of metal atoms, their ions, and electrons. A burst of helium gas at room temperature is often squirted through the sample holder to cool the metal plasma. Helium does not react, even with hot metal atoms, but through thousands of collisions it cools the metal vapor to near room temperature. As they cool, metal atoms recombine and condense to form solid material again. It is this recombination of atoms that produces the clusters. If allowed to, the recombining metal vapor would plate out again on the surface from which it came. However, the flowing helium gas pushes the small metal particles out as an expanding spray into the vacuum chamber. This gaseous spray, known as a *molecular beam*, is where cluster experiments are conducted. For example, if the cluster molecular beam is sprayed into a mass spectrometer, the masses, and therefore the sizes of the clusters formed, can be measured. In a mass spectrum of metal clusters produced by this kind of experiment, the various molecules formed with two, three, four, ... atoms are called *dimers, trimers, tetramers,* and so on. If the cluster spray is intersected by another laser beam, the absorption spectrum of the clusters can be measured to determine their structures (overall shape, bond lengths, bond angles). Other experiments with lasers and mass spectrometers measure the energy required to ionize clusters of different sizes or to break the chemical bonds that hold cluster atoms together.

Although clusters of virtually any metal can be formed and studied, these molecules are not "stable" in the usual sense of the word. Aggregated atoms in a cluster are not as stable as more conventional molecules, but they are far more stable than the same atoms separated, and so clusters do not spontaneously fly apart. But it is usually not possible to draw simple bonding schemes such as Lewis dot formulas for these kinds of molecules. One exception occurs for "dimers" of alkali metals, where the atoms have a single s electron in their valence shell just like hydrogen. Like diatomic hydrogen, H_2, the molecules Li_2, Na_2, K_2, etc. are held together by a two-electron covalent single bond. The same is true for the coinage metal dimers, Ag_2, Cu_2, and Au_2, which have a filled d electron shell not involved in the bonding and a single s electron in the valence shell which forms the covalent bond.

Large particles of sodium, silver, and copper are well-known conductors of electricity, but interestingly the diatomic particles cannot conduct because they contain no free electrons. Transition metals have more complex bonding schemes.[2] For example, a chromium atom has five d electrons and one s electron in its valence shell, and all of these are used in Cr_2, which has a *sextuple* bond! Iron has six d electrons and two s electrons, but Fe_2 has only a single bond. The "unused" electrons in Fe_2 give it "unsatisfied" bonding capacity, and it is therefore an extremely reactive molecule. When clusters larger than dimers are considered, the chemical bonding becomes too complex for any simple description. Some of the electrons are used to hold the atoms together in chemical bonds, and some are delocalized over the surface and volume of the cluster as it begins to look like a piece of solid metal. Even with the best computer models available today, it is extremely difficult to predict what chemical bonding scheme will be found for clusters with 10–20 atoms.

Closely related to the chemical bonding in clusters are their geometric structures. Although solid metals usually have their atoms arranged in orderly crystal structures, this is not necessarily the case for clusters. Many solid metals have their atoms arranged in close-packed hexagonal or cubic networks in which each atom has 12 nearest neighbors. If these structural patterns are drawn for clusters, they end up with sharp edges and flat faces. In reality, effects such as surface tension combine with chemical bonding forces to give clusters a smoother exterior. One structure that occurs frequently as an especially stable arrangement of cluster atoms is the 13 atom *icosahedron*. This structure resembles close-packed lattices in that it has one atom surrounded by 12 nearest neighbors, but it has an essentially spherical exterior. Such spherical and elliptical geometries are common in cluster structures. It is interesting to note that 12 out of the 13 atoms in the icosahedral structure are on the surface of the particle. Even in a cluster with 100 atoms, about 80 are on the surface. This high surface area, where other molecules can stick and react, makes clusters all the more interesting as potential catalysts. One interesting twist is that chemical bonds in clusters are often weak, making the bonding network easy to disrupt. When this begins to occur, cluster bonds may break and reform rapidly as if the particle were a liquid droplet instead of a solid!

This behavior, which is like melting in larger particles, requires a characteristic temperature depending on the size of the cluster and the stability of the bonding arrangement. If they are heated enough, as occurs with laser excitation, clusters may "evaporate" and lose atoms.

Experimental section

Molecular beam experiments which produce metal clusters are extremely complex and expensive, and are beyond the scope of an undergraduate laboratory. However, it has been shown that many metals can produce clusters by laser vaporization of neat samples without the rare gas for collisional cooling.[7] Silver, antimony, and bismuth are such metals, and we examine clusters of one or more of these metals in this experiment.

Prepare a sample of silver oxide powder in a hollowed-out solid sample probe such as that used for C_{60} or MALDI experiments. The hollowed-out probe has a shallow hole about $\frac{1}{8}$ inch in diameter drilled into its face. Pack this hole tightly with the metal or metal oxide powder, and insert it into the mass spectrometer. Alternatively, metal oxide powder can be rubbed into the surface of a flat-ended probe tip using a "Scotch-Brite" pad. Use the 355 nm wavelength from the Nd:YAG laser with a pulse energy of 1–5 mJ/pulse, and focus this light onto the sample with a lens located outside the vacuum system. Measure the mass spectrum for as many clusters as you can. Adjust the laser focus and position on the sample probe to optimize the signal.

Clusters of silver are interesting because this metal has two naturally occurring isotopes, ^{107}Ag and ^{109}Ag. These two isotopes have roughly equal natural abundance, and so both are clearly evident in mass spectra. When silver atoms form clusters, each mass peak has a distribution of isotopes determined by the relative probability of forming each isotopomer. For example, the silver dimer mass

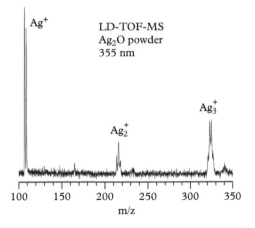

Figure 12.4 *The mass spectrum of silver atom clusters produced by laser ablation of silver oxide powder. The isotopic mass peaks are evident for the atom, dimer, and trimer.*

peak is composed of the following masses: ^{107}Ag-^{107}Ag (mass 214), ^{107}Ag-^{109}Ag (mass 216), and ^{109}Ag-^{109}Ag (mass 218). The two peaks for silver atom and the three peaks for the silver dimer can be seen in the mass spectrum in Figure 12.4. The silver trimer has a pattern of four masses, which is just barely resolved in the figure. Larger silver atom clusters have an even wider mass range. As described earlier, such isotopic abundances can be modeled statistically with various programs available online, such as IsoPro.[8] Obtain one of these programs and model the isotopic patterns for Ag$_2$ and Ag$_3$.

Questions for consideration

1. Laser vaporization of pure silver metal powder does not produce silver clusters as efficiently as vaporization of silver oxide powder. Why is this true?

2. It is usually true in silver cluster mass spectra that the odd-numbered cluster peaks (e.g., Ag$_3$$^+$, Ag$_5$$^+$, etc.) are more intense in the mass spectrum than the even-numbered cluster peaks. Why does this occur?

REFERENCES

1. M. A. Duncan and D. A. Rouvray, "Microclusters," *Sci. Am.* **161**, 110 (1989).
2. M. D. Morse, "Clusters of transition-metal atoms," *Chem. Rev.* **86**, 1049 (1986).
3. M. A. Duncan, "Spectroscopy and photochemistry in clusters and organometallic complexes of silver," *Advances in Metal and Semiconductor Clusters*, Vol. I, M. A. Duncan, ed., JAI Press, Inc., Greenwich, CT, 1993, p. 123.
4. H. Haberland, ed., *Clusters of Atoms and Molecules I: Theory, Experiment, and Clusters of Atoms*, Springer-Verlag, Berlin, 1995.
5. R. L. Johnston, *Atomic and Molecular Clusters*, Taylor & Francis, New York, 2002.
6. M. A. Duncan, "Laser vaporization cluster sources," *Rev. Sci. Instrum.* **83**, 041101 (2012).
7. S. W. McElvaney, H. H. Nelson, and A. P. Baronavski, "FTMS studies of mass-selected, large cluster ions produced by direct laser vaporization," *Chem. Phys. Lett.* **134**, 214 (1987).
8. These isotopic patterns for silver clusters can be modeled with the software "IsoPro," available free of charge via the Internet at https://sites.google.com/site/isoproms/.

Laser vaporization to produce sulfur and phosphorus clusters

Sulfur

Elements such as sulfur or phosphorus can also form atomic clusters and laser vaporization is a good way to make and detect these species with a mass spectrometer. Lower laser powers are generally required, but otherwise the methodology is the same as that used for metal atoms.

Figure 12.5 shows the mass spectra measured for cations (bottom) and anions (top) when sulfur powder is vaporized with the laser at 355 nm. Sulfur cluster ions of the form S_n^+, for $n = 2$–8, are observed. The signal for S^+ is extremely weak, but the S_2^+ and S_5^+ mass peaks are prominent. The spectrum extends out to S_8^+, but the mass peaks corresponding to larger cluster ions, S_9^+ and S_{10}^+, are much less intense. The anion spectrum is noticeably different from that for the cations. Only the mass peaks for $n = 1$ –4 are observed, with S_3^- by far the most intense. These observations agree with previous results reported by Johnson and coworkers.[1]

The interpretation of these mass spectra should stimulate speculation about which exact atomic or molecular species are desorbed, whether there is subsequent growth of larger species, and whether or not there is fragmentation in the ionization process. It is well known that solid sulfur primarily contains the S_8 allotrope.[2,3] The vapor above heated sulfur samples has also been shown to contain neutral S_8 molecules.[4,5] It is therefore possible that laser vaporization desorbs molecular S_8 directly into the gas phase, and that the ions observed represent fragmentation products from this molecular species. Another possibility is that desorbed sulfur atoms recombine to form these clusters, and that the growth ends near the S_8 species. Unfortunately, there is no way of determining from mass spectra alone which is the predominant mechanism of cluster production. However, either mechanism suggests the likely importance of eight-membered rings in the gas-phase cations. These results for sulfur cations vary noticeably from corresponding experiments reported by Martin and coworkers.[5] In their work, clusters formed by heating the powder in an oven and ionizing it with electron impact produced multiples of the S_8 species. It is understandable that different vapor production methods and ionization processes lead to different mass spectra. [6] provides a more detailed discussion of cluster growth under different conditions and the role of ionization conditions on the appearance of mass spectra.

It is also interesting to note the differences between anions and cations. This suggests that the number of electrons present in these species plays an important role in their stability. The details of ion formation and cluster growth in laser vaporization processes are not well understood, and therefore speculation about these processes is entirely acceptable. Students should be encouraged to propose structures and bonding configurations for both the prominent cation and anion species to attempt to rationalize why certain clusters are produced more

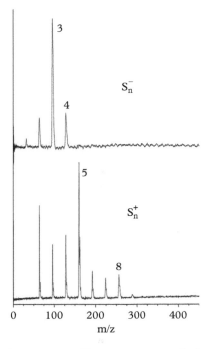

Figure 12.5 *The mass spectrum of sulfur anions and cations produced by laser vaporization at 355 nm. (Reprinted with permission from J. Chem. Ed. 91, 291 (2014). Copyright 2014 American Chemical Society.)*

than others. In particular, they should question why the abundant clusters here (S_3^-, S_2^+, and S_5^+) all have an *odd* number of electrons. It is particularly useful to discuss what information can and cannot be derived with certainty from mass spectra. *Ab initio* theory on these systems is possible to elucidate cluster ion structures, and this kind of work is also recommended for more advanced students (see Chapter 15). The student could also calculate ionization potentials or electron affinities for sulfur clusters to see how these properties vary with cluster size.

Phosphorus

A mass spectrum obtained by the vaporization of red phosphorus powder is presented in Figure 12.6. Previous laser desorption experiments have obtained similar results.[7–9] Unlike sulfur, phosphorus forms much larger clusters in these experiments. The most interesting details in this spectrum are the prominent P_3^+, P_4^+, P_5^+, and P_7^+ ions and the preference for odd-numbered cations that begins after n = 11.

As in the case of sulfur, the mechanism of cluster growth is fascinating in this system. Again, we consider growth from vaporized atoms versus desorption of larger intact molecular species. Because red phosphorus is composed of polymeric chains,[10] it is tempting to conclude that clusters are produced by direct desorption of these, perhaps followed by additional fragmentation. However, it is difficult to imagine that chains could be desorbed efficiently, as they must necessarily be entangled in the solid and have significant interactions (van der Waals) with their surroundings. Desorption of atoms or small molecules would require the breaking of fewer bonds, but it is difficult to understand how such

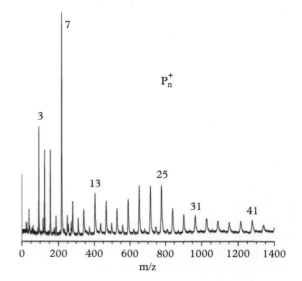

Figure 12.6 *The mass spectrum of phosphorus cations produced by laser vaporization at 355 nm. (Reprinted with permission from J. Chem. Ed. 91, 291 (2014). Copyright 2014 American Chemical Society.)*

large clusters could grow on the timescale of the vaporization without a collision gas present. However, larger phosphorus clusters were seen previously by Martin and coworkers using an oven source,[11] and they were interpreted to grow from atomic vapor. Additionally, McElvaney and coworkers have shown that large clusters of other elements can grow without an added collisional gas.[12] If enough elemental density can be produced in the vaporization process, self-collisions can cool the atomic vapor and stabilize growing clusters.

Another interesting observation is the preference for odd-numbered cations, where P_7^+ is the most prominent. These odd-numbered clusters all have an overall even number of electrons, perhaps explaining their stability.[11] However, an additional feature is the prominence of the P_3^+, P_5^+, and P_7^+ ions compared to other odd-numbered species. Electron counting rules found in the Polyhedral Skeletal Electron Pair Theory, also known as "Wade's Rules," may provide further insight into this.[13,14] These electron counting rules are frequently applied to atomic clusters in the same way that the 18-electron rule is applied to transition metal–ligand complexes. They relate the number of skeletal electrons in clusters to specific polyhedral structures with delocalized electron clouds in their interiors that provide enhanced stability. Clusters that are electron deficient can share electrons more effectively in this way, and some are believed to possess three-dimensional aromaticity.[13,14] The bonding in these clusters only involves valence p electrons, and an N-atom cluster achieves stability when there are $2N + 2$, $2N + 4$, or $2N + 6$ skeletal electrons, representing the "closo," "nido," and "arachno" polyhedral structures, respectively. P_3^+, P_5^+, and P_7^+, with 8, 14, and 20 valence electrons respectively, meet the criteria for the "closo," "nido," and "arachno" structures, perhaps partially explaining their stability. Theoretical research has investigated the stability of phosphorus cluster cations,[7–9] but no experiments have yet been able to measure their structures. More discussion of this experiment and these mass spectra is contained in a recent article.[15]

....

REFERENCES

1. A. K. Hearley, B. F. G. Johnson, J. S. McIndoe, and D. G. Tuck, "Mass spectrometric identification of singly-charged anionic and cationic sulfur, selenium, tellurium and phosphorus species produced by laser ablation," *Inorg. Chim. Acta* **334**, 105 (2002).

2. B. Meyer, ed. *Elemental Sulfur*, John Wiley & Sons, New York, 1965.

3. R. Steudel and B. Eckert, "Elemental sulfur and sulfur-rich compounds II," *Top. Curr. Chem.* **230**, 1 (2003).

4. J. Berkowitz and J. R. Marquart, "Equilibrium composition of sulfur vapor," *J. Chem. Phys.* **39**, 275 (1963).

5. T. P. Martin, "Cluster beam chemistry – from atoms to solids," *Angew. Chem. Int. Ed.* **25**, 197 (1986).

6. M. A. Duncan, "Laser vaporization cluster sources," *Rev. Sci. Instrum.* **83**, 041101 (2012).

7. M. D. Chen, J. T. Li, R. B. Huang, L. S. Zheng, and C. T. Au, "Structure prediction of large cationic phosphorous clusters," *Chem. Phys. Lett.* **305**, 439 (1999).

8. A. V. Bulgakov, O. F. Bobrenok, and V. I. Kosyakov, "Laser ablation synthesis of phosphorous clusters," *Chem. Phys. Lett.* **320**, 19 (2000).

9. M. D. Chen, R. B. Huang, L. S. Zheng, Q. E. Zhang, and C. T. Au, "A theoretical study for the isomers of neutral, cationic and anionic phosphorus clusters P_5, P_7, P_9," *Chem. Phys. Lett.* **325**, 22 (2000).

10. D. E. C. Corbridge, *The Structural Chemistry of Phosphorus*, Elsevier Scientific Publishing Co., Amsterdam, 1974.

11. T. P. Martin, "Compound clusters," *Z. Phys. D - Atoms, Molecules and Clusters* **3**, 211 (1986).

12. S. W. McElvaney, H. H. Nelson, and A. P. Baronavski, "FTMS studies of mass-selected, large cluster ions produced by direct laser vaporization," *Chem. Phys. Lett.* **134**, 214 (1987).

13. K. Wade, "Structural and bonding patterns in cluster chemistry," *Adv. Inorg. Chem. Radiochem.* **18**, 1 (1976).

14. J. D. Corbett, "Homopolyatomic ions of the post-transition elements - Synthesis, structure, and bonding," *Prog. Inorg. Chem.* **21**, 129 (1976).

15. T. M. Ayers, S. T. Akin, C. J. Dibble, and M. A. Duncan, "Laser desorption time-of-flight mass spectrometry of inorganic nanoclusters: An experiment for physical chemistry or advanced instrumentation laboratories," *J. Chem. Educ.* **91**, 291 (2014).

Laser vaporization to produce bismuth oxide clusters

In the same way described already for silver, sulfur, and phosphorus, laser vaporization without any collision gas present can produce bismuth oxide clusters.[1] The sample in this case is pure bismuth metal that has been exposed to ambient air for a few days and has become darkened by this exposure. It has an oxide layer on its surface. Pure stoichiometric bismuth oxide (with formula Bi_2O_3) is a white powder, and does not work well for this experiment.

Figure 12.7 shows a mass spectrum of bismuth oxide cluster cations (bottom) and anions (top) produced from partially oxidized bismuth metal powder. Similar mass spectra were reported previously for these systems, and also for corresponding antimony oxides.[2,3] The fascinating observation here is that the metal oxide stoichiometries are not purely random. Instead, for each number of bismuth atoms there is one main oxide stoichiometry (e.g., $Bi_3O_4^+$). This preference must arise from either the geometric or electronic stability of these clusters. Additional insight is provided by a comparison of the corresponding cations and anions. If geometry alone is important, clusters containing the same number of bismuth atoms should exhibit the same oxide stoichiometry regardless of their

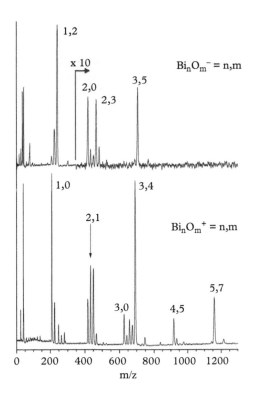

Figure 12.7 *The mass spectrum of bismuth anions and cations produced by laser vaporization at 355 nm. (Reprinted with permission from J. Chem. Ed. 91, 291 (2014). Copyright 2014 American Chemical Society.)*

charge. However, if the number of bonding electrons is more important, clusters containing the same number of bismuth atoms would have different stoichiometries, as they do (e.g., $Bi_3O_4^+$ versus $Bi_3O_5^-$). Electronic stabilization is therefore apparently the primary consideration. Students should be encouraged to explain these oxide patterns in terms of specific bonding configurations for each of these clusters. As discussed previously,[2] reasonable configurations can be obtained in terms of Bi-O-Bi-O network structures having localized two-electron bonds. There is no need to invoke delocalized electron configurations as we did for phosphorus, because there is a better balance in the number of electrons between bismuth and oxygen.

In addition to the atomic clusters described here, many other systems can be studied with this methodology. The Duncan lab at Georgia has also examined pure carbon clusters, clusters of tin and lead, aluminum and aluminum-sulfide clusters, and metal-C_{60} or metal-PAH clusters formed from mixed powder samples.[4]

....

REFERENCES

1. S. W. McElvaney, H. H. Nelson, and A. P. Baronavski, "FTMS studies of mass-selected, large cluster ions produced by direct laser vaporization," *Chem. Phys. Lett.* **134**, 214 (1987).
2. M. R. France, J. W. Buchanan, J. C. Robinson, S. H. Pullins, J. L. Tucker, R. B. King, and M. A. Duncan, "Antimony and bismuth oxide clusters: A new family of magic number clusters," *J. Phys. Chem. A* **101**, 6214 (1997).
3. M. Kinne, T. M. Bernhardt, B. Kaiser, and K. Rademann, "Formation and stability of antimony and bismuth oxide clusters: A mass spectrometric investigation," *Int. J. Mass Spectrom. Ion Processes* **167/168**, 161 (1997).
4. T. M. Ayers, B. C. Westlake, and M. A. Duncan, "Laser plasma production of metal and metal-compound complexes with PAH's," *J. Phys. Chem. A* **108**, 9805 (2004).

MALDI mass spectrometry of insulin

The mass spectra of many biomolecules, such as proteins, enzymes, DNA, etc., can be measured using MALDI.[1,2] In the MALDI experiment, the molecule of interest is suspended as a minor component in a matrix material, which is usually an organic acid (dihydroxy benzoic acid, sinapinic acid, etc.). UV laser excitation, usually at very low pulse energies, causes absorption by the acid film, subsequent explosive plume formation, and ejection of the molecule of interest into the gas phase. Ionization is believed to occur by proton transfer from the organic acid to the analyte in the plume chemistry. Therefore, the ion detected has an M+1 mass peak, and because the energy is absorbed by the matrix, there is virtually no fragmentation. Insulin (5734 amu) is typical for these experiments, but other popular examples include cytochrome c (12,361 amu), myoglobin (17,568 amu), and ubiquitin (8564 amu). The mass spectrometry measurement is similar to others described here, differing only in the sample preparation and laser conditions used. Insulin obtained from any chemical manufacturer (e.g., Sigma-Aldrich) can be combined with any one of several matrix species for this measurement.

Sample preparation

Mix a few milligrams of solid insulin together with a 10^3 excess of matrix material (e.g., sinapinic acid) and dissolve the mixture in a small amount of methanol. Use the minimum amount of solvent necessary to dissolve the mixture. Use a pipet, or micro-pipet if available, to deposit the solution drop-wise onto the flat metal surface of a sample probe. Let each drop dry before depositing a subsequent drop. Two or three drops of solution should be sufficient. Let the sample dry in air, and then insert it into the mass spectrometer inlet port.

Focus the laser to be used onto the sample probe tip, but use an attenuator or aperture to limit the laser power at the sample. For the sinapinic acid matrix, ultraviolet light is required, such as that from a nitrogen laser (337 nm) or the third harmonic of a Nd:YAG laser (355 nm). If a Nd:YAG laser such as the Continuum Minilite is to be used, the low-power setting is recommended. If other higher power Nd:YAG lasers are employed, the beam must be attenuated by using a reflection off a glass or quartz plate instead of the main beam. Measure the mass spectrum and optimize the signal with the laser power and the focus. Experiments with several sample preparations may be required to achieve optimum concentration for the measurement.

If higher acceleration voltages are available in the mass spectrometer, these should be turned up from their usual level (e.g., 3–5 kV) to a higher value of 10–15 kV. This adjustment is necessary to obtain optimum signal intensity for all high molecular weight molecules. When the mass spectrum is obtained, calculate the mass with some reasonable choice for the reference mass. Figure 12.8 shows a spectrum obtained for insulin. The same experiment can be repeated

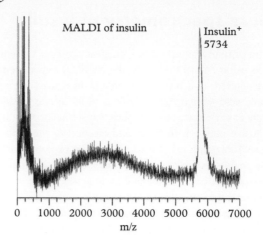

Figure 12.8 *The MALDI-TOF mass spectrum measured for insulin with the 355 nm YAG laser.*

with other proteins or matrix materials if time allows. Other experiments using MALDI mass spectrometry in undergraduate chemistry labs have been described in the literature.[3–10]

Questions for consideration

1. Why are higher acceleration voltages required for high molecular weight molecules?
2. Are mass peaks corresponding to the matrix molecule produced?

REFERENCES

1. G. Hillenkamp and J. Peter-Katalinić, eds., *MALDI MS*, Wiley-VCH, Weinheim, Germany, 2007.
2. R. B. Cole, ed., *Electrospray and MALDI Mass Spectrometry*, John Wiley, Hoboken, NJ, 2010.
3. D. C. Muddiman, R. Bakhtiar, S. A. Hofstadler, and R. D. Smith, "Matrix-assisted laser desorption/ionization mass spectrometry," *J. Chem. Educ.* **74**, 1288 (1997).
4. A. E. Counterman, M. S. Thompson, and D. E. Clemmer, "Identifying a protein by MALDI–TOF mass spectrometry," *J. Chem. Educ.* **80**, 177 (2003).

5. O. A. Moe, W. A. Patton, Y. K. Kwon, and M. G. Kedney, "Ladder sequencing of a peptide using MALDI-TOF mass spectrometry," *Chem. Educator* **9**, 272 (2004).

6. N. C. Dopke and T. N. Lovett, "Illustrating the concepts of isotopes and mass spectrometry in introductory courses: A MALDI-TOF mass spectrometry laboratory experiment," *J. Chem. Educ.* **84**, 1968 (2007).

7. I. J. Arnquist and D. J. Beussman, "Incorporating biological mass spectrometry into undergraduate teaching labs, Part 2: Peptide identification via molecular mass determination," *J. Chem. Educ.* **86**, 382 (2009).

8. J. C. Albright, D. J. Dassenko, E. A. Mohammed, and D. J. Beussman, "Identifying gel-separated proteins using in-gel digestion, mass spectrometry and database searching," *Biochem. & Molec. Bio. Educ.* **37**, 49 (2009).

9. C. W. Harmon, S. A. Mang, J. Greaves, and B. J. Finlayson-Pitts, "Identification of fatty acids, phospholipids and their oxidation products using matrix-assisted laser desorption ionization mass spectrometry and electrospray ionization mass spectrometry," *J. Chem. Educ.* **87**, 186 (2010).

10. M. Eibisch, B. Fuchs, J. Schiller, R. Süβ, and K. Teuber, "Analysis of phospholipid mixtures from biological tissues by matrix-assisted laser desorption and ionization time-of-flight mass spectrometry (MALDI-TOF MS): A laboratory experiment," *J. Chem. Educ.* **88**, 503 (2011).

13

Multiphoton Ionization Mass Spectrometry of Metal Carbonyls

Introduction

Metal carbonyls represent a family of transition metal complexes used in synthetic inorganic chemistry and noted for their interesting photochemistry.[1–5] Some of the commonly available members of the family include: $Fe(CO)_5$ (iron pentacarbonyl); $Cr(CO)_6$ (chromium hexacarbonyl); $Ni(CO)_4$ (nickel tetracarbonyl); and $Mo(CO)_6$ (molybdenum hexacarbonyl). The stabilities of these complexes are usually explained using the so-called 18-electron rule, which counts the valence electrons around the metal and those donated into empty metal orbitals from the CO ligands. A complex with 18 electrons has a rare gas configuration, explaining its stability. Each CO donates 2 electrons, and so a metal like chromium (with a d^5s^1 configuration) achieves an 18-electron configuration with 6 CO ligands. Such stability trends explain why different metals coordinate different numbers of CO ligands. Metal carbonyls are also popular systems for the study of vibrational spectroscopy (IR and/or Raman).[6] The CO stretching frequency is sensitive to the amount of charge transfer between the metal and CO, causing the C–O stretching frequency to vary significantly in these complexes compared to the gas-phase vibration of an isolated CO molecule (2143 cm^{-1}).[7] In recent research, these effects have been studied for unsaturated metal carbonyls and their ions.[8,9]

All of the metal carbonyls are photochemically active. If left exposed to light in the solid or liquid phase, they decompose and rearrange to other carbonyl complexes. For example, iron pentacarbonyl rearranges to form di-iron nonacarbonyl, $Fe_2(CO)_9$. In the gas phase, photoexcitation causes elimination of carbonyl ligands, e.g.,

$$Fe(CO)_5 + h\nu \rightarrow Fe(CO)_4 + CO$$

Additional elimination steps may occur if the radical intermediates absorb more light before encountering another unreacted molecule. The gas-phase chemistry can eventually produce the same kind of species as the condensed-phase chemistry, but the time for this to occur is slower because of the reduced density. Because of their tendency to dissociate via ligand elimination, metal carbonyls are used in the microelectronics industry in the process of photochemical vapor deposition. In this process, UV excitation of metal carbonyls near a surface eliminates the ligands, leaving behind a patterned deposit of pure metal, which can be used, e.g., as wires to connect other components in circuitry.

Nickel carbonyl is a highly toxic gas, iron carbonyl is a toxic liquid with high vapor pressure, and both chromium and molybdenum are white solids which can sublime easily into a mass spectrometer at room temperature. We focus this experiment on these latter two species which are safer to handle than their more volatile counterparts.

If metal carbonyls are excited with a laser, both photochemistry and photoionization may occur. The ionization energy of selected metal carbonyl complexes are given in Table 13.1.[10] As shown, the values for these ionization energies are well above the photon energy of the wavelengths available from a typical Nd:YAG laser (532 nm: 2.33 eV; 355 nm: 3.49 eV; 266 nm: 4.66 eV). Therefore, absorption of UV light produces excited states of the neutral complex, which can either dissociate directly or absorb additional photons because of the high peak power of the pulsed laser. Dissociation causes elimination of carbonyl ligands, and additional light absorption causes further elimination and ionization steps. The mass spectrum, which is measured in this experiment, reflects the final result of a complex sequence of fragmentation + ionization and/or ionization + fragmentation events. The ions resulting at the end of these processes are detected with a time-of-flight mass spectrometer. This experiment therefore provides one example of laser ionization mass spectrometry, which has become a widely used and valuable analytical technique.[11] In this experiment, chromium or molybdenum carbonyls are injected into the mass spectrometer as a vapor and excited with the ultraviolet light from the Nd:YAG laser. The 355 nm wavelength is recommended, but experiments may also be conducted at the 266 nm UV wavelength, or even at the 532 nm visible green wavelength. Both chromium and molybdenum have more than one naturally occurring isotope, and these facilitate the assignment of the mass spectra.

Table 13.1 *Ionization energies of selected metal carbonyl complexes.[10]*

Complex	Ionization energy (eV)
$Fe(CO)_5$	7.897
$Cr(CO)_6$	8.142
$Mo(CO)_6$	8.227
$Ni(CO)_4$	8.3
$W(CO)_6$	8.242

Procedure

Chromium or molybdenum hexacarbonyls are available from chemical manufacturers such as Sigma-Aldrich. Load a small amount of powder into a sample tube connecting into the gas addition inlet on the mass spectrometer. A time-of-flight mass spectrometer such as that discussed in Chapter 12 is recommended for these experiments. It must have windows allowing a laser beam to pass between the first and second plates that establish the acceleration fields, as shown in Figure 13.1.

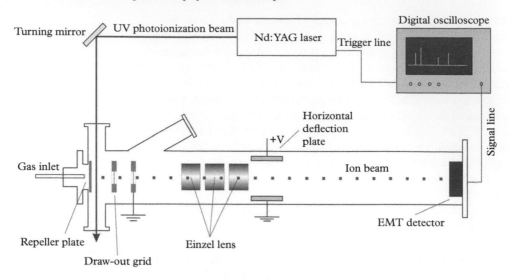

Figure 13.1 *Schematic diagram of the photoionization time-of-flight mass spectrometer.*

A needle valve should be situated between the sample and the vacuum system to control the partial pressure of the carbonyl vapor. Open the needle valve until the ionization gauge pressure increases to about 5×10^{-6} Torr. Adjust the Nd:YAG laser to operate at either of the 355 or 266 nm wavelengths. Close the curtains around the laser table, and then switch the laser into the high power mode of operation to photoionize this vapor and measure a mass spectrum. To do this, the oscilloscope must be triggered with the synch-out or Q-switch-out pulse from the laser so that the time sweep begins exactly when the pulse of laser light forms the ions in the mass spectrometer. Another option for triggering is to use a photodiode which detects the light output from the laser. Collect the mass spectrum by averaging the signal from several laser shots and then analyze the resulting spectrum to determine the masses produced, as discussed in Chapter 12.

Figure 13.2 shows a mass spectrum obtained from multiphoton ionization of molybdenum carbonyl at the 355 nm wavelength. The only ion detected is Mo^+; the multiple peaks seen are those from the isotopes of molybdenum. This pattern is attractive for tuning up the mass spectrometer grid voltages to optimize the mass resolution. This isotopic pattern for molybdenum can be modeled with the software "IsoPro," available free of charge via the Internet at https://sites.google.com/site/isoproms/. Such modeling is useful in the case of ions detected in a mass spectrometer containing multiple elements with multiple isotopes. It can also be used to estimate the resolution of the mass spectrometer.

It is clear from the detection of only the atomic ion that extensive fragmentation has occurred in the absorption/ionization process. This behavior was

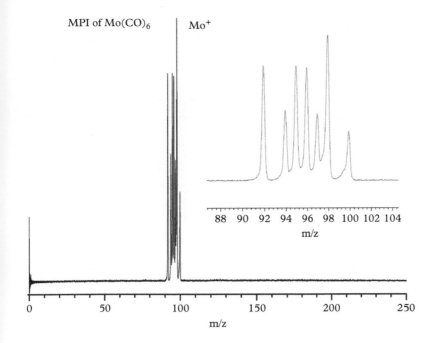

Figure 13.2 *MPI mass spectrum ob-tained from molybdenum hexacarbonyl at the 355 nm wavelength. The inset shows an expanded view of the* Mo$^+$ *peak with its isotopes.*

documented previously for several metal carbonyl systems.[12] It is tempting to try to investigate the mechanism for this fragmentation and to determine whether it involves fragmentation followed by ionization of fragments as opposed to ionization followed by fragmentation of ions. Because it necessarily involves multiphoton absorption, the mechanism could be investigated with studies of the laser power dependence of the signal. Similar investigations have been described for many transition metal or organometallic complexes, including, for example, a study of uranium hexafluoride.[13] In the case of UF$_6$ the main ions observed were U^{n+} with n = 1, 2, 3, 4 and a "collective excitation" process was suggested as the ionization mechanism. In a true multiphoton process, the power dependence could reveal the number of photons involved in generating the signal, e.g., a two-photon process would vary with the square of the laser power. However, the present system involves a sequence of several allowed single-photon steps which are likely to have very different cross-sections. This scenario significantly limits the information that can be obtained from power dependence.

　　Metal carbonyl complexes and their ions can be produced in the gas-phase environment of molecular beams using the method of laser vaporization discussed in Chapter 12. Infrared spectroscopy combined with quantum chemistry calculations has provided the structures for many unusual systems not seen before in the condensed phase.[9]

Questions for consideration

1. What sequence of photochemistry and photoionization steps can explain the mass spectrum observed? Is it possible to distinguish between neutral absorption and fragmentation followed by ionization of the metal, as opposed to ionization of the carbonyl followed by carbonyl ion fragmentation?

2. What is the "mass resolution" (defined as $\Delta m/m$) at the most intense peaks measured?

3. Use the IsoPro program to simulate the Mo^+ ion mass spectrum in order to estimate the mass resolution.

REFERENCES

1. N. A. Beach and H. B. Gray, "Electronic structures of metal hexacarbonyls," *J. Am. Chem. Soc.* **90**, 5713 (1968).

2. M. Wrighton, "Photochemistry of metal carbonyls," *Chem. Rev.* **74**, 401 (1974).

3. C. R. Bock and E. A. Koerner von Gustorf, "Primary photoprocesses of organo-transition metal compounds," *Adv. Photochem.* **10**, 221 (1977).

4. E. A. Koerner von Gustorf, L. H. C. Leenders, I. Fischler, and R. Perutz, "Aspects of organo-transition-metal photochemistry and their biological implications," *Adv. Inorg. Chem. Radiochem.* **19**, 65 (1976).

5. G. Frenking and N. Fröhlich, "The nature of the bonding in transition-metal compounds," *Chem. Rev.* **100**, 717 (2000).

6. K. Nakamoto, *Infrared and Raman Spectra of Inorganic and Coordination Compounds*, John Wiley, New York, 1997.

7. K. P. Huber and G. Herzberg, *Molecular Spectra and Molecular Structure IV. Constants of Diatomic Molecules*, Van Nostrand Reinhold Co., New York, 1979.

8. M. Zhou, L. Andrews, and C. W. Bauschlicher, Jr., "Spectroscopic and theoretical investigations of vibrational frequencies in binary unsaturated transition-metal cations, neutrals and anions," *Chem. Rev.* **101**, 1931 (2001).

9. A. M. Ricks, Z. E. Reed, and M. A. Duncan, "Infrared spectroscopy of mass-selected metal carbonyl cations," *J. Mol. Spec.* **266**, 63 (2011).

10. E. P. Hunter and S. G. Lias, NIST Standard Reference Database 69: *NIST Chemistry WebBook*, http://webbook.nist.gov.

11. D. H. Parker, in *Ultrasensitive Laser Spectroscopy*, D. S. Kliger, ed., Academic Press, New York, 1983, p. 234.

12. M. A. Duncan, T. G. Dietz, and R. E. Smalley, "Efficient multiphoton ionization of metal carbonyls cooled in a pulsed supersonic beam," *Chem. Phys.*
44, 415 (1979).

13. D. M. Armstrong, D. A. Harkins, R. N. Compton, and D. Ding, "Multiphoton ionization of uranium hexafluoride," *J. Chem. Phys.* **100**, 28
(1994).

Part IV

Laser Experiments for Quantum Chemistry and Spectroscopy

Optical Spectroscopy

14

Introduction

Optical spectroscopy has a long history in chemistry and physics, and remains a broad area of widespread research activity. Many advances have been made possible by the incorporation of lasers and laser-based techniques, and new applications are steadily emerging. The interaction of electromagnetic radiation with samples in the gaseous or condensed phases, including optical absorption, nuclear magnetic resonance (NMR), X-ray diffraction, etc., makes it possible to investigate the physical properties of molecules in unprecedented detail. Optical methods include microwave (pure rotational), infrared and Raman (rotational/vibrational), and UV-visible (electronic) spectroscopy. Although there are interesting new applications of microwave measurements, infrared, Raman, and electronic spectroscopy are more widely applicable in modern research and analysis, and these are the areas of focus in the present text. There are many previous textbooks on the theory of spectroscopy in these regions of the spectrum.[1] The three-volume set of monumental texts by Gerhard Herzberg [2–4] has served as the primary source of information in this field and has played an important role in the development of modern molecular spectroscopy. Herzberg was awarded the Nobel Prize in Chemistry in 1971 due, in part, to these books. The latest (fourth) volume by Huber and Herzberg [5] tabulates many of the molecular constants which are used in this textbook. Terahertz spectroscopy, which covers frequencies between the far infrared and microwave regions, represents an emerging new area of science, but will not be covered in this book.

Infrared (IR) and Raman spectroscopy provide the most common ways to investigate ground-state molecular vibrations and rotations. These two methods can be described by reference to Figure 14.1. Upon excitation, molecules can rotate and vibrate with characteristic frequencies which depend upon their masses and bond lengths, as well as the force constants involved. The total number of vibrational degrees of freedom for a molecule containing N atoms is 3N–6 if the molecule is nonlinear and 3N–5 if it is linear. This is easily understood if one considers that the total degrees of freedom for N free atoms is 3N (each atom can move in x, y, z directions). Since the atoms are connected with bonds, the molecule as a whole (its center of mass) can move in three dimensions giving 3N–3 non-rotational degrees of freedom. The molecule can also rotate about the center of mass in three dimensions. Therefore subtracting the three translational and the three rotational degrees of freedom gives a total of 3N–6

Laser Experiments for Chemistry and Physics. First Edition. Robert N. Compton and Michael A. Duncan.
© Robert N. Compton and Michael A. Duncan 2016. Published in 2016 by Oxford University Press.

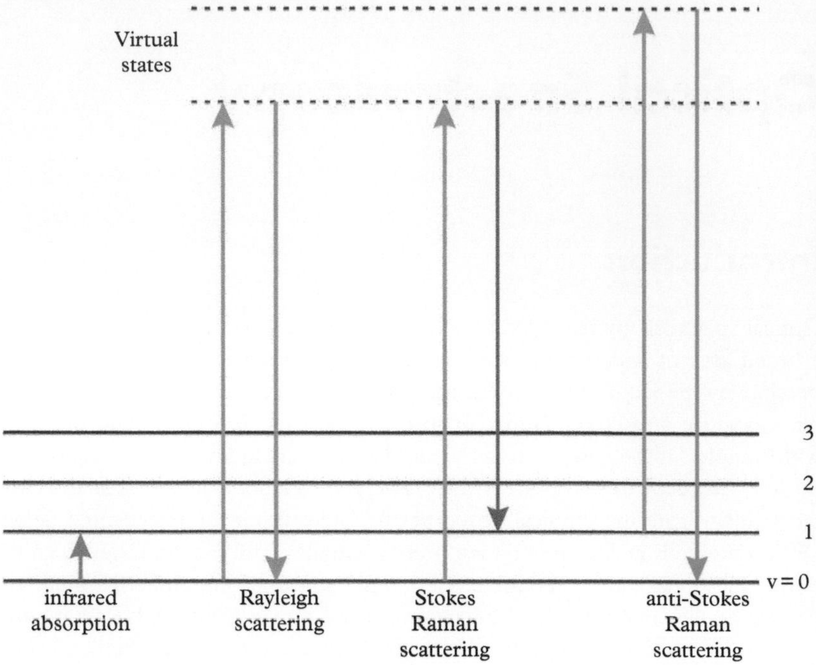

Figure 14.1 *Illustration of infrared and Raman spectroscopy. IR involves direct absorption between vibrational states of the molecule, whereas Raman involves two-photon scattering steps through a virtual intermediate state. The intermediate can also be a real state, in which case the process is called resonance-Raman scattering.*

vibrational degrees of freedom. If the molecule happens to be linear (e.g., CO_2 or HCCH), rotations about the molecular axis do not count (no angular momentum) and the total degrees of freedom for a linear molecule is 3N–5. Thus, H_2O has 3 vibrational degrees of freedom, CO_2 has 4, benzene has 30, and Buckminsterfullerene (C_{60}) has 174 vibrational degrees of freedom. Although we focus here on optical excitation, molecular vibrations can also be excited (or relaxed) via collisions with other molecules, surfaces, electrons, or neutrons. The energies of light corresponding to rovibrational spectroscopy fall in the infrared region and are usually discussed in cm^{-1} units, where the typical range is 500–4000 cm^{-1}. The conversion of cm^{-1} to other common units is given in Table 1.1.

Infrared and Raman are complementary to each other because of the different selection rules governing the physical processes involved. Rovibrational transitions are IR active if the dipole moment of the molecule changes during the vibration. Likewise, a rovibrational transition is Raman active if the polarizability of the molecule changes during the vibration. Immediately, one can see that a homonuclear diatomic molecule cannot undergo IR excitation because vibration of the two atoms cannot produce a dipole moment change. However the polarizability of the homonuclear diatomic does change upon vibration, giving rise to Raman excitation. Generally speaking, symmetric vibrations are Raman active and asymmetric vibrations are IR active. Additionally, if a molecule possesses a center of symmetry, only asymmetric modes will be excited in IR spectroscopy, whereas only symmetric modes will be excited in Raman spectroscopy. This is known as the *mutual exclusion rule* and can be nicely illustrated by the CO_2 molecular vibrations shown below.

$\leftarrow$ O - C - O $\rightarrow$ ν_1 symmetric stretch vibration, Raman allowed

O - C - O ν_2 asymmetric bend vibration, IR allowed

$\leftarrow$ O - C $\leftarrow$ O ν_3 asymmetric stretch vibration, IR allowed

The ν_2 vibration is "doubly-degenerate" (i.e., counted twice) since it can be in the plane of the paper or perpendicular to it, giving a total of four vibrations. Another rather vivid demonstration of the mutual exclusion rule is given by the C_{60} molecule, also known as Buckminsterfullerene. Of its 174 vibrational modes, only four (triply-degenerate) are IR active and ten are Raman active. The prediction and observation of these vibrations was the first clear evidence for the soccer ball (icosahedral) structure of the C_{60} molecule (see Chapter 12). Its geodesic dome shape prompted the name of Buckminsterfullerene after the American architect Buckminster Fuller or "Buckyball" (because it has the same structure as a soccer ball).

Electronic spectroscopy involves the excitation of electrons from their lowest ground atomic or molecular orbitals into any one of the higher energy orbitals of the system. In the case of atoms, the general pattern of orbitals is illustrated by hydrogen, as shown in Figure 14.2. The single electron in the 1*s* atomic orbital can be excited into a variety of other familiar *s*, *p*, *d*, *f*, etc., orbitals as shown in the figure. The diagram in Figure 14.2 is called a Grotrian diagram, and the energies are relative to the ionization potential, defined by convention as the zero of the binding energy. More negative energies on this scale correspond to more stable electronic configurations. The energies for the hydrogen atom depend only on the principle quantum number n and are given by the Rydberg formula,

$$E(n) = R\left(\frac{1}{n_1^2} - \frac{1}{n_2^2} \right) \tag{14.1}$$

where R is the Rydberg constant ($R = 109{,}677 \text{ cm}^{-1}$), which is equivalent to the ionization potential of the hydrogen atom (13.6 eV). Levels are shown here for the different *s*, *p*, *d*, *f* states corresponding to different angular momentum states, indicated by the L quantum number. In the hydrogen system, these are all at the same energy for a given value of n, except for extremely small correction terms unimportant in chemistry but of considerable interest in atomic physics.

In multielectron atoms, the electron–electron repulsion causes levels of the same n to split out to different energies. For example, as discussed in Chapter 16,

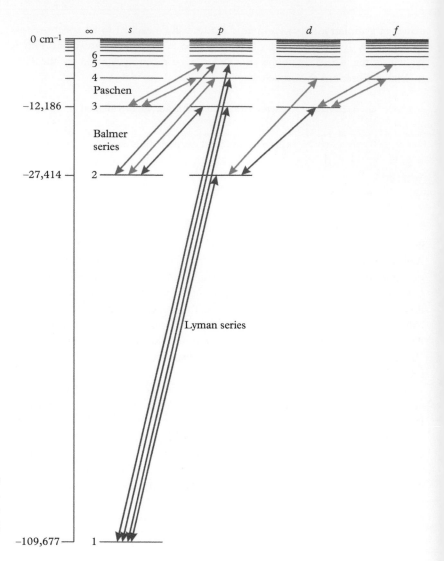

Figure 14.2 *Grotrian diagram for the levels of atomic hydrogen. Selected transitions allowed in either absorption or emission are indicated with colored arrows.*

the energy levels of the "one-electron" alkali atoms are shown to be represented by a Rydberg series similar to that described by equation 14.1. However, the n quantum number is modified to have different values for different angular momentum states. The quantum number n is replaced by $n-\delta_\ell$ where δ_ℓ is the "quantum defect" for a given angular momentum quantum number ℓ. In essence the outer electron in its orbit penetrates the core electrons and therefore the interaction depends upon its angular momentum. The *s* electrons penetrate more than *p*, *p* penetrates more than *d*, and so on, and therefore $\delta_s > \delta_p > \delta_d$. In addition to the electron–electron interactions, multielectron atoms can have states with different multiplicities $(2s + 1)$, i.e. singlets, doublets, triplets, etc., whereas hydrogen can only exist as a doublet. With these two caveats, the pattern of states shown for hydrogen in Figure 14.2 is representative of that for other atoms.

Ionization energies of atoms are given in electron volts, and they typically fall in the range of 5–15 eV (see Table 1.4). The first excited states of atoms are usually more than halfway in energy to the ionization potential, with successively higher states becoming more closely spaced because of the inverse n^2 dependence of the energies. Transitions between these states induced by light absorption and emission must conserve angular momentum. Because the photon has one unit of angular momentum, its absorption or emission requires a corresponding change of one unit of angular momentum in the atom, i.e., that $\Delta L = \pm 1$. Therefore, $s \to s$, $p \to p$, $d \to d$, $s \to d$, etc., transitions are forbidden and $s \to p$, $p \to s$, $p \to d$, $d \to p$, etc., transitions are allowed. Depending on the atom, the transitions involving ground to excited state absorption may occur at higher energies (e.g., the vacuum ultraviolet for the Lyman system of hydrogen), or much lower energies for atoms with lower ionization potentials (e.g., visible yellow for the $3s \to 3p$ transition in sodium). Transitions between excited states of atoms can be seen in the emission from electrical discharges or flames, and these fall at lower energies in the visible or even infrared regions. This is the case for the Balmer and Paschen series of hydrogen (with lower levels of n = 2 and n = 3 respectively). Figure 1.12 shows an example of the visible Balmer series lines generated by a discharge lamp. These emission lines are the basis for Atomic Absorption (AA) spectroscopy, a mainstay in elemental analysis. Transitions between very high n Rydberg levels can also occur in the microwave region. In multielectron atoms, the transitions seen usually involve the excitation or relaxation of the single outermost valence electron. Excitation of core electrons can also occur, but this requires much higher energies outside the UV-visible region. Atomic spectroscopy experiments based on these principles are described in Chapters 11, 16, 24, 25, and 26.

Electronic spectroscopy of molecules is similar in many ways to that of atoms. Ionization potentials generally fall at higher energies for small molecules and lower energies for larger molecules, but the excited electronic states usually begin to be found at energies about halfway to ionization. Higher excited states become more closely spaced, approaching the ionization limit, and following an approximate inverse n^2 energy dependence. From the perspective of an electron in a highly excited orbital near the ionization limit, the core of either an atom or a molecule both appear to be a small positive particle, and therefore the highest electronic states for both follow a similar pattern. The highly excited levels for molecules are therefore also called Rydberg states. Transitions from the ground state to such Rydberg levels lie at high energies in the vacuum ultraviolet wavelength region, and are not often studied. The most common absorption transitions for molecules also involve the excitation of the single outermost valence electron from its ground state in its "highest occupied molecular orbital" (HOMO) to its next-lowest energy state, the "lowest unoccupied molecular orbital" (LUMO). Figure 14.3 shows the

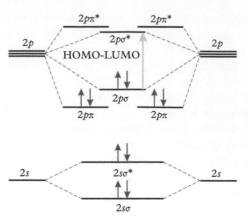

Figure 14.3 *The molecular orbital diagram for the nitrogen molecule, showing the valence electrons in the levels occupied in the ground state. In this example, the HOMO is the $2p\sigma$ orbital and the LUMO is the $2p\sigma^*$.*

molecular orbitals involved for the valence electrons of the nitrogen molecule, N_2 and the HOMO → LUMO transition. Absorption and emission transitions between higher excited states are also possible in electrical discharges. All transitions must conserve angular momentum, but the *s, p, d,* etc., symbols indicating the L values for atoms change over to σ, π, δ, etc., indicating the Λ values for diatomics and linear polyatomics. Nonlinear polyatomic molecules have symmetry-based selection rules which are discussed in detail elsewhere.[1]

Most stable molecules have closed-shell electronic configurations like the rare gas atoms, giving them relatively higher ionization energies, and corresponding higher energy excited states. The N_2 molecule shown in Figure 14.3 is such an example, with an ionization energy of 15.6 eV and a HOMO → LUMO transition lying at 8.45 eV, which falls in the vacuum ultraviolet wavelength region. Similar energetics apply for other small molecules. Larger molecules have more complex patterns of molecular orbitals and smaller HOMO–LUMO energy gaps, but most electronic spectroscopy for these systems usually falls in the ultraviolet wavelength region. An exception to this occurs for molecules composed of heavy multielectron atoms, in which the valence electrons are far removed from the nucleus, and then smaller HOMO–LUMO gaps are possible. An example of this behavior is the iodine molecule, I_2, explaining why iodine is popular for laser experiments in this book (Chapters 17, 18, 28). However, all of the simple, stable diatomic and triatomic molecules composed of first- and second-row elements (e.g., N_2, O_2, H_2, CO, CO_2, Cl_2, HF, HCl) have no low-energy absorption spectra and therefore these systems are not good candidates for electronic spectroscopy with available tunable lasers. Absorption spectra for small molecules can occur at low energies if the system is open shell, which occurs for radicals or ions. UV-visible lasers are therefore well suited to study species such as OH, CH, NH, CO^+, N_2^+, O_2^+, CO_2^+, etc. Indeed, the hydroxyl radical has a well-known electronic spectrum near 290 nm that is employed to detect this species in the atmosphere and in hydrocarbon flames, where it plays major roles in both atmospheric and combustion chemistry. UV-vis spectroscopy is intrinsically more sensitive than infrared or Raman measurements (the ν^3 factor in the Einstein A coefficient makes extinction coefficients larger in this region; see Chapter 3), and is therefore preferred for measurements on ions or radicals, which can only be produced in very low concentrations. Polyatomic radicals and ions also have lower energy spectra and can often be detected at visible or near-UV wavelengths with available tunable lasers. As shown later in Chapters 16 and 18, if single-photon allowed transitions are not accessible at low energies, multiphoton methods can also be used to probe higher energy states.

The importance of spectroscopy

Optical spectroscopy is commonly employed to study the structures of molecules as well as for the analysis of the composition of materials or multicomponent mixtures. In addition to its many laboratory applications, such spectroscopy is also essential for studies of remote environments, such as the upper atmosphere or even interstellar gas clouds. For example, the light from the Sun can be employed to investigate the IR absorbing molecules in the Earth's atmosphere. With concerns about global warming caused by gases such as carbon dioxide and methane, this subject has become critical for the preservation of the Earth as we know it. Thus, before discussing the details of spectroscopic transitions, we present selected examples of the application of spectroscopy to atmospheric processes. Undergraduate and graduate students at the University of Tennessee have been involved in studies of IR absorption in the Earth's atmosphere for the past 15 years and we use examples from this work. Figure 14.4 shows an atmospheric absorption spectrum using the Sun as the light source. This spectrum was recorded by Dr. Stewart Hager, Dr. Jeffrey Steill, and RNC at the University of Tennessee using an ABBomem DA-8 Fourier Transform Infrared Spectrometer (FT-IR). The dips seen in all of the spectra shown are from rovibrational lines of atmospheric gases. All of the lines shown are caused by real absorption and not "noise" as illustrated in further expansion of the bands in other figures below.

The spectrum in Figure 14.4 shows the IR absorption mainly due to water, carbon dioxide, and nitrous oxide. Svante August Arrhenius (1859–1927), one of the fathers of physical chemistry, was the first to point out the importance of these and other gases for heating of the earth. In 1896 he made the analogy of a greenhouse to the warming of the Earth's atmosphere.[6] Infrared light being reradiated from the Earth warms the atmosphere due to absorption by "greenhouse gases." Increasing amounts of these gases, notably CO_2, have been implicated to contribute to the increasing average temperature of the Earth. Shortly after Arrhenius published his paper he received criticism by some, noting that the CO_2 lines appear to be saturated. Thus addition of more CO_2 into the atmosphere would not affect the energy balance. Arrhenius responded by pointing out that

Figure 14.4 *The black-body (T ~5700 K) solar emission spectra incident upon the Earth exhibits absorption "dips" from atmospheric gases. The spectrum shows a myriad of absorption lines. Saturated absorption is seen in regions corresponding to CO_2 and H_2O. Absorptions due to N_2O and O_2 are also indicated.*

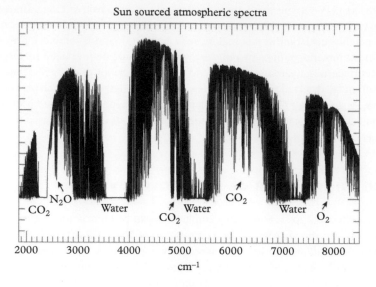

Sun sourced atmospheric spectra

Figure 14.5 *Spectra showing the saturated absorption profile of CO_2 for sunlight coming nearly straight down on the Earth at 1 pm (black), compared to that at different angles at 9 (blue) and 6 am (red). The 6 am spectrum clearly shows broader IR absorption and saturation due to the greater path length.*

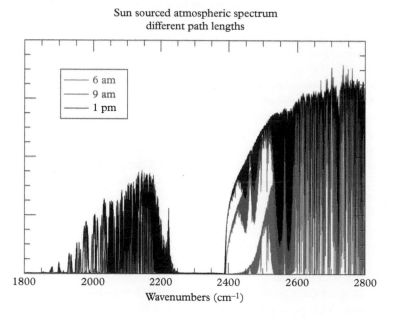

Sun sourced atmospheric spectrum
different path lengths

the line center may be saturated, however adding more CO_2 causes the band to broaden due to increased participation of the unsaturated rotational lines away from its center, in addition to collisional broadening, thus increasing the effective absorption. This is nicely illustrated in Figure 14.5, where spectra of the CO_2 band are shown at different times of the day for sunlight passing through different "path lengths" of the atmosphere. The absorption feature is clearly seen to broaden with increasing path lengths.

It is important to note that the CO_2 absorption band shown in Figures 14.4 and 14.5 above is not the one most responsible for atmospheric heating. Arrhenius used the Earth's reflection from the Moon to show that the doubly degenerate ν_2 bending vibration of CO_2 at ~ 667 cm^{-1} (see Chapter 21) is at the center of the spectrum of the Earth's ~300 K black-body radiation, and therefore it is this band that contributes most to global warming. In addition to the broadening of the absorption due to the addition of CO_2, Arrhenius also pointed out the importance of greater abundance of CO_2 to the radiative balance in the upper atmosphere. However, it is certainly true that because of this saturation the contribution of CO_2 *per molecule* to global warming is less than that of other molecules, such as methane, for which the vibrational bands are not saturated.

Provided that the interested student has an IR spectrometer with direct access to sunlight, metal mirrors can be used to introduce sunlight into this instrument to observe the IR absorption from these greenhouse gases. The spectra shown in Figure 14.4 were obtained using computer-driven mirrors to continuously correct for the position of the Sun, i.e., a "suntracker." The clever student could record an atmospheric absorption spectrum by manual adjustment of mirrors to direct the light into the spectrometer. Many absorption lines in the solar spectrum are due to atoms and molecules and their excited states in the chromosphere of the Sun. These features are called Fraunhofer lines, named for their discoverer, Joseph von Fraunhofer (1787–1826). Such lines are normally seen in the visible spectrum, but Fraunhofer lines due to absorption of Ca, Si, and H atoms can also be found in the IR, as shown in Figure 14.6.

In addition to IR absorption, it is also relatively straightforward to observe near-IR electronic absorption in the atmosphere using the sun as the light source. Figure 14.7 shows the rovibrational structure due to the $X\,^3\Sigma_g^- \to a\,^1\Delta_g$ transition of molecular oxygen measured in this way. As shown in Figure 14.4 and in more detail in Figure 14.7, this transition occurs near 7874 cm^{-1} (1.27 μm). This transition is totally forbidden under dipole selection rules (singlet $\leftrightarrow$ triplet, $\Sigma \leftrightarrow \Delta$, g $\leftrightarrow$ g are all forbidden) and only magnetic dipole (M1) and electric quadrupole (E2) transitions are allowed.[7] The long path afforded by the Earth's atmosphere allows easy access to this otherwise exceedingly weak absorption. However, this is not the only way to observe this transition. In the terrestrial laboratory, Miller et al. used the sensitive technique known as cavity ring down spectroscopy (CRDS) [8] to measure this spectrum, obtaining an extremely small Einstein A_{10} coefficient (2.3×10^{-4} s^{-1}). This group also determined an exceedingly long radiative lifetime of ~72 min (4300 $\pm$ 400 sec), resulting from the extreme forbiddenness of the $X\,^3\Sigma_g^- \to a\,^1\Delta_g$ transition.

The spectrum in Figure 14.7 also has a broad dip for sunlight passing through the longer path length atmosphere. As the sun declines in the west and the light path through the atmosphere increases, this dip becomes more pronounced. It is attributed to collisions between O_2 and surrounding molecules, which make the

Figure 14.6 *IR absorption spectra of both atmospheric N_2O and species in the Sun using the solar spectrum of light as a light source. The spectral region near the atmospheric N_2O $v_1 + 2v_2$ absorption band shows absorptions due to atoms and ions in the chromosphere of the sun such as Ca, Si, and the $n=4$ to $n=5$ transition of hydrogen. The hydrogen line is both pressure- and Doppler-broadened.*

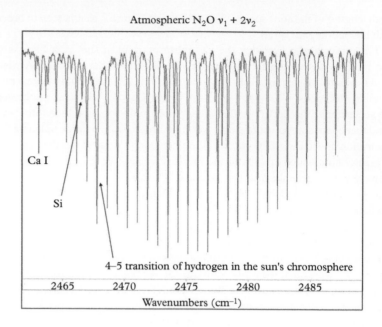

Figure 14.7 *The spectrum of the $X\ ^3\Sigma_g^- \rightarrow a\ ^1\Delta_g$ transition of molecular oxygen measured by passing sunlight through the Earth's atmosphere into a FT-IR spectrometer situated in the Science and Energy Research Facility (SERF) at the University of Tennessee. The lower trace was recorded when the sunlight was passing straight down (~noon). The upper trace was recorded late in the afternoon when the light probed a longer path through the atmosphere. Many of the rotational lines are saturated and there is a broad "dip" appearing around 1.27 μm.*

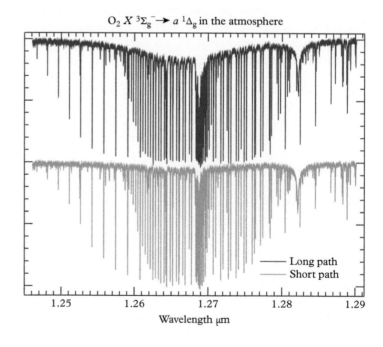

transition more allowed. Interestingly, if two O_2 $(a\,^1\Delta_g)$ molecules happen to collide, emission at about twice the energy of the $a\,^1\Delta_g$ state can be observed. This emission was first observed by Arnold et al. in a discharge flow in oxygen gas and has been called "dimol" emission.[9] Two emission bands are observed according to:

$$O_2\left(a^1\Delta_g\right) + O_2\left(a^1\Delta_g\right) \rightarrow O_2\left(X^3\Sigma_g^-\right)(v=0) + O_2\left(X^3\Sigma_g^-\right)(v=0) + h\nu\,(\lambda = 635\,\text{nm})$$

$$\rightarrow O_2\left(X^3\Sigma_g^-\right)(v=0) + O_2\left(X^3\Sigma_g^-\right)(v=1) + h\nu\,(\lambda = 703\,\text{nm})$$

Molecules in the $a\,^1\Delta_g$ state are called "singlet-delta" oxygen. Among its many applications, singlet-delta oxygen is employed in photodynamic therapy to kill cancer cells. Unfortunately, the $a\,^1\Delta_g$ state can adversely affect normal human cells and is also believed to induce cancer. A large research effort has been devoted to this subject. Due to collisions with atmospheric gases (notably other O_2 molecules), the lifetime of the $a\,^1\Delta_g$ state in air is reduced to 100 to 200 ms. Dr. Michael Kasha, one of the early investigators of singlet-delta oxygen, once suggested to a co-author of this book (RNC) that one should wait approximately 20 sec before opening a microwave oven to avoid breathing any O_2 $(a\,^1\Delta_g)$ that might have been produced.

Singlet-delta oxygen is also a component in one of the most powerful lasers in existence. The reaction of $a\,^1\Delta_g$ O_2 with molecular iodine produces excited $^2P_{1/2}$ iodine atoms, which are stimulated to radiate to the lower $^2P_{3/2}$ fine-structure state, producing laser light at the 1.315 μm wavelength. Light from this powerful Chemical Oxygen-Iodine Laser (COIL) can pass through the atmosphere without absorption from common atmospheric gases. Following the discovery that fullerenes and oxygen in the presence of light (even solar light) can produce copious amounts of singlet-delta oxygen, many groups, especially those in the military, have been working on the development of a high powered Fullerene-Oxygen-Iodine Laser (FOIL).

Rovibrational spectroscopy

Following this introduction to the importance of spectroscopy, we discuss the basic physics of the subject. The vibrational energy levels of a molecule can be investigated using infrared spectroscopy, in which IR light is passed through a gaseous sample and the dips in transmitted light intensity are recorded as a function of its frequency, as shown in Figure 14.4. The dips correspond to absorption of light, inducing transitions from the lower (v″) to upper (v′) vibrational energy levels of the molecule, as shown in Figure 14.1. Raman measurements access these same quantum states. To fully describe such molecular vibrations it is necessary to introduce the *anharmonic oscillator* potential function. In a first course in quantum mechanics, a student is introduced to the *harmonic oscillator* in which the potential of interaction between the two atoms of a diatomic molecule is given by $V(\Delta r) = (k/2)(\Delta r)^2$, where k is the force constant and Δr represents the distance between the two atoms about their equilibrium position, r_e, i.e., $\Delta r = r - r_e$. The vibrational energy levels for this fictitious molecule are equally spaced by a constant energy of $h\nu_0$. As a result of the uncertainty principle, the lowest energy level for the harmonic oscillator is $(\frac{1}{2})h\nu_0$. This quantum mechanical "jiggle" energy is called the zero-point energy. The potential of interaction for a real molecule is anharmonic (the force constant changes with stretching) and the bond will break upon stretching beyond the "elastic limit." Many empirical forms for this potential have been proposed. A commonly used form for the interaction between two atoms that becomes stiffer as they are compressed and softer as they are forced apart is the Morse potential, given by:

$$V(r) = D_e\left[1 - e^{-\beta(r-re)}\right]^2 \tag{14.2}$$

D_e is the depth of this potential well, r_e is the equilibrium position, and β is a measure of the curvature of the potential at the bottom of the well at $r = r_e$.

The solutions to the Schrödinger equation for the energy levels and wavefunctions for the Morse potential are given in most textbooks on quantum mechanics. The wavefunctions involve the Laguerre polynomials. The exact solutions of the Schrödinger equation for the allowed energy levels for this potential are,[1]

$$E_v = \omega_e\,(v + \tfrac{1}{2}) - \omega_e x_e\,(v + \tfrac{1}{2})^2 + \omega_e y_e\,(v + \tfrac{1}{2})^3 \tag{14.3}$$

where ω_e is the *vibrational constant* and $\omega_e x_e$ and $\omega_e y_e$ are the anharmonicity corrections to the frequency. Below we neglect the term involving $\omega_e y_e$ because its contribution is much smaller than the resolution of typical experiments. The relationships between the various parameters D_e, β, and r_e can be easily derived. At an internuclear distance equal to the equilibrium, i.e., $r = r_e$, the first and second derivatives of the Morse function are

$$dV(r)/dr = 0 \quad \text{and} \quad d^2V(r)/dr^2 = 2\beta^2 D_e \equiv k \tag{14.4}$$

where k is the force constant of the bond defined through

$$\omega_e = (1/2\pi c)\sqrt{(k/\mu)} = \beta\sqrt{(hD_e/2\pi^2 c\mu)} \tag{14.5}$$

It can also be shown that

$$\omega_e x_e = h\beta^2/8\pi^2 c\mu \tag{14.6}$$

Thus, from measurements of ω_e, $\omega_e x_e$, and r_e one can use these relationships to construct a Morse potential for a diatomic molecule. ω_e and $\omega_e x_e$ can be determined from measurements of IR spectroscopy for polar (heteronuclear) diatomic molecules or from Raman spectroscopy for non-polar (homonuclear) diatomics.

The energies of transitions from the ground state ($v'' = 0$) to a higher level, v', can be calculated from

$$E_v - E_0 = \omega_e v - \omega_e x_e v(v + 1) \tag{14.7}$$

An excitation from $v = 0$ to $v = 1$ is called the *fundamental* and transitions to higher v are called *overtones* (i.e., $\Delta v = 0 \to 2$ is the first overtone; $\Delta v = 0 \to 3$ is the second overtone, etc.). Because of the anharmonicity in the potential, the intervals between energy levels decrease as v increases. Eventually, increasing v by one unit causes the bond to break. The total energy (sum of all of the vibrational energies plus the last step) required to break the bond is called the dissociation energy, D_0. The energy from the very bottom of the well to the bond-breaking point is unfortunately also called the dissociation energy, but designated D_e. D_e can be calculated, but it cannot be measured because molecules do not exist at this energy. The lowest energy at which they can exist is the $v = 0$ level, and therefore D_0 is the quantity that can be measured. If the vibrational constants are known, and the Morse potential is adequate to describe the system, D_e and D_0 can be related to each other. D_0 is equal to the energy from the bottom of the well, D_e, minus the energy of the molecule in its zero point energy, E_0:

$$D_0 = D_e - \omega_e \,(\tfrac{1}{2}) + \omega_e x_e (\tfrac{1}{2})^2 = D_e - (\tfrac{1}{2})\, \omega_e + (\tfrac{1}{4})\, \omega_e x_e \tag{14.8}$$

In addition to their vibrational motions, molecules can rotate. If we confine a molecule so that the bond length does not change with rotational state, we can solve the Schrödinger equation again to obtain the rotational energy levels of the rigid rotor:

$$E_J = (h^2/8\pi^2 I)J\,(J + 1) = BJ\,(J + 1) \tag{14.9}$$

where $I = \mu r_e^2$ is the moment of inertia, μ is the reduced mass (for a diatomic molecule $\mu = m_1 m_2/(m_1 + m_2)$), and the rotational constant is $B = h^2/8\pi^2 I$. Typically the rotational constant is given in units of cm^{-1} or $B = h/8\pi^2 c I$. Combining vibrations with rotations, we can now write the energy for a molecule in a specific vibrational and rotational state. As a molecule is excited to higher vibrational levels, the internuclear separation of the atoms increases. If we let B_v indicate that r_e changes with vibrational level v (i.e., $B_v = h^2/8\pi^2 \mu r_v^2$), the energy levels of a vibrating and rotating molecule can be written as

$$E_{v,J} = \omega_e\,(v + \tfrac{1}{2}) - \omega_e x_e (v + \tfrac{1}{2})^2 + B_v J\,(J + 1) \tag{14.10}$$

A simplified picture of these energy levels is presented in Figure 14.8. v_0 is the energy difference between the two vibrational levels without rotation ($J'' = 0 \to J' = 0$). Other transitions in which $J'' = J'$ all pile up at v_0 are called the Q-branch. It is easy to show from equation 14.10 that the energy difference between adjacent rotational energy levels is equal to $2B_v$, i.e.,

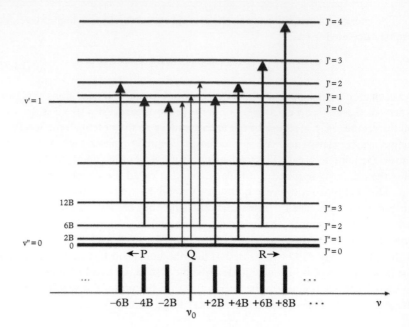

Figure 14.8 *Energy levels of the rotational states for the* v" = 0 *and* v' = 1 *states of a diatomic molecule. The vertical lines along the bottom show the energies of the transitions between these levels.*

$E(J = n) - E(J = n \pm 1) = \pm 2B_v$. Transitions in which J increases by one ($\Delta J = +1$) belong to the R-branch and those transitions in which J decreases by one ($\Delta J = -1$) belong to the P-branch.

Real molecules do not behave like rigid rotors and can expand upon rotation. As the molecule rotates, the bond length increases due to the centrifugal force. The rotational energy decreases by an amount $-D_e[J(J+1)]^2$ with increasing rotational states, giving the total rovibrational energy level as

$$E_{vJ} = \omega_e (v + \tfrac{1}{2}) - \omega_e x_e (v + \tfrac{1}{2})^2 + B_v J (J + 1) - D_e [J (J + 1)]^2 \qquad (14.11)$$

It can be shown that $D_e = 4B_e^3/\omega_e^2$. Note that the D_e here is the centrifugal distortion constant—not the dissociation energy. Although this is an unfortunate historical practice, the constants for these two quite different physical quantities use the same symbol. D_e is a small constant and its effect can only be detected with high-resolution measurements.

The vibrational energy levels for a molecule in a Morse potential are best illustrated for the simplest neutral molecule, H_2. Figure 14.9 is a plot of the Morse potential for hydrogen along with the vibrational energy levels and the wavefunction up to the dissociation limit. The square of the wavefunction gives the probability of finding the molecule at a particular separation R. Thus it is clear that as the vibrational level increases the molecule spends more time at the farthest extreme of its vibrational motion. This is typical of a classical oscillator and

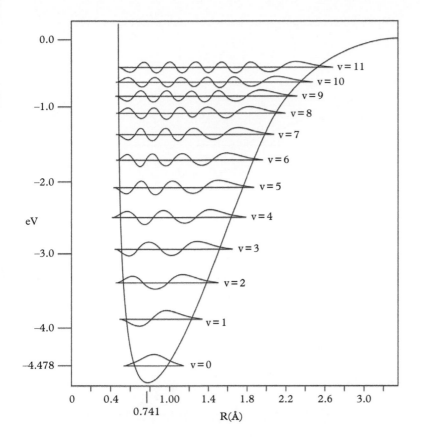

Figure 14.9 *Morse potential, vibrational energy levels, and wavefunctions for the hydrogen molecule, H_2. The levels were calculated using $r_e = 0.74144$ Å, $\omega_e = 4401.21$ cm^{-1}, and $\omega_e x_e = 121.33$ cm^{-1}. The dissociation energy, D_0, is 4.4781 eV. β is determined from equation 14.5.*

the condition in which a quantum system approaches that predicted by classical mechanics is called the correspondence principle.

Going back to equation 14.3, it is customary to equate E(v) with G(v) in the literature. Thus we write

$$G(v) = \omega_e (v + \tfrac{1}{2}) - \omega_e x_e (v + \tfrac{1}{2})^2 \qquad (14.12)$$

Taking the difference between G(v + 1) and G(v) gives

$$G(v + 1) - G(v) = \omega_e - 2\omega_e x_e (v + 1) \qquad (14.13)$$

Thus a plot of G(v + 1) − G(v) versus (v + 1) gives a straight line with a slope of −2$\omega_e x_e$ and an intercept at (v + 1) = 0 equal to ω_e. This is known as a Birge–Sponer plot.

To illustrate the above analysis for determining ω_e and $\omega_e x_e$ for a diatomic molecule, we again use data for hydrogen. Of course H_2 is inactive in the IR

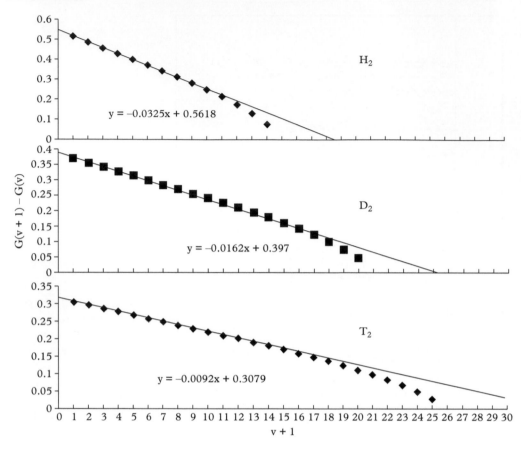

Figure 14.10 *Birge–Sponer plots of* $G(v + 1) - G(v)$ *in electron volts versus* $(v + 1)$ *for the homonuclear isotopomers* H_2, D_2, *and* T_2.

but we employ accurately calculated vibrational levels for H_2, D_2, and T_2 from a report by Fantz and Wunderlich.[10] We have plotted such curves for three isotopes of the hydrogen molecule containing hydrogen (H), deuterium (D), and tritium (T) as summarized in Figure 14.10.

The values for ω_e and $\omega_e x_e$ obtained from the intercept and slope in Figure 14.10, after converting from eV to cm^{-1}, are in good agreement with the accepted values [5] for H_2 ($\omega_e = 4401.21$ cm^{-1}; $\omega_e x_e = 121.34$ cm^{-1}), D_2 ($\omega_e = 3115.50$ cm^{-1}; $\omega_e x_e = 61.82$ cm^{-1}), and T_2 ($\omega_e = 2546.47$ cm^{-1}; $\omega_e x_e = 41.23$ cm^{-1}).

Before leaving molecular hydrogen and its isotopomers, we emphasize that infrared absorptions do not occur for homonuclear diatomic molecules like these. Raman measurements are possible, as noted before, but there are also other methods with which to determine vibrational frequencies for these systems. Most notable are the recent measurements of Dickenson et al. [11] who employed two-photon electronic excitations from the $X\,^1\Sigma_g{}^+$, $v = 0, 1$ states to the $EF\,^1\Sigma_g{}^+$, v = 0 level in a molecular beam (i.e., Doppler free) to determine very accurate (to 2×10^{-4} cm^{-1}) vibrational fundamentals for H_2, HD, and D_2. The values obtained are H_2 (4161.16632$_{18}$), HD (3632.16054$_{24}$), and D_2 (2993.61713$_{17}$). These results are in excellent agreement with the predictions of *ab initio* calculations which include relativistic and quantum electrodynamic (QED) effects.

Modern infrared spectroscopy is generally accomplished by employing a Fourier Transform Infrared (FT-IR) spectrometer, an instrument found in essentially every chemistry department and many physics departments throughout the world. Although lasers are employed for calibration purposes, the FT-IR method does not *directly* involve a laser. Students are encouraged to employ departmental FT-IR spectrometers to record IR spectra for gas- and liquid-phase molecules. A standard physical chemistry laboratory experiment is to record the IR spectrum for a heteronuclear molecule such as HCl to obtain ω_e, $\omega_e x_e$, r_e, and other molecular constants. The physical chemistry textbook by Sime provides a description of this experiment.[12]

Using such absorption measurements, we have recorded IR spectra for a number of gas-phase molecules (HF, NO, and acetylene) using a commercial ABBomen DA-8 FT-IR. These spectra illustrate the patterns found for such systems. Figures 14.10 and 14.12 show the regions of the fundamental vibration, v" = 0 → v' = 1, and the first overtone, v" = 0 → v' = 2, for gas-phase hydrogen fluoride, HF. The experimental parameters used to record the spectra are shown in the figures. The data for the first overtone was recorded under lower resolution and at a slightly higher pressure to obtain good signal to noise. Because listing these transitions can be cumbersome, special transition labels have been developed which are used in these figures. If the J in R(J) and P(J) represents its value in the lower state, then the transition from the ground rotational state J" = 0 to J' = 1 is designated as the R(0) line, and the transition from J" = 1 to J' = 0 is designated as the P(1) line. If there is no Q-branch, the R(0) line is the first at higher frequency from the band center, and P(1) is the first to lower frequency.

Because the potential describing a real molecule is anharmonic, the equilibrium bond length is a function of the vibrational state and increases with increasing vibrational quantum number. This requires that the rotational energy levels become more closely spaced with increasing v. This vibrational state dependence of the rotational energy constant is given by the vibration–rotation coupling constant α_e through the relation

$$B_v = B_e - \alpha_e \, (v + \tfrac{1}{2}) \qquad (14.14)$$

where $B_e = h/8\pi^2\mu r^2$. This trend is clearly present in the two spectra for HF (Figures 14.11 and 14.12) where the R-branch rotational lines are closer together for the overtone spectrum. We can also write

$$B_0 = B_e - \alpha_e \, (\tfrac{1}{2}) \qquad (14.15)$$

$$B_1 = B_e - \alpha_e \, (3/2), \ \text{etc.} \qquad (14.16)$$

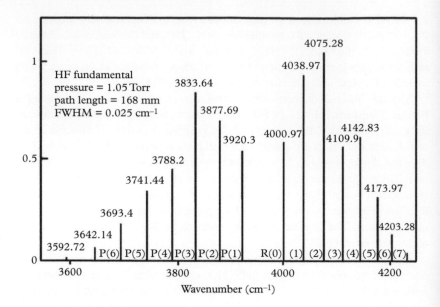

Figure 14.11 *P- and R-branches of the fundamental vibration* ($v'' = 0 \rightarrow v' = 1$) *of the HF molecule. The FWHM of each line in the spectra is 0.025 cm⁻¹.*

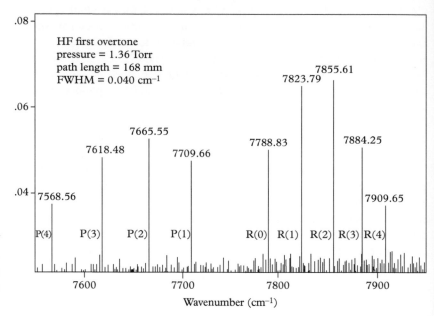

Figure 14.12 *P- and R-branches of the first overtone* ($v'' = 0 \rightarrow v' = 2$) *vibration of the HF molecule. The FWHM of each line in the spectra is 0.04 cm⁻¹.*

Therefore $\alpha_e = B_0 - B_1$ and $\alpha_e = B_n - B_{n-1}$. Including terms containing the vibrational constant, the total energy levels become

$$E_{v,J} = \omega_e (v + \tfrac{1}{2}) - \omega_e x_e (v + \tfrac{1}{2})^2 + B_e J (J + 1) - D_e J^2 (J + \tfrac{1}{2})^2 - \alpha_e (v + \tfrac{1}{2}) J (J + 1) \tag{14.17}$$

Applying equation 14.14 to the data for the fundamental vibration (Figure 14.11) gives $B_0 = 20.4771$ cm^{-1} and $B_1 = 19.707$ cm^{-1}, with a standard deviation of 0.32%. Similar analysis for the first overtone data gives $B_2 = 18.99$ cm^{-1} with a standard deviation of 0.16%. Combining the B_v values, we calculate $r_e = 0.9223$ Å.

The determination of ω_e and $\omega_e x_e$ is best done from an analysis of many vibrational transitions. However, reasonable values can be determined from the measurements of the $v = 0 \rightarrow 1$ and $v = 0 \rightarrow 2$ transitions alone. Because the Q-branch is forbidden for the HF molecule, these energy levels can be taken from halfway between the R(0) and P(1) energies in both graphs. Solutions of two equations and two unknowns gives $\omega_e = 4133.82$ cm^{-1} and $\omega_e x_e = 86.13$ cm^{-1}.

Looking at the transitions for $\Delta J = +1$, the energies of the lines in the R-branch are

$$\tilde{\nu}_R = (v' - v'')\omega_e - \omega_e x_e [(v' - v'')(v' + v'' + 1)]$$
$$+ (B_{v'} + B_{v''})(J'' + 1) + (B_{v'} - B_{v''})(J'' + 1)^2 - 4D_e(J'' + 1)^3 \tag{14.18}$$

Similarly, the energies of the lines in the P-branch ($\Delta J = -1$) are

$$\tilde{\nu}_P = (v' - v'')\omega_e - \omega_e x_e [(v' - v'')(v' + v'' + 1)] - (B_{v'} + B_{v''})J'' + (B_{v'} - B_{v''})J''^2 - 4D_e J''^3 \tag{14.19}$$

Measurement of the lines in the P- and R-branches allows one to calculate the molecular constants $B_{v'}$, $B_{v''}$, D_e, ω_e, and $\omega_e x_e$. Consider the case in which both the R- and P-branch originate from $J'' = J$; then the two lines on each side of the band center are P(1) and R(0). Proper combination of equations 14.16 and 14.17 leads to so-called "combination difference" relationships between the molecular parameters $B_{v'}$, $B_{v''}$, D_e, ω_e, and $\omega_e x_e$ and the P- and R-lines:[1,2]

$$\frac{R(J) + P(J)}{4(J + \tfrac{1}{2})} = B_{v'} - 2D_e(J^2 + J + 1) \tag{14.20}$$

$$\frac{R(J-1) - P(J+1)}{4(J + \tfrac{1}{2})} = B_{v''} - 2D_e(J^2 + J + 1) \tag{14.21}$$

$$\frac{R(J) + P(J+1)}{2} = (v' - v'')\omega_e [1 - x_e(v' + v'' + 1)] + (B_{v'} - B_{v''})(J + 1)^2 \tag{14.22}$$

From equation 14.18 one can see that $B_{v'}$ and D_e are obtained from the intercept ($J = 0$) and slope ($-2D_e$) of a plot of $[R(J) + P(J)]/(J + 1)] [4(J + \tfrac{1}{2})]$

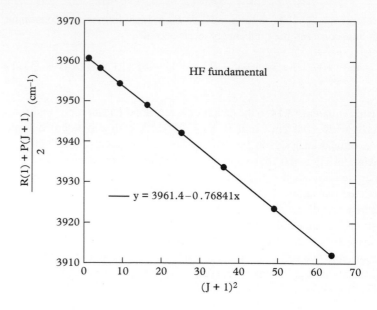

Figure 14.13 *Plot of {R(J) + P(J + 1)}/2 versus (J + 1)² for the vibrational fundamental of gas-phase HF.*

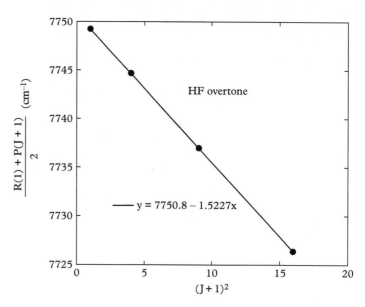

Figure 14.14 *Plot of {R(J) + P(J + 1)}/2 versus (J + 1)² for the first overtone of gas-phase HF.*

versus $[J^2 + J + 1]$. Likewise from equation 14.19, a plot of $[R(J–1) - P(J + 1)]/[4(J + \frac{1}{2})]$ versus $[J^2 + J + 1]$ gives a straight line of slope $-2D_e$ and intercept $B_{v''}$. Of particular interest from equation 14.20, the slope of a plot of $[R(J) + P(J + 1)]/2$ versus $[(J + 1)^2]$ gives $B_{v'} - B_{v''}$. Such plots are shown for the fundamental and first overtone transitions of HF in Figures 14.13 and 14.14.

Table 14.1 *Combination differences [1,2] for the HF fundamental transition. The band origin is given by the intercept of the line of best fit.*

P(J)	E_J (cm^{-1})	$\omega_e(1 - 2x_e) - 2JB_e - J(J - 2)\alpha_e + 4J^3 D_e$
P(1)	3920.30	$\omega_e(1 - 2x_e) - 2B_e + \alpha_e + 4D_e$
P(2)	3877.69	$\omega_e(1 - 2x_e) - 4B_e + 32D_e$
P(3)	3833.64	$\omega_e(1 - 2x_e) - 6B_e - 3\alpha_e + 108D_e$
P(4)	3788.20	$\omega_e(1 - 2x_e) - 8B_e - 8\alpha_e + 256D_e$
P(5)	3741.44	$\omega_e(1 - 2x_e) - 10B_e - 15\alpha_e + 500D_e$
P(6)	3963.40	$\omega_e(1 - 2x_e) - 12B_e - 24\alpha_e + 864D_e$
P(7)	3644.14	$\omega_e(1 - 2x_e) - 14B_e - 35\alpha_e + 1372D_e$
P(8)	3593.72	$\omega_e(1 - 2x_e) - 16B_e - 48\alpha_e + 2048D_e$

Specific values used in the HF fundamental plot and their relationships to the molecular constants are given in Tables 14.1 and 14.2. The intercept of this line at J = 0 gives the band origin, ν_0.

$$\nu_0 = (v' - v'') \omega_e [1 - x_e (v' + v'' + 1)] \qquad (14.23)$$

In the case for $v'' = 0$ and $v' = 1$, the band origin occurs at an energy given by $\omega_e - 2\omega_e x_e$.

A comparison of these derived molecular constants with accurately known values can be seen in the caption for Figure 14.15, which shows a Morse potential plot and vibrational levels for HF. The literature values were taken from Huber and Hertzberg.[5] The error for ω_e and r_e is less than 1%, where $\omega_e x_e$ differs by 4%. The value for α_e is only 3.5% lower than the accepted value.

Table 14.2 *Line positions in the P- and R-branches for the fundamental and first overtone transitions of HF.*

J	P(J) (cm^{-1})	R(J) (cm^{-1})
0	-	4000.97
1	3920.30	4038.87
2	3877.69	4075.28
3	3833.64	4109.91
4	3788.20	4124.83
5	3741.44	4173.97
6	3693.40	4203.28
7	3644.14	4230.76
8	3593.72	-

J	P(J) (cm^{-1})	R(J) (cm^{-1})
0	-	7788.83
1	7709.66	7823.79
2	7665.55	7855.61
3	7618.48	7884.25
4	7568.56	7909.65

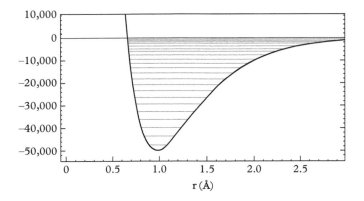

Figure 14.15 *Plot of the Morse potential ($V(r) = D_e[1 - e^{-\beta(r-r_e)}]^2$) for the HF molecule using D_e = 49378 cm^{-1} and r_e = 0.91680 Å. β is determined from equation 14.5 using ω_e = 4138.32 cm^{-1} and $\omega_e x_e$ = 89.88 cm^{-1}. D_0(HF) = 5.869 eV (D_e here is the dissociation energy).*

Upon inspection of the vibrational levels for H_2 in Figure 14.9 and for HF in Figure 14.15, an interesting question arises: How many vibrational levels exist in the two potential wells? An exact answer to this question requires precise knowledge of the long-range potential and its deviation from the Morse potential (equation 14.2). It is easily seen that the extrapolation of the Morse potential for H_2 or its deuterated and triton derivatives in Figure 14.9 gives an overestimate for the highest vibrational level, v_{max}. An estimate of v_{max} can be derived by setting E $(v +1) - E(v)$ equal to zero, i.e.

$$\omega_e (v + 3/2) - \omega_e x_e(v + 3/2)^2 - \omega_e (v + \tfrac{1}{2}) + \omega_e x_e (v + \tfrac{1}{2})^2 = 0 \qquad (14.24)$$

and solving for v_{max} which results in $v_{max} = 1/(2x_e) - 1$. Another method of estimating v_{max} is to set the derivative of $\omega_e (v + \tfrac{1}{2}) + \omega_e x_e (v + \tfrac{1}{2})^2$ to zero. This procedure results in $v_{max} = 1/(2x_e) - \tfrac{1}{2}$. Using the measured values of $\omega_e x_e$ and ω_e for HF gives $x_e = 0.02172$ and $v_{max} \sim 22$. For H_2 $x_e = 0.0275674$ and $v_{max} \sim 16$. Note that the Birge–Sponer extrapolation in Figure 14.10 gives $v_{max} \sim 18$. The student should find the approximate v_{max} values for D_2 and T_2 and compare these estimates with the extrapolated results from Figure 14.10.

Finally, we present rovibrational infrared spectra for the nitric oxide molecule in Figure 14.16. Unlike most diatomics, the extra electron in NO has one unit of angular momentum about the molecular axis and thus NO has a $^2\Pi$ ground state. As a result of this odd electron, the IR spectrum of NO exhibits a Q-branch. A greatly expanded view of this Q-branch is shown in Figure 14.17.

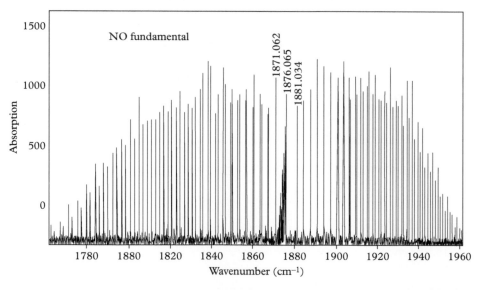

Figure 14.16 *Infrared absorption spectrum for the fundamental of NO. The center Q-branch is expanded in Figure 14.17.*

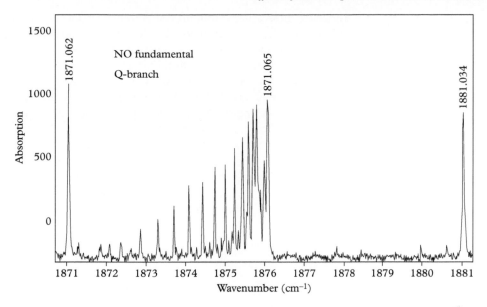

Figure 14.17 *Expanded view of the Q-branch for the fundamental band of NO shown in Figure 14.16.*

The HF and NO examples presented here show typical rovibrational spectra of diatomic gases and how they are analyzed. Similar examples and analyses can be found in many laboratory texts on physical chemistry for other more commonly studied species such as HCl. The IR spectra reported here via FT-IR absorption can also be measured with infrared lasers using direct absorption or other related techniques. Chapter 4 describes infrared diode lasers and infrared OPOs that can be used for these experiments. In Chapter 27 we show how stimulated electronic Raman scattering can be used to produce another kind of infrared laser device. Infrared spectroscopy using such lasers is described in Chapters 23 and 27.

Effects of nuclear spin and statistics on molecular spectra

The spectra of many symmetric molecules are found to exhibit a regular alternation of rotational line intensities and sometimes there are even *missing* rotational lines. This alternation may be simple in the case of diatomic molecules, or more complex as in the case of nonlinear polyatomic molecules.[1–4] The alternation of intensities is a direct consequence of the exchange symmetry (i.e., statistics) of the molecule and the statistical weights attached to the various rotational levels. The intensity alternations and missing lines were first observed by Rasetti [13] in the diatomic gases hydrogen, nitrogen, and oxygen. Mecke [14] had earlier associated these patterns with molecules containing identical nuclei. The full explanation involves a quantum description of the wavefunction and its dependence on the nuclear spins (Bose or Fermi particles) of the constituent atoms. We consider the effects of spin and statistics for a homonuclear diatomic molecule, but the treatment can easily be generalized to polyatomics.

According to the Pauli Principle, wavefunctions describing aggregates of particles (molecules) must be well-behaved mathematical functions when space and spin coordinates are interchanged (i.e., when equivalent particles exchange places). In fact one finds that the total wave function is either unchanged or merely changed in sign. If the wavefunction is unchanged, the particles obey what is called Bose–Einstein statistics. If the wavefunction changes sign, the particles obey Fermi–Dirac statistics. This is illustrated below upon exchange of arbitrary particles m and n in the wavefunction

$$\Psi(q_a, q_b, \ldots q_m, \ldots q_n, \ldots q_z) =$$
$$+ \Psi(q_a, q_b, \ldots q_n, \ldots q_m, \ldots \ldots q_z) \qquad \text{Bose–Einstein} \qquad (14.25)$$

$$\Psi(q_a, q_b, \ldots \ldots q_m, \ldots q_n, \ldots \ldots q_z) =$$
$$- \Psi(q_a, q_b, \ldots q_n, \ldots q_m, \ldots q_z) \qquad \text{Fermi–Dirac} \qquad (14.26)$$

Particles having zero or integer nuclear spins are called Bose particles, or "bosons," and their total wavefunction must remain unchanged upon exchange of nuclei. Particles with half-integer spins are called Fermi–Dirac particles, or "fermions," and their wavefunctions must change sign upon interchange of nuclei. Ehrenfest and Oppenheimer [15] first showed that nuclei containing an odd number of nucleons obey Fermi–Dirac statistics and those with an even number of nucleons obey Bose–Einstein statistics. This follows from the fact that nucleons, protons, or neutrons are half-integer spin particles. This exchange symmetry has profound effects on the rotational properties of symmetric molecules, as we demonstrate below for homonuclear diatomics.

We write the total wavefunction for a diatomic molecule as a product of the electronic, ψ_e, vibrational, ψ_v, rotational, ψ_r, and nuclear spin ψ_{ns} wavefunctions.

$$\Psi = \psi_e \psi_v \psi_r \psi_{ns} \qquad (14.27)$$

Now consider the exchange symmetry of each wavefunction separately:

ψ_e: The ground electronic states of most homonuclear diatomic molecules are symmetric with respect to exchange of nuclei. A notable exception is the ground $X\,^3\Sigma_g^-$ state for molecular oxygen.

ψ_v: ψ_v is always symmetric with respect to interchange of particles since the vibrational wavefunction depends only on the magnitude of the internuclear separation.

ψ_r: The rotational symmetry depends upon the rotational quantum number J. In spherical coordinates exchanging nuclei changes θ to $\pi - \theta$ and φ to $\pi - \varphi$. The solution to the rigid rotor gives the Legendre polynomials $P_J[\cos(\pi - \theta)]$. Since

$$P_J[\cos(\pi - \theta)] = (-1)^{J-m} P_J[\cos(\theta)] \tag{14.28}$$

and

$$e^{im(\pi+\varphi)} = (-1)^m e^{im\varphi}$$

We have

$$\psi_r = (-1)^J \psi_{r(\theta,\varphi)}$$

Thus ψ_r is symmetric for even J (J = 0, 2, 4, ...) and antisymmetric for odd J (J = 1, 3, 5 ...) rotational states.

ψ_{ns}: ψ_{ns} can be either symmetric or antisymmetric and will combine with even J or odd J rotational functions in just such a manner to make the total wavefunction obey the proper statistics. This is illustrated below:

	$\Psi = \psi_e \psi_v$	ψ_r	ψ_{ns}
Bose–Einstein Statistics		s even J	s
		a odd J	a
Fermi–Dirac Statistics		a odd J	s
		s even J	a

If I is the nuclear spin, there are $(2I+1)^2$ possible nuclear spin states of which $(2I+1)(I+1)$ are symmetric and $(2I+1)I$ are antisymmetric. Therefore the symmetric nuclear spin states accompany the even J rotational levels for bosons (and conversely for fermions). Thus for Bose–Einstein statistics the statistical weight of even J levels to that of odd J levels, N_s/N_a, is equal to the statistical weight of symmetric spin states to antisymmetric spin states or $(I+1)/I$. For Fermi–Dirac statistics, the converse is true or $N_s/N_a = I/(I + 1)$. Summarizing,

$$N_s/N_a = (I + 1)/I \quad \text{Bose–Einstein Statistics} \tag{14.29}$$

$$N_s/N_a = I/(I + 1) \quad \text{Fermi–Dirac Statistics} \tag{14.30}$$

This now provides an explanation for the alternating intensities and missing lines in rotational spectroscopy for homonuclear diatomic molecules.

For molecular hydrogen, the H atom has an intrinsic spin of ½ and is a fermion. Thus the even J levels combine with antisymmetric nuclear spins and the odd J states combine with even nuclear spin functions, giving a ratio of intensities of $I/(I + 1) = \frac{1}{2}/(\frac{1}{2} + 1) = 3$. Indeed, a 3:1 alternation in intensity is observed in the rotational spectrum of H_2 (and those for other molecules such as H_2O or CH_2O having two equivalent hydrogens).

For bosons in which $I = 0$, N_s/N_a is infinite and the asymmetric rotational lines (J = odd) are missing. This is true for most molecules with $I = 0$, however, there is one important exception. The ground state of molecular oxygen is $X\,^3\Sigma_g^-$ and is antisymmetric upon exchange of O atoms. Thus, the *symmetric* rotational states are missing. This means that the $^{16}O_2$ (and $^{18}O_2$) molecule cannot rotate with even J, i.e., these levels are missing in nature! Also, there are no missing rotational lines (or alternating intensities) for $^{18}O^{16}O$ molecules since the molecule is no longer symmetric. Steinfeld [1] has pointed out that ^{17}O was first discovered as a weak line from a spectrum of molecular oxygen. It is important to point out that many of the first nuclear spins were inferred from alternating or missing intensities of the rotational spectrum.

Finally, the above analysis can be generalized to a linear molecule with a center of symmetry like O=C=O or HC≡CH. For CO_2 (bosons) the odd rotational lines are missing and for C_2H_2 (fermions) one observes a 3:1 ratio of odd to even intensities. In 1931 Houston and Lewis [16] observed Raman shifts in the rotational spectrum of CO_2 which could only be explained by assuming that the odd rotational lines were missing. This is precisely what is expected if CO_2 is linear and symmetrical.

The 3:1 ratio is seen in the spectrum for acetylene in Figure 14.18. Note that the 3:1 ratio is only seen at the "wings" of the spectrum where the absorption is weaker since there is saturation occurring in the central part of the P- and R-branches.

The hydrogen atom (nuclear spin = $\frac{1}{2}$) is a fermion while the deuterium atom (nuclear spin = 1) is a boson. Consequently, the rotational state populations for H_2 and D_2 are restricted by nuclear spin statistics and the relative intensities of rotational lines can be calculated according to equations 14.29 and 14.30 above. Likewise, the two naturally occurring carbon isotopes ^{12}C (nuclear spin = 0) and ^{13}C (nuclear spin = $\frac{1}{2}$) are bosons and fermions, respectively, and similar restrictions on the rotational levels of the $^{12}C_2$ and $^{13}C_2$ homonuclear diatomic molecules will occur. Similar considerations apply to the rotationally resolved spectra of any molecule with atoms that are equivalent by theory. In addition to acetylene, organic molecules such as ethane, ethylene, methane, benzene, etc., are also well-known examples.[17]

A particularly extreme example of equivalent atoms is found for the C_{60} molecule. A number of theoretical studies have considered the interesting case of the effects of nuclear spin and statistics on the rovibrational states of $^{12}C_{60}$ and $^{13}C_{60}$. We discuss here the treatment of Saito, Dresselhaus, and Dresselhaus.[18] C_{60} has icosahedral, I_h, symmetry and the ground-state electronic structure of C_{60} is of A_g symmetry. Therefore, the total wavefunction is symmetric with respect to the permutation of any two ^{12}C atoms in $^{12}C_{60}$ or antisymmetric with exchange

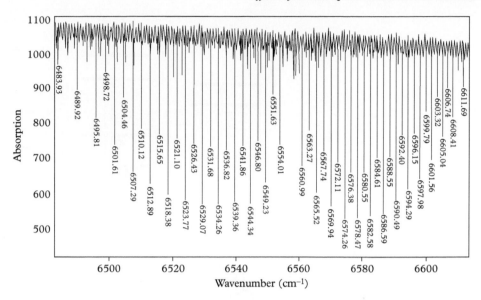

Figure 14.18 *Infrared spectrum of acetylene in the C–H overtone region showing the even–odd alternation in line intensities. In the wing regions of the spectrum where the absorptions are not experiencing saturation, an approximate 3:1 intensity ratio is seen.*

of any two ^{13}C in ^{13}C$_{60}$. In the case of ^{12}C$_{60}$, Ψ_{el}, Ψ_{vib}, and Ψ_{ns} all transform as A$_g$ species. This requires that the rotational wavefunctions also transform as A$_g$. Saito et al. [18] showed that rotational motion is restricted to only those J values which contain the irreducible representation A$_a$ in the I$_h$ point group. Table 1 of [18] shows that for the ground vibrational state the possible J values are restricted to J = 0, ... 6, ... 10, ... 12, ... 15, 16, ... Likewise a different set of allowed rotational states is found for the upper vibrational states, giving rise to an interesting rovibrational spectrum. Sogoshi et al. [19] have summarized the rotational levels for both the ground and upper vibrational levels for ^{12}C$_{60}$ showing the expected P, Q, and R transitions as reproduced in Figure 14.19.

The study by Sogoshi et al. [18] also presented a simulated calculation for the expected gas-phase P, Q, R infrared absorption spectrum for the A$_g$ → F$_{1u}$(3) transition of ^{12}C$_{60}$ at T = 5.0 and 2.5 K using a rotational constant of $B = 0.0028$ cm^{-1}. In this simulated spectrum the branches each cover only 1 cm^{-1} and contain ~80 lines over this interval. The separation between lines is only 0.05 cm^{-1} and it therefore requires extremely good instrumental resolution and rotationally cold gas-phase molecules to confirm these predictions. In this regard, Sogoshi et al. [19] were unable to observe the boson-exchange restrictions on the rotations of C$_{60}$ in para-hydrogen, p-H$_2$/C$_{60}$, as a result of complications due to ^{13}C isotopic contributions, p-H$_2$ matrix effects and possible hindered rotation.

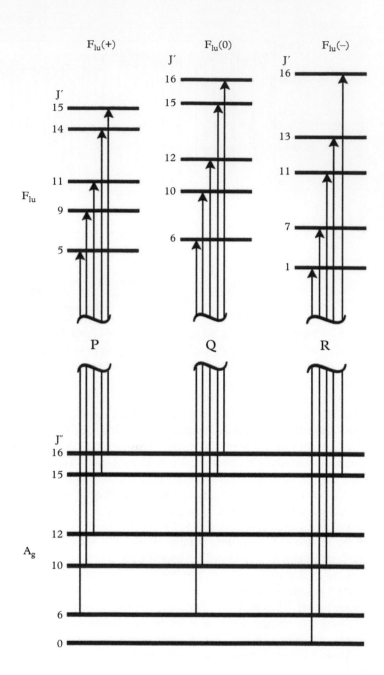

Figure 14.19 *Illustration of the expected* P, Q, *and* R *transitions for rovibrational states of* $^{12}C_{60}$ *between the lower* A_g *and upper* F_{1u} *states. Notice the missing rotational lines due to Bose exchange symmetry restrictions (adapted from Sogoshi et al. [19]).*

Electronic spectroscopy

As noted above, electronic transitions involve the excitation or relaxation of electrons between two electronic states. In most absorption measurements, the electron is in the HOMO and is excited to the LUMO. The ground electronic state has the same structure and energy levels that would be investigated with an infrared or Raman experiment. These include the 3N–6 (or 3N–5 for a linear system) vibrational modes, each with their respective vibrational constants, and the three rotational constants describing the distribution of mass about the framework of the system. Because the excited electronic state has a different orbital occupancy, it generally has a slightly different structure, with slightly different bond strengths. This in turn leads to rotational constants and vibrational frequencies that are different from those in the ground state. Because electronic transitions generally occur in the UV-vis wavelength region, their energies are in the 10,000–50,000 cm^{-1} range. Vibrational energies are in the range of 500–4000 cm^{-1}, and rotational intervals are on the order of 0.1–10.0 cm^{-1}. Therefore, as an experiment scans through the UV-vis energy range, it is possible to excite not only the electronic transition but also the vibrational and rotational excitations that go along with it. The name sometimes used for this kind of "all-state" investigation is *rovibronic* spectroscopy. It is rare for even the best spectroscopists to get the information about all the vibrational modes of the system in its ground and excited states, but electronic spectroscopy can get some of this information. A full description of rovibronic spectroscopy for polyatomic molecules is beyond the scope of this discussion. The interested reader is referred to the spectroscopy texts cited earlier.[1] Here, we use the example of a diatomic system to illustrate the principles involved in this kind of work. We focus on the electronic and vibrational, i.e., *vibronic*, transitions, since the full rovibronic patterns can also be quite complex.

Figure 14.20 shows a schematic level diagram depicting the potential energy curves, quantum levels, and transitions for the electronic spectroscopy of a diatomic molecule. By convention, prime and double prime superscripts are used to indicate parameters for the excited and ground state, respectively. As discussed earlier, the ground-state potential is characterized by a Morse function, with a bond distance r_e'', and the vibrational levels in this potential are described by the vibrational constants ω_e'' and $\omega_e x_e''$. The depth of the well provides the dissociation energy, which can be defined relative to the bottom of the well (D_e'') or relative to the v = 0 level (D_0''). This ground state can be investigated with either Raman or IR, depending on whether or not the system is homonuclear. The excited state will usually have a slightly different bond length (r_e'), well depth (D_0'), and vibrational constants (ω_e' and $\omega_e x_e'$) because of its different electronic structure. In the present example, the excited state has a longer bond length and lower dissociation energy than the ground state. The energy difference between the bottom of the ground-state potential and that of the excited-state potential is defined as T_e. This is the pure electronic-state energy difference.

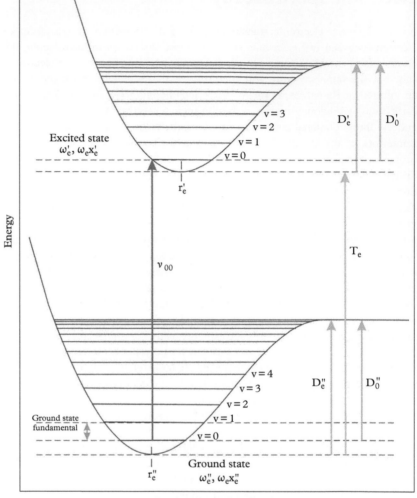

Figure 14.20 *The ground and excited potentials for discussion of the electronic spectroscopy of a diatomic molecule. The vertical transition (arrow) from the ground state* v = 0 *level accesses an excited state with a longer bond distance.*

Internuclear seperation (r)

The absorption of light therefore takes the system from the ground state to the excited state, assuming that the transition is allowed by selection rules. Diatomic electronic transitions are allowed if they conserve angular momentum, which is indicated by the capital Greek letter term symbols used to indicate states. Thus, Σ, Π, and Δ states have 0, 1, 2 units of electronic (orbital) angular momentum. Because photons have one unit of angular momentum, their absorption or emission indicates that the angular momentum should change by 1, i.e., that $\Sigma \to \Pi$, $\Pi \to \Sigma$, and $\Pi \to \Delta$ transitions are allowed. Transitions that would require two units of angular momentum are thus forbidden (e.g., $\Sigma \to \Delta$). However, it is the *total* angular momentum that must be conserved, which must also include that from rotations. Therefore, $\Sigma \rightarrow \Sigma$, $\Pi \rightarrow \Pi$, etc., transitions are also allowed because the angular momentum can be conserved by changes in rotational states.

Vibronic transitions in a diatomic do not involve angular momentum and are thus allowed from any ground-state vibrational level to any excited-state vibrational level. However, the most probable transitions will have the greatest line intensities. The intensity of the vibronic transition is given by the dipole moment integral, which should properly be expressed with ground- and excited-state wavefunctions that are products of vibrational and electronic wavefunctions, as shown on the left side of equation 14.31. However, according to the Franck–Condon Principle, electronic excitation proceeds so much more rapidly than nuclear movement (i.e., vibration), that the two motions can be considered independently, as shown on the right side of equation 14.31.

$$ \int \int \psi_v' \psi_e' \mu_e \psi_e'' \psi_v'' \, d\tau_e \, d\tau_n \approx \int \psi_e' \mu_e \psi_e'' \, d\tau_e \int \psi_v' \psi_v'' \, d\tau_n \qquad (14.31) $$

Electronic transitions are then assumed to occur when nuclei are essentially frozen in their ground-state structure, corresponding to a *vertical* transition in Figure 14.20 (i.e., one in which r does not change). An example of such a transition is indicated in the figure. The probability of such a transition is given by the electronic part of the dipole moment integral in a product with the vibrational part, where the vibrational part $\int \psi_v^* \psi_v \, d\tau_n$ is known as the Franck–Condon factor connecting a ground-state v'' level with an excited v' level. This is also referred to as the vibrational overlap integral, because it expresses the product of wavefunctions in the ground and excited state at the most probable r'' value of the v'' state.

If the system is at relatively low temperature, only the lowest vibrational state in the ground electronic state (v'' = 0) will be populated thermally, and this provides the starting point for a series of transitions to different v' levels known as a "progression." The v'' = 0 → v' = 0 transition is called the origin band. The spacings between successive members in such a progression provide the excited state vibrational level intervals. The progression energies are given by

$$ G\,(v'', v') = E(v'' = 0 \rightarrow v') = T_e + G(v') - G(v'' = 0) \qquad (14.32) $$

where G(v') is given by equation 14.12. If the sample is hotter, or if the vibrational frequency is low (the case for I_2 in Chapter 17 and 18), transitions may originate in v'' levels other than v'' = 0, producing more complex spectral patterns. Such transitions that begin in excited v levels are called "hotbands."

Figure 14.21 shows an excited-state progression in the electronic spectrum measured for the Ag-Kr diatomic molecule.[20] Silver atoms produced by laser vaporization were combined with krypton gas in a supersonic molecular beam, producing ultracold conditions, so that this unusual molecule could be formed. It was detected with resonant two-photon ionization spectroscopy (also known as

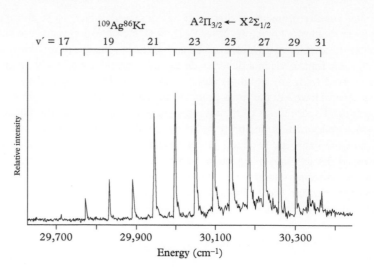

Figure 14.21 *A progression in the electronic spectrum of diatomic Ag-Kr measured with resonance enhanced ionization detection. (Reprinted with permission from L. R. Brock, M. A. Duncan, "Photoionization spectroscopy of Ag–rare gas van der Waals complexes," J. Chem. Phys. 103, 9200 (1995). Copyright 1995, AIP Publishing.)*

(1+1) REMPI; see Chapter 18), in which a single photon produced the excited state and another ionized those excited complexes. The intensity of the AgKr$^+$ ion was measured with a mass spectrometer as a function of the laser wavelength to record the spectrum. In this system, the molecule is extremely cold (about 5 K) so that only v" = 0 molecules are present initially. The excited-state bond distance is much shorter than that in the ground state, so that the best Franck–Condon overlap is for transitions from v" = 0 to v' = 20–30 (quantum numbers were assigned by isotopic studies). The level spacing decreases toward higher v' levels because of the excited state anharmonicity. A Birge–Sponer plot was employed to analyze the excited-state potential and vibrational constants. The vibrational frequency turned out to be ω_e' = 121.6 cm^{-1} and the excited state dissociation energy was 2286 cm^{-1}. Because of the heavy atoms and very small rotational constants, rotational structure was not resolved with the available laser linewidth (0.5 cm^{-1}).

Electronic spectroscopy can be measured with a variety of techniques including direct absorption (see Chapter 17), resonant ionization (see Chapter 16 and 18), or laser-induced fluorescence (see Chapter 25). Ionization and fluorescence detection schemes are much more sensitive than direct absorption methods and are used more often in modern research when samples in low density media are studied (e.g., low-pressure gases, molecular beams). These techniques measure a signal produced as a result of light absorption rather than absorption itself, and are two examples of so-called "action" spectroscopies. We introduce these methods here with a discussion of laser-induced fluorescence (LIF).

Laser-induced fluorescence is by now a well-established method that has been employed for many years. Credit for its invention is generally given to Professor Richard Zare (Stanford University).[21] Figure 14.22 shows a level

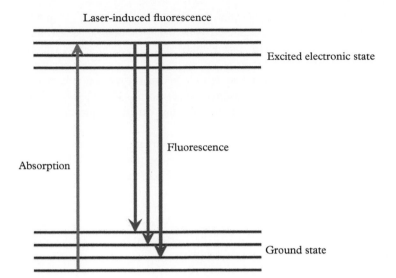

Figure 14.22 *Levels diagram showing absorption versus fluorescence, and the different wavelengths of fluorescence produced from a single* v″ → v′ *excitation transition.*

diagram illustrating how absorption can occur on a single v″ → v′ resonance, and then fluorescence can occur from the v′ state excited to any of several v″ ground-state levels depending on the favorable Franck–Condon Factors. The key to the sensitivity of LIF and other action spectroscopy methods is that the transmitted intensity of the light source is not what is measured. When low density samples are studied, their concentrations are not great enough for absorption to attenuate the transmitted light to any measurable degree. The fraction of the laser light absorbed is actually much less than the inherent noise level in the laser intensity. Amplification of the transmitted intensity does not help, as this also amplifies the noise. Instead, LIF detects the fluorescence that the excited molecules produce. As shown in the typical experimental setup in Figure 14.23, this fluorescence comes out in all directions from the excited molecules. In a direction orthogonal from the excitation laser, there should be no signal other than this fluorescence. A PMT detector mounted here can detect even a low level of light; and because there is in principle

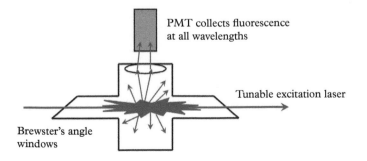

Figure 14.23 *Experimental configuration used to detect laser-induced fluorescence excitation spectra. All light is detected regardless of its color as the excitation laser scans. The spectrum is recorded as the yield of total fluorescence versus the excitation wavelength.*

no other background, amplification improves the signal level. In practice there is always a small level of scattered light that produces a background. However, pulsed dye lasers and time gating can be used to avoid scattered light from the excitation laser, improving sensitivity further. Fluorescence only occurs following absorption, and so its yield versus the excitation laser wavelength can be used to map out the absorption spectrum. This technique works well for molecules with efficient fluorescence, but can fail if the quantum yield is too low.

Another configuration of LIF is the measurement of dispersed fluorescence, as shown in Figure 14.24. If fluorescence is efficient, the excitation laser can be fixed on a particular v" → v' transition, and the different wavelengths of light produced can be analyzed with a scanned monochromator. This provides a progression with ground state vibrational intervals (see right side of Figure 14.22). This method can be used to investigate the ground-state potential when other methods fail. Wavelength-specific fluorescence can be used to distinguish one component molecule in a mixture, such as hydroxyl radicals present in a flame environment or in the atmosphere.

The concepts demonstrated by LIF and dispersed fluorescence are reproduced in different ways in other forms of action spectroscopy. In resonance ionization methods (Chapter 16 and 18), the production of ions is measured as a function of the excitation laser wavelength. Mass analysis of these ions with a mass spectrometer provides selectivity like that of dispersed fluorescence. Photodissociation spectroscopy detects fragments produced when neutral or ionized molecules are excited with lasers.[22] Again, mass spectrometry can be used to achieve selectivity. In photoacoustic spectroscopy (Chapter 23) or optogalvanic spectroscopy (Chapter 24), sound or discharge current variations are caused by laser absorption. All of these new techniques, which can provide exquisite sensitivity

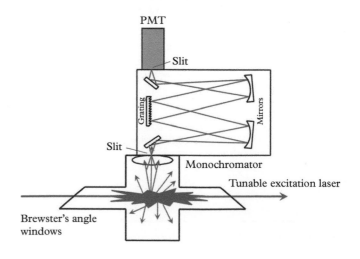

Figure 14.24 *Experimental configuration used to detect dispersed fluorescence spectra. The excitation laser is fixed on a particular v" → v' transition and the monochromator scans to analyze the fluorescence wavelengths produced.*

and selectivity for absorption spectroscopy, are only possible when spectroscopy employs tunable lasers.

....

REFERENCES

1. See for example: J. I. Steinfeld, *Molecules and Radiation: An Introduction to Modern Molecular Spectroscopy*, The MIT Press, Cambridge, MA, 1985; W. S. Struve, *Fundamentals of Molecular Spectroscopy*, John Wiley and Sons, New York, 1989; D. C. Harris and M. D. Bertolucci, *Symmetry and Spectroscopy: An Introduction to Vibrational and Electronic Spectroscopy*, Dover Publications, Inc., New York, 1978; P. F. Bernath, *Spectra of Atoms and Molecules*, Oxford University Press, 1995; J. M. Hollas, *High Resolution Spectroscopy*, second edition, John Wiley & Sons, Chichester, UK, 1998.

2. G. Herzberg, *Molecular Spectra and Molecular Structure I. Spectra of Diatomic Molecules*, D. van Nostrand Company, Inc., New York, 1950.

3. G. Herzberg, *Molecular Spectra and Molecular Structure II. IR and Raman Spectra of Polyatomic Molecules*, D. van Nostrand Company, Inc., New York, 1945.

4. G. Herzberg, *Molecular Spectra and Molecular Structure III. Electronic Spectra and Electronic Structure of Polyatomic Molecules*, D. van Nostrand Company, Inc., New York, 1966.

5. K. P. Huber and G. Herzberg, *Molecular Spectra and Molecular Structure IV. Constants of Diatomic Molecules*, D. van Nostrand Company, Inc., New York, 1979.

6. S. Arrhenius, "On the influence of carbonic acid in the air upon the temperature of the ground," *Phil. Mag. and J. of Sci.* **41**, 237(1896).

7. I. E. Gordon, S. Kassi, A. Campargue, and G. C. Toon, "First identification of the $a^1\Delta_g - X\ ^3\Sigma_g^-$ electric quadrupole transitions of oxygen in solar and laboratory spectra," *J. Quant. Spec. Rad. Trans.* **111**, 1174 (2010).

8. H. C. Miller, J. E. McCord, J. Choy, and G. D. Hager, "Measurement of the radiative lifetime of $O_2\ (a^1\Delta_g)$ using cavity ring down spectroscopy," *J. Quant. Spec. Rad. Trans.* **69**, 305 (2001).

9. S. J. Arnold, E. A. Ogryzlo, and H. Witzke, "Some new emission bands of molecular oxygen," *J. Chem. Phys.* **40**, 1769 (1964).

10. U. Franz and D. Wünderlich, "Franck-Condon factors, transition probabilities and radiative lifetimes for hydrogen molecules and their isotopomers," *Atomic Data Nucl. Data Tables* **92**, 853 (2006).

11. G. D. Dickenson, M. L. Niu, E. J. Salumbides, J. Komasa, K. S. E. Eikema, K. Pachucki, and W. Ubachs, "Fundamental vibration of molecular hydrogen," *Phys. Rev. Lett.* **110**, 193601 (2013).

12. R. J. Sime, *Physical Chemistry: Methods, Techniques, and Experiments*, Saunders College Publishing, Philadelphia, 1990.

13. F. Rasetti, "On the Raman effect in diatomic gases," *Proc. Natl. Acad. Sci.* **15**, 234 (1929); F. Rasetti, "On a fluorescence spectrum of oxygen," *Proc. Natl. Acad. Sci.* **15**, 411 (1929).

14. R. Mecke, "Zum Aufbau der Bandenspektra," *Z. Physik* **32**, 823 (1925).

15. P. Ehrenfest and J. Oppenheimer, "Note on the statistics of nuclei," *Phys. Rev.* **37**, 333 (1931).

16. W. V. Houston and C. M. Lewis, "Rotational Raman spectrum of CO_2," *Proc. Natl. Acad. Sci.* **17**, 229 (1931).

17. E. B. Wilson, "The statistical weights of the rotational levels of polyatomic molecules, including methane, ammonia, benzene, cyclopropane and ethylene," *J. Chem. Phys.* **3**, 276 (1935).

18. R. Saito, G. Dresselhaus, and M. S. Dresselhaus, "Hindered rotation of solid $^{12}C_{60}$ and $^{13}C_{60}$," *Phys. Rev. B* **50**, 5680 (1994).

19. N. Sogoshi, Y. Kato, T. Wakabayashi, T. Momose, S. Tam, M. E. DeRose, and M. E. Fajardo, "High-resolution infrared absorption spectroscopy of C_{60} molecules and clusters in para-hydrogen solids," *J. Phys. Chem. A* **104**, 3733 (2000).

20. L. R. Brock and M. A. Duncan, "Photoionization spectroscopy of Ag-rare gas van der Waals complexes," *J. Chem. Phys.* **103**, 9200 (1995).

21. R. N. Zare, "My life with LIF: A personal account of developing laser-induced fluorescence," *Annu. Rev. Anal. Chem.* **5**, 1 (2012).

22. M. A. Duncan, "Frontiers in the spectroscopy of mass-selected molecular ions," *Int. J. of Mass Spectrom.* **200**, 545 (2000).

Quantum Chemistry Calculations

15

In this chapter we introduce the student to several modern methods used to perform quantum chemical calculations on a desktop computer. Most of the molecular properties discussed in this text can be calculated using computational methods now readily available to non-experts. Once solely in the hands of quantum chemists, these powerful methods can now be employed at the undergraduate level on a personal computer. In fact, quantum chemistry calculations are even finding their way into some high school curricula (see http://chemistry.ncssm.edu). A number of commercial and open-source software packages are available to perform such calculations. Some of these programs have been developed to the point that, once a molecular geometry has been estimated and the atomic coordinates entered into the program, any student can calculate molecular properties at different levels of electronic structure theory. Virtually every chemistry or physics graduate student who performs experimental measurements employs these programs to predict/guide/interpret their experiments. In some cases the measurements are sufficiently difficult or imprecise that one could perhaps trust the calculations over the measurements. Quantum chemical calculations, when understood and applied correctly, are valuable tools in the experimentalist's toolbox. They can be used to help understand experimental results, and provide the only method for evaluating certain systems where experiments are unavailable. An excellent example is the calculation of reaction transition states or barriers to molecular rearrangements, where experiments generally cannot access the relevant energetics directly.

Many excellent textbooks are devoted to the subject of quantum chemistry (see [1–9]). [1] and [2] are particularly helpful for the application of available quantum chemistry programs. These two books provide descriptions of the methods employed to calculate physical properties of molecules such as electronic/vibrational energies, IR/Raman spectra, optical rotary dispersion (ORD), circular dichroism (CD), and multipole moments (dipole, quadrupole, etc.). Students should refer to these books for in-depth treatments of approximate solutions to the Schrödinger equation. The purpose of this chapter is to summarize the basic methods and especially to introduce the student to the *mechanics* of employing the available programs. Many colleges and universities have site licenses for the commercial programs such as Gaussian (http://www.gaussian.com), SPARTAN (http://www.wavefun.com), Q-CHEM (http://www.q-chem.com),

Laser Experiments for Chemistry and Physics. First Edition. Robert N. Compton and Michael A. Duncan.
© Robert N. Compton and Michael A. Duncan 2016. Published in 2016 by Oxford University Press.

or Turbomole (http://www.turbomole-gmbh.com); other programs such as GAMESS (http://www.msg.ameslab.gov/GAMESS.html), NWCHEM (http://www.nwchem-sw.org/index.php/Main_Page), and PSI4 (http://www.psicode.org) are open-source and available free of charge. Also, the HyperChem (http://www.hyper.com) program is easy to use and relatively inexpensive. All of these programs can run on Windows PCs or workstations. This chapter focuses on the use of the Gaussian program.[10] In addition to the Gaussian manual, Foresman and Frisch have published an essential book describing the use of the Gaussian suite of programs.[11]

As a part of the graduate and undergraduate courses in quantum chemistry and spectroscopy, as well as in the advanced undergraduate laboratories at the Universities of Tennessee and Georgia, students make measurements of IR, Raman, UV-vis, ORD, or CD spectra for a chosen molecule and carry out quantum chemical calculations using Gaussian to compare with the experiment. This exercise serves as a good introduction to experimental methods for the more theoretically inclined students, and provides a jump-start for graduate research in other areas of chemistry, physics, and materials science.

In the interest of completeness, a brief introduction to quantum chemical calculations follows. Students should refer to the many available texts (e.g., [1–9]) for a more comprehensive treatment. To outline the basic methods of electronic structure calculations, we begin by writing down the time-independent Schrödinger equation for a molecule.

$$\hat{H}\Psi = E\Psi \tag{15.1}$$

$\hat{H}$ is the electronic Hamiltonian operator, which operates on the eigenfunction Ψ, to give the eigenvalue or electronic energy, E. The Hamiltonian operator of a system is frequently referred to as simply the "Hamiltonian" of the system. Since we focus on calculating time-independent quantum mechanical properties of molecules, all mention of the Schrödinger equation in this chapter refers exclusively to its time-independent form. The more general form of the equation is the time-dependent Schrödinger equation, which applies to both eigenstates and non-eigenstates of a system. If we neglect relativity (a good approximation for low Z elements) the total Hamiltonian for a polyatomic molecule containing M nuclei and N electrons can be written as

$$\hat{H} = -\frac{\hbar^2}{2m_e}\sum_{i=1}^{N}\nabla_i^2 - \frac{\hbar^2}{2}\sum_{A=1}^{M}\frac{1}{M_A}\nabla_A^2 - \sum_{i=1}^{N}\sum_{A=1}^{M}\frac{Z_A e^2}{4\pi\varepsilon_0 r_{iA}}$$
$$+ \sum_{i=1}^{N}\sum_{j>i}^{N}\frac{e^2}{4\pi\varepsilon_0 r_{ij}} + \sum_{A=1}^{M}\sum_{B>A}^{M}\frac{Z_A Z_B e^2}{4\pi\varepsilon_0 R_{AB}} \tag{15.2}$$

where m_e is the electron mass, ε_0 is the permittivity of free space, $\hbar$ is Planck's constant divided by 2π, e is the electron charge, Z_A and Z_B are the atomic numbers of atoms A and B, M_A is the mass of A, R_{AB} refers to the separation between nuclei A and B, r_{ij} denotes the separation between electrons i and j, and r_{iA} is the distance between electron i and nucleus A.

Almost all quantum chemistry programs employ atomic units for their calculations. The atomic unit of energy is the hartree (E_h), defined as $E_h = e^2/4\pi\varepsilon_0 a_0 = 27.211\,385\,05(60)$ eV, and the atomic unit of length is the bohr (a_0), defined as $a_0 = 4\pi\varepsilon_0\hbar^2/m_e e^2 = 5.2917721092(17) \times 10^{-11}$ m. In atomic units $\hbar$, e, m_e, and $4\pi\varepsilon_0$ are all equal to 1, and the energy output for the Gaussian calculations is in hartrees. However, the output for the dipole moment, is in Debye instead of in atomic or ea_0 units ($ea_0 = 8.47836 \times 10^{-30}$ C·m, 1 debye $= 3.33564 \times 10^{-30}$ C·m $= 0.393430$ ea_0).

Converting the Hamiltonian into atomic units is straightforward and results in

$$\hat{H} = -\sum_{i=1}^{N}\frac{1}{2}\nabla_i^2 - \sum_{A=1}^{M}\frac{1}{2M_A}\nabla_A^2 - \sum_{i=1}^{N}\sum_{A=1}^{M}\frac{Z_A}{r_{iA}} + \sum_{i=1}^{N}\sum_{j>i}^{N}\frac{1}{r_{ij}} \tag{15.3}$$

$$+ \sum_{A=1}^{M}\sum_{B>A}^{M}\frac{Z_A Z_B}{R_{AB}}$$

If we invoke the *Born–Oppenheimer approximation* (the electrons are so much lighter and faster than nuclei that the nuclei can be assumed to be stationary during electronic motion) the wavefunction Ψ can be written as a product of an electronic wavefunction, ψ_e, times a nuclear wavefunction, ψ_n, i.e.,

$$\Psi = \psi_e\psi_n \tag{15.4}$$

The product form implies that there is no coupling between electronic and nuclear degrees of freedom. The electronic wavefunction is a function of each electron's spatial and spin coordinates, and depends parametrically on the nuclear positions, which are treated as classical point charges in this approximation. Denoting the spatial coordinates of electron i as r_i and the spin as α or β, there are K spatial orbitals $\varphi(r)$ and 2K spin orbitals, $\varphi(r)\alpha$ plus $\varphi(r)\beta$ (technically, this applies only to restricted wavefunctions; unrestricted wavefunctions can have different spatial orbitals, as explained later). If we let x_i denote the combination of spatial and spin coordinates, and $\chi(x)$ represents a product of space and spin wavefunctions, we can write

$$\psi_e(x_1, x_2, \ldots\ldots x_N) = \chi_1(x_1)\,\chi_2(x_2)\ldots\ldots\ldots\ldots\chi_N(x_N) \tag{15.5}$$

Electrons are fermions, and therefore all electronic wavefunctions must be antisymmetric upon exchange of any two electrons. This postulate of quantum mechanics gives rise to the Pauli exclusion principle: no two electrons in an atom can have the same four quantum numbers, n, l, m_l, and m_s. Fortunately such antisymmetric wavefunctions can be conveniently written as a Slater determinant, named after John C. Slater who introduced the concept in 1930.

$$\psi_e(x_1, x_2, \ldots \ldots x_N) = (1/\sqrt{N!}) \begin{vmatrix} \chi_i(x_1) & \chi_j(x_1) & \ldots & \chi_K(x_1) \\ \chi_i(x_2) & \chi_j(x_2) & \ldots & \chi_K(x_2) \\ \vdots & \vdots & \ddots & \vdots \\ \chi_i(x_N) & \chi_j(x_N) & \ldots & \chi_K(x_N) \end{vmatrix} \tag{15.6}$$

This determinant has precisely the properties needed to represent the antisymmetric electronic wavefunction. Firstly, the value of the determinant changes sign upon interchange of any two rows or any two columns. Secondly, the determinant is zero if any two rows or two columns are the same. These properties satisfy both exchange antisymmetry and the Pauli Principle.

The Schrödinger equation can only be solved exactly for "hydrogen-like" or "hydrogenic" atoms (i.e., those atoms or atomic ions with one electron) or positronium (electron/positron pair). It is the goal of quantum chemistry to develop increasingly accurate yet tractable approximations to the many-electron Schrödinger equation, and solve them to obtain excellent approximations to the wavefunctions and energies of molecules. We begin by making the *Hartree–Fock* (HF) approximation to solve the Schrödinger equation, and introduce the HF wavefunction again as a Slater determinant conveniently written in the Dirac "bra-ket" notation as

$$\Psi_{HF} = | \chi_1(x_1) \chi_2(x_2) \ldots \ldots \ldots \chi_K(x_N) \rangle \tag{15.7}$$

HF is a "mean-field" approximation, because each electron feels the average electrostatic field generated by the remaining $N-1$ electrons.

At this point the concept of a one-electron basis set is considered. As before, the HF wavefunction can be written as a spatial part times a spin function. Furthermore, the spatial part is written as a summation of one-electron functions, Φ_j (i.e., a linear combination of basis functions),

$$\varphi_i = \sum_j c_{i,j} \Phi_j \tag{15.8}$$

The choice of one-electron functions Φ_j is termed a "basis set" as discussed below. The expansion coefficients $c_{i,j}$ are determined from the variation theorem, which rigorously shows that the true ground-state energy, E_0, of a system described by the Hamiltonian, $\hat{H}$, will always be less than or equal to the calculated average energy (or expectation value of energy) obtained using a normalized trial wavefunction, Ψ, for the system, i.e.,

$$E_0 \leq \langle \Psi | \hat{H} | \Psi \rangle \tag{15.9}$$

In many cases for ground-state molecules there is an even number of electrons (an exception is NO) and one can assume that the electrons are spin-paired (an exception is O_2). Thus the spatial degrees of freedom of paired spin-up and spin-down electrons can be considered as a single orbital. This is referred to as the *Restricted Hartree–Fock Approximation* and the energy can be written as

$$E_{RHF} = 2 \sum_{i=1}^{\frac{N}{2}} h_{ii} + \sum_{i=1}^{\frac{N}{2}} \sum_{j=1}^{\frac{N}{2}} \left(2J_{ij} - K_{ij} \right) \tag{15.10}$$

The sum here is taken over N/2 because we have used the restricted HF constraint of having two electrons occupying a single spatial orbital.

In equation 15.10, h_{ii}, a one-electron integral, is the sum of the kinetic energy associated with the electron in spatial orbital φ_i and the Coulomb interaction energy between electron i and the nuclei,

$$h_{ii} = \left\langle \varphi_i(1) \left| -\frac{1}{2} \nabla_1^2 - \sum_{A=1}^{M} \frac{Z_A}{r_{1A}} \right| \varphi_i(1) \right\rangle \tag{15.11}$$

where J_{ij}, a two-electron integral, represents the average Coulomb energy of interaction between electrons in spatial orbitals φ_i and φ_j,

$$J_{ij} = \left\langle \varphi_i(1)\varphi_j(2) \left| \frac{1}{r_{12}} \right| \varphi_i(1)\varphi_j(2) \right\rangle \tag{15.12}$$

and K_{ij}, also a two-electron integral, is an "exchange" integral which accounts for the fact that electrons having the same spin quantum number in spatial orbitals φ_i and φ_j avoid each other.

$$K_{ij} = \left\langle \varphi_i(1)\varphi_j(2) \left| \frac{1}{r_{12}} \right| \varphi_j(1)\varphi_i(2) \right\rangle \tag{15.13}$$

The exchange integral has no physical interpretation except through the Pauli exclusion principle, and has no classical counterpart. Without the presence of the exchange integral, K_{ij}, the Coulomb integral, J_{ij}, would over-compute the repulsion energy. The beauty of the Slater determinant is that the exchange integral correctly accommodates the Pauli principle (notice that the exchange contribution is negative and the Coulomb contribution is positive, see equation 15.10). Typically, the exchange integrals are roughly 25% of the magnitude of the Coulomb integrals. The evaluation of the two-electron integrals 15.12 and 15.13 is usually the most computationally expensive part of a HF calculation.

Thus far we have only considered molecules having closed shells (even-electron systems in which all electron spins are paired). The *Unrestricted Hartree–Fock* (UHF) method, as the name implies, does not restrict electron pairs to the same spatial orbital, and treats the spin functions separately. Likewise there exists a method known as the *Restricted Open Shell Hartree–Fock* (ROHF) approximation, which restricts the paired electrons in an open-shell molecule to the same spatial orbital, but allows different spatial orbitals for the unpaired electrons. For example, the NO molecule (15 electrons) would be treated as an "alkali molecule" with the odd electron outside a "shell of 14 spin-paired electrons." The main advantage of the ROHF method over the UHF method is that, unlike the UHF wavefunction, the ROHF wavefunction is an eigenfunction of the total spin operator $(\hat{S}^2)$ and therefore avoids spin contamination.

The above Hartree–Fock equations must be solved using a self-consistent field (SCF) procedure. One assumes a set of orbitals, based on a "trial" wavefunction (expanded in the space spanned by the basis set) in order to calculate the external potential felt by each electron in the presence of the nuclei and the other electrons. This external potential is then used to construct a new Hamiltonian operator for each electron, and the resulting Schrödinger equation is solved to obtain new eigenfunctions (expressed as new coefficients, c_i, for basis functions) for each electron. The procedure is repeated iteratively, whereby these eigenfunctions are used to calculate modified external potentials for each electron, and the resulting new Hamiltonian operators produce revised electron eigenfunctions, etc., until a "self-consistent" set of single-electron wavefunctions, i.e. orbitals, is obtained. As mentioned earlier, the final many-electron wavefunction is comprised of a single Slater determinant of the self-consistent orbitals. The point at which the derived orbitals no longer improve (lower) the electronic energy approaches the *Hartree–Fock limit*. The HF limit is the exact solution of the HF equation, i.e. using the complete basis set limit.

The HF limit is the best approximation to the energy that can be obtained without explicitly considering *electron correlation*, i.e., the tendency of electrons to avoid each other in space and to thus lower their repulsion energy. Electron correlation energies already show up in the terms resulting from the single Slater determinant calculation. The Hartree–Fock approximation accounts for correlation between electrons of the same spin, but neglects correlation between interacting electrons of opposite spin. The single Slater determinant does not take into account these other electron correlation terms. A further designation for this electron correlation was put forward by Per-Olov Löwdin in 1955 and expresses the fact that a single Slater determinant cannot account for the full electron correlation effects.[12] In essence, the correlated motion of electrons acts to keep the electrons as far apart from each other as possible. HF lacks a complete description of this because it is a mean-field approximation. A simple definition of the electron correlation energy, E_{corr}, can be stated as the difference between the "true" energy, E_{true}, and the Hartree–Fock energy, E_{HF},

$$E_{corr} = E_{true} - E_{HF} \qquad (15.14)$$

Thus electron correlation effects are not physical observables. They essentially account for the errors inherent in HF many-electron theory. Interested students should refer to the paper by Raghavachari and Anderson [13] for estimates of the magnitude of correlation energies: "a frequently used rough rule of thumb is that correlation effects contribute ~1 eV (~23.06 kcal/mol) for a pair of electrons in a well-localized orbital.[14] In the case of many pairs of electrons in close proximity, correlation effects become sizeable. For example, correlation effects contribute more than 100 kcal/mol to the bond energy in N_2."

The straightforward approach to correct for electron correlation is to employ *configuration interaction* (CI) theory, in which electrons are promoted into virtual orbitals in higher energy states to form an excited Slater determinant (i.e., excited electronic configurations). Thus, the CI wavefunction will consist of a summation of the HF wavefunctions plus a linear combination of excited Slater determinants or configurations. For example, the lithium atom electronic configuration is $1s^2 2s$. In performing an exact calculation of the ground-state energy level for $Li(1s^2 2s)$ one would include $1s^2 2s + 1s2s^2 + 1s^2 3p + \ldots$ Unfortunately, it is not practical to include all possible configurations and a truncation of the configuration space is necessary. Allowing the wavefunction to be a linear combination of many-electron configurations, one can include single (S), double (D), triple (T), quadruple (Q), or higher excitations, where the expansion coefficients are determined by an iterative procedure as before.

There are many electronic-structure methods available in the Gaussian program suite [11] which include electron correlation. In addition to the CI techniques mentioned above, *Møller–Plesset* (MP) perturbation theory can be used to improve upon the Hartree–Fock approximation. First-order Møller–Plesset theory is identical to Hartree–Fock, whereas MP2, MP3, and MP4 designate the second, third, and fourth orders of perturbation theory. The calculations are designed to improve as the order increases at the expense of complexity and computational time (CPU time).

Coupled Cluster (CC) theory is another method in which a nonlinear excitation expansion is used (contrary to the linear expansion of CI). An approximate CC method is superior to the corresponding CI method, although the computational scaling is the same. CC and CI are usually more expensive than MP methods (at the same level of excitation) because they are iterative, but they are also more accurate. As of this writing, the so-called "gold standard" in calculations is the CCSD(T) method, in which single and double excitations are included in the coupled cluster expansion and triple excitations are added in as a perturbation.

Of particular importance to the experiments described in this book is the *Density Functional Theory* (DFT) developed by Pierre Hohenberg and Walter Kohn.[15] Kohn [16] and John A. Pople shared the Nobel Prize in 1998 for their individual contributions to quantum chemistry. The Gaussian program was the initial creation of Pople. DFT replaces the detailed many-electron wavefunction with an electron density $\rho(x,y,z)$ to calculate the molecular properties of the ground electronic state of the system, such as energies, geometries, vibrational frequencies, multipole moments, etc. Excited electronic-state properties can be calculated using time-dependent DFT (TDDFT). Details of this theory can be found in essentially all of the references in this chapter, as well as in the book by

Parr and Yang.[17] DFT proves that the ground-state energy of a molecule is a function of its ground-state electron density, which is in turn a function of position in space. A function of a function is a "functional." However, since the exact energy functional is not known, DFT is applied to molecules by calculating an approximate functional with the best possible description of exchange and correlation effects, using an appropriate basis set.

Hybrid functionals also have been introduced as further treatments. In 1993 Becke [18] constructed a hybrid between DFT and HF approaches to treat complex molecules. One of the most popular DFT functionals is the B3LYP (Becke, 3-parameter, Lee-Yang-Parr), which includes the Becke exchange functional and the correlation functional of Lee, Yang and Parr. This method is employed to support most of the experiments described in this book.

Before we provide the details of using the Gaussian program, it is necessary to introduce the basis sets employed for quantum chemical calculations. Detailed information on basis sets is also contained in essentially all of the references cited in this chapter. Students are further referred to the guide to basis set selection for molecular calculations by Davidson and Feller.[19] At the time of this review (1986) there were close to 100 basis sets from which to choose and this number has grown since. An online database of basis sets for atoms in the periodic table is available at https://bse.pnl.gov/bse/portal.

Slater-type orbital (STO) and Gaussian-type orbital (GTO) basis sets

In 1930 J. C. Slater proposed the use of the "Slater-Type Orbitals" for constructing a molecular orbital,

$$\mathrm{STO}_{nlm_l(r,\theta,\varphi)} = N_{nl}\mathrm{r}^{n-1}\mathrm{e}^{-\zeta r}\mathrm{Y}_1^{m_l}(\theta,\varphi) \qquad (15.15)$$

where the normalization factor N_{nl} is

$$N_{nl} = (2\zeta)^{n+\frac{1}{2}}/[(2n)!]^{\frac{1}{2}} \qquad (15.16)$$

Explanations of the terms r and Y(l,m) are required for equation 15.15. The radial part of this equation is similar to the radial solution to the Schrödinger equation for the hydrogen-like atom, except that ζ is taken to be a variable and not Z/n as for the hydrogen atom, where Z is the nuclear charge and n is the principal quantum number. Combinations of Slater orbitals can be employed as well. For example, a helium atom can be described by a sum of two Slater 1s orbitals using two different values of ζ. Such an orbital is called a *double-zeta* orbital (see for example pages 442–443 in [2]) and can lead to very good results for small systems.

The use of Slater orbitals for polyatomic molecules involves multicentered integrals which are difficult to handle computationally. To better facilitate the calculations, F. Boys introduced the use of Gaussian-type orbital (GTO) wave-functions of the form

$$\text{GTO}_s(r, \alpha) = \left(\frac{2\alpha}{\pi}\right)^{\frac{3}{4}} e^{-\alpha r^2} \tag{15.17}$$

$$\text{GTO}_{sx}(r, x, \alpha) = \left(\frac{128\alpha^5}{\pi^3}\right)^{\frac{1}{4}} x e^{-\alpha r^2} \tag{15.18}$$

where replacing x with y and z in equation 15.18 generates functions dependent on y and z coordinates, respectively. The differences between the STO and the GTO are seen at small r where the STO exhibits a "cusp" at $r = 0$. Although not as accurate as the STOs, the GTOs reduce the computational complexity and time required for convergence to the SCF result. For this reason it is well recognized that atom-centered Gaussian basis sets are most useful for quantum chemical calculations. Often linear combinations of Gaussian orbitals are constructed to "mimic" the Slater orbitals at low r but especially at large r.

A *minimal basis set* is one which contains only as many orbitals as are needed to account for the electrons of the neutral atoms in the molecule to retain spherical symmetry. Thus H has $1s$ or one orbital, Li to Ne have ($1s$, $2s$, $2p_x$, $2p_y$, $2p_z$) or five orbitals and Na to Ar have ($1s$, $2s$, $2p_x$, $2p_y$, $2p_z$, $3s$, $3p_x$, $3p_y$, $3p_z$) or nine possible orbitals. This minimal basis set is designated as STO-3G. STO-3G indicates that each orbital is a linear combination of three Gaussian functions.

Improvements in molecular orbital calculations over the minimal basis set come about by (1) making the basis set larger (so-called split valence basis set) and (2) including polarization functions. Split-valence basis sets add two more "sizes" designated by a prime for the orbitals involved, i.e. H:$1s$ becomes H:$1s$, $1s'$ and C:$1s$, $2s$, $2p$ becomes C:$1s$, $2s$, $2s'$, $2p$, $2p'$, where p implies p_x, p_y, p_z. The split valence basis set allows for only a change in orbital size, not the "shape" of the orbitals. The addition of d functions allows the orbitals to be "polarized," i.e., to shift an orbital away from the center of the nucleus. Polarization is accommodated by mixing an orbital with functions containing one unit of higher angular momentum. To account for polarization, the computation includes six d functions (x^2, y^2, z^2 as well as xy, xz, and yz) for all but the H atoms. It is easily visualized that an s orbital can polarize in one direction if mixed with a p orbital and a p orbital can polarize if mixed with d orbitals, etc.

It is often necessary to build so-called *diffuse functions* into the molecular orbitals. These are Gaussian functions with small exponents, i.e., small α in equations 15.17 and 15.18. As one might expect, diffuse functions are essential to properly describe negative ions as well as excited states and many other molecular properties such as optical rotary dispersion (ORD). It is conventional to use the + sign to indicate a diffuse orbital. For example, a diffuse carbon atom would be designated by $1s$, $2s$, $2p$, $2s+$, $2p+$ where again p implies p_x, p_y, p_z. Adding diffuse functions along with the polarization functions gives a complicated construction for an atomic orbital. For example, including the polarization functions up to five $3d$ orbitals for the carbon atom gives $1s$, $2s'$, $2p'$, $2s''$, $2p''$, five $3d$ orbitals, $2s+$ and $2p+$ for a total of 18 basis functions! (the primes and double primes indicate a split-valence basis set; see [2], p. 622–627) Realizing that this is for only one of the atoms, one can easily see how complicated a quantum chemistry computation can become.

The Gaussian program employs all of the split-valence basis set functions discussed above. The nomenclature for the description of such functions can be stated as P-XYG, where P indicates the number of primitive Gaussian (G) functions for each core atomic-orbital basis function. X and Y indicate that the valence orbitals are made of two individual basis functions each having a linear combination of X and Y primitive functions. An asterisk (*) or double asterisk (**) after the G indicates the inclusion of one or two polarization functions, respectively, and a + before the G indicates the presence of diffuse functions. In some cases the type of polarization function added is placed in parenthesis, e.g. (p,d). As an example, consider the molecular orbital for a molecule made of atoms from H to Zn in the periodic table. The designation 6–31+G* signifies that six d orbitals are added to the 6-31G function for the atoms Li through Ca, ten f orbitals are added for atoms Sc through Zn, and diffuse functions are also included.

Mechanics of using the Gaussian program

Performing a calculation using the Gaussian program is significantly simplified by using a molecular modeling program to determine the approximate positions of each atom in the molecule as a starting geometry before the computations begin. This can be done with a number of free or relatively inexpensive software packages, available on the Internet, that allow one to calculate the geometry of molecules. Most colleges and universities have a site license for such programs. A popular program available free of charge is Avogadro (http://avogadro.cc.wiki.main_page). Relatively inexpensive programs are also available such as Hyperchem program (http://www.hyper.com/) or PC Model by Serena Software (http://www.serenasoft.com/).

One also needs a program to interface with Gaussian that will properly display the Gaussian data output, as well as molecular structures, vibrational modes, and IR or Raman spectra. The Chemcraft program (http://www.chemcraftprog.com/) is a graphical program for displaying results of quantum chemical calculations and especially for preparing new jobs for the computation. Many colleges and universities have site licenses for Chemcraft. Also, GaussView is a graphical user interface that allows one to prepare input for submission to Gaussian and to graphically display its output (Version 5, Roy Dennington, Todd Keith, and John Millam, Semichem Inc., Shawnee Mission, KS, 2009).

As an example, we use the *cis-* and *trans-*dichloroethane molecules to be investigated in Chapter 29. First the geometry of *cis*-dichloroethane is generated using say, PC Model or Avogadro. After input into Chemcraft the coordinates are displayed in terms of the atomic number and X, Y, Z position as

6	0.651890000	0.895842000	−0.382405000
6	−0.651893000	0.895843000	0.382408000
17	1.741827000	−0.471377000	0.081092000
1	1.197366000	1.814517000	−0.161934000
1	0.488769000	0.823850000	−1.455966000
17	−1.741827000	−0.471378000	−0.081093000
1	−0.488759000	0.823840000	1.455966000
1	−1.197366000	1.814516000	0.161930000

These coordinates are then pasted into the Gaussian program Job Entry under the Molecule Specification. One then chooses the Charge and Multiplicity. For any molecular species, one chooses the appropriate charge based on whether it is neutral or ionic, and the spin multiplicity of $2s + 1$, where s denotes the number of unpaired electrons. If there is a mistake, the program immediately identifies this upon trying to do the calculation. However, the program cannot distinguish between allowed multiplicities, for instance 1, 3, 5, etc. The multiplicity of the lowest energy state is determined by calculation. Next, one chooses the level of computation under the Route Section, e.g., #B3LYP/6–311+G*, indicating that it is a DFT calculation using the B3LYP exchange-correlation functional. From the earlier discussion on basis sets, it should be clear that polarization and diffuse functions are included. Vibrational frequencies and the infrared spectrum are produced by default. However, if a Raman spectrum is desired, the additional input line of "OPT FREQ=raman" must be added after the basis set. The Gaussian computation takes a few minutes depending on the capabilities of the computer, and the energy output will appear as

#B3LYP/6–311+G* OPT FREQ=raman
cis-dichloroethane
Charge = 0 Multiplicity = 1

Stoichiometry C2H4Cl2
Framework group C1[X(C2H4Cl2)]
Deg. of freedom 18
Full point group   C1 NOp 1
Largest Abelian subgroup   C1 NOp 1
Largest concise Abelian subgroup C1 NOp 1

|HF=−999.0936382|
|Dipole=−0.0407703,−1.1561388,0.0002885|
Job cpu time: 0 days 0 hours 7 minutes 2.0 seconds.

File lengths (MBytes): RWF = 16 Int = 0 D2E = 0 Chk = 5 Scr = 1
Normal termination of Gaussian 03 at Tue May 19 13:26:00 2009.

An identical calculation for the *trans*-dichloroethane molecule gives

#B3LYP/6–311+G* OPT FREQ=raman
trans-dichloroethane
Charge = 0 Multiplicity = 1

Stoichiometry   C2H4Cl2
Framework group   CI[X(C2H4Cl2)]
Deg. of freedom   9
Full point group   CI   NOp   2
Largest Abelian subgroup   CI   NOp   2
Largest concise Abelian subgroup   CI   NOp   2

|HF = –999.0963372|
|Dipole = 0.,0.,0.|
Job cpu time: 0 days 0 hours 2 minutes 51.0 seconds.
File lengths (MBytes): RWF = 16 Int = 0 D2E = 0 Chk = 5 Scr = 1
Normal termination of Gaussian 03 at Tue May 19 12:14:32 2009.

Notice that the energy of the *trans*-dichloroethane is 0.0026989 au more negative than that of the *cis*-dichloroethane. Multiplying by 27.21 eV/au gives 0.0734 eV. Thus the *trans*-dichloroethane is 0.0734 eV (1.69 kcal/mole) more stable than *cis*-dichloroethane.

The calculation also gives a printout of the multipole moments of the molecules. In the case of the *cis*-dichloroethane, the dipole moment was 2.88 Debye and the traceless quadrupole moment Q_{zz} was +1.05 Debye-Angstrom. For *trans*-dichloroethane the dipole moment was zero and Q_{zz} was +2.295 Debye-Angstrom.

It is important to note the inclusion of OPT FREQ=raman in the Route Section. This is a command that directs the geometry optimization followed by a force constant calculation to determine vibrational frequencies and intensities for both IR and Raman modes, as well as thermochemistry at 298 K. See Chapter 29 for examples of the Raman spectra output for *cis*- and *trans*-dichloroethane.

To calculate relative energies for different structures of a molecule, or for ionization potentials or electron affinities, it is necessary to correct for the respective zero point energies (ZPE). The definition of the adiabatic IP is the energy difference between the $v = 0$ level of the neutral and the $v = 0$ level of the ion, and the electron affinity is defined in a similar way between the anion and the neutral. It is therefore necessary to calculate the vibrational frequencies for the neutral and the corresponding ion and then to take the difference between these to get IPs and EAs. Within the harmonic approximation, the ZPE is given by ZPE = $\frac{1}{2} \sum h\nu$, or one-half the sum of all harmonic vibrational frequencies. It is important to ensure that none of the computed vibrational frequencies are imaginary. An imaginary frequency indicates that the structure obtained is not a true minimum in energy on the potential energy surface. The ZPE for a molecule or its ion can be rather large. However, because the difference between the two energies is used to obtain the relative energies between structures or to get EA or IP values, the correction will usually be small.

The utilization of programs such as Gaussian has become so convenient that almost anyone can now perform such computations. However, there are many pitfalls in the applications of these methods, and care must be taken to produce *meaningful* results. The most basic result of computational studies is the structure of a molecule of interest. If a closed-shell molecule composed of lighter main group elements is investigated, most programs will get the qualitative structure correct, although different methods may produce variations in subtle aspects of bond distances and angles. Structures are more challenging for open-shell species or for those composed of heavier elements such as transition metals, and different methods may produce different results. Only experienced computational chemists can select the most appropriate and reliable methods for these kinds of systems. Even with relatively simple closed-shell molecules, the *absolute* energies may vary substantially from one method to another, and the *relative* energies of two or more structures may change their order, depending on the method and basis set employed. The best energetics require the largest calculations, such as those at the CI or coupled cluster level with large basis sets, but these calculations are severely demanding on computer time. As noted above, DFT is an attractive alternative to these methods, but it is known to be less reliable for absolute energies, and it has special problems with open-shell species, transition (or heavier) metals, and non-covalent or electrostatic interactions.[20–23]

Computational chemistry can investigate almost any physical property of molecules that can be measured. In the theoretical chemistry community, the focus is often on the absolute energies of molecules at a particular level compared to the result at the basis set limit for that method. In this way computations can be evaluated even if relevant experimental results are not available. However, it is also true that the experimental results of molecular energetics are often not reliable. Traditional thermochemical measurements are among the most difficult in physical science, because they require the measurement of small temperature changes in the lab. In some cases spectroscopic results can be used to derive thermochemical data, and then the resulting quantities are likely to be more accurate.

In the experiments in this laboratory text, the focus is on laser spectroscopy and other optical measurements. In particular, the results from infrared and Raman spectroscopy experiments are used to investigate molecular structures. In rotationally resolved spectra, such as those described in Chapter 14, the level spacings in spectra can be used to derive rotational constants for molecules.

These can in turn be compared directly to the same constants resulting from a structure obtained in a computational study. However, in the case of vibrational frequencies measured in infrared, Raman or UV-vis spectra, the raw results of theory cannot be compared directly to those from an experiment. The vibrational frequencies resulting from computations are almost always higher than the experimentally measured values because theory neglects the effects of anharmonicity on the vibrations. All of the standard methods discussed above produce results in the "harmonic approximation." Methods exist for including anharmonicity explicitly, but these are so computationally demanding that they can only be applied to the smallest systems (those with fewer than 5–6 atoms). More commonly, the effects of anharmonicity are taken into account by multiplying the computed vibrational frequencies by an empirical "scaling" factor determined from prior experience on similar molecules. The scaling factor varies with the particular level of theory employed and the basis set, and recommended values for many combinations of these have been determined through systematic study.[24,25] A typical value for such a scaling factor would be about 0.96, consistent with the physical reality that anharmonicity tends to lower frequencies by a few percent. Such scaling can account for the anharmonic shift to lower frequencies in an individual vibrational mode, but it does not allow the calculations to reproduce other anharmonic effects such as overtones, combination bands, or Fermi resonances. The vibrational dynamics in systems containing these effects, or of large molecules with multiple isomers or conformations, continue to be challenging for both theory and experiment.

. . . .

REFERENCES

1. I. N. Levine, *Quantum Chemistry*, seventh edition, Prentice Hall, Upper Saddle River, NJ, 2013.
2. D. A. McQuarrie, *Quantum Chemistry*, University Science Books, Sausalito, CA, 2008.
3. T. Engel, *Quantum Chemistry and Spectroscopy*, third edition, Pearson, Upper Saddle River, NJ, 2013.
4. J. P. Lowe, *Quantum Chemistry*, second edition, Academic Press, San Diego, CA, 1993.
5. C. Cramer, *Essentials of Computational Chemistry*, second edition, John Wiley, Chichester, UK, 2002.
6. T. Schlick, *Molecular Modeling and Simulation*, Springer, Berlin, 2002.
7. A. Szabo and N.S. Ostlund, *Modern Quantum Chemistry: Introduction to Advanced Electronic Structure Theory*, MacMillan Publishing Co., New York, 1989.
8. W. Hehre, J. Yu, P. E. Klunzinger, and L. Lou, *A Brief Guide to Molecular Mechanics and Quantum Chemical Calculations*, Wavefunction Inc.,1998.

9. D. Rogers, *Computational Chemistry Using the PC*, Wiley Interscience, Hoboken, NJ, 2003.

10. M. J. Frisch, G. W. Trucks, H. B. Schlegel, G. E. Scuseria, M. A. Robb, J. R. Cheeseman, J. A. Montgomery, Jr., T. Vreven, K. N. Kudin, J. C. Burant, J. M. Millam, S. S. Iyengar, J. Tomasi, V. Barone, B. Mennucci, M. Cossi, G. Scalmani, N. Rega, G. A. Petersson, H. Nakatsuji, M. Hada, M. Ehara, K. Toyota, R. Fukuda, J. Hasegawa, M. Ishida, T. Nakajima, Y. Honda, O. Kitao, H. Nakai, M. Klene, X. Li, J. E. Knox, H. P. Hratchian, J. B. Cross, V. Bakken, C. Adamo, J. Jaramillo, R. Gomperts, R. E. Stratmann, O. Yazyev, A. J. Austin, R. Cammi, C. Pomelli, J. W. Ochterski, P. Y. Ayala, K. Morokuma, G. A. Voth, P. Salvador, J. J. Dannenberg, V. G. Zakrzewski, S. Dapprich, A. D. Daniels, M. C. Strain, O. Farkas, D. K. Malick, A. D. Rabuck, K. Raghavachari, J. B. Foresman, J. V. Ortiz, Q. Cui, A. G. Baboul, S. Clifford, J. Cioslowski, B. B. Stefanov, G. Liu, A. Liashenko, P. Piskorz, I. Komaromi, R. L. Martin, D. J. Fox, T. Keith, M. A. Al-Laham, C. Y. Peng, A. Nanayakkara, M. Challacombe, P. M. W. Gill, B. Johnson, W. Chen, M. W. Wong, C. Gonzalez, and J. A. Pople, Gaussian, Inc., Wallingford CT, 2004.

11. J. B. Foresman and A. Frisch, *Exploring Chemistry with Electronic Structure Methods: A Guide to Using Gaussian*, Gaussian, Inc., 1996.

12. P.-O. Löwdin, "Quantum theory of many-particle systems. III. Extension of the Hartree–Fock Scheme to include degenerate systems and correlation effects," *Phys. Rev.* **97**, 1509 (1955).

13. K. Raghavachari and J. B. Anderson, "Electron correlation effects in molecules," *J. Phys. Chem.* **100**, 12960 (1996).

14. A. C. Hurley, *Electron Correlation in Small Molecules*, Academic Press, New York, 1976.

15. P. Hohenberg and W. Kohn, "Inhomogeneous electron gas," *Phys. Rev. B* **136**, B864 (1964); W. Kohn and L. J. Sham, "Self-consistent equations including exchange and correlation effects," *Phys. Rev.* **140**, 1133 (1965).

16. W. Kohn, Nobel Prize lecture, http://www.nobelprize.org/nobel_prizes/chemistry/laureates/1998/kohn-lecture.pdf.

17. R. G. Parr and W. Yang, *Density-Functional Theory of Atoms and Molecules*, Oxford University Press, 1989.

18. A. D. Becke, "A new mixing of Hartree-Fock and local density-functional theories," *J. Chem. Phys.* **98**, 1372 (1993).

19. E. Davidson and D. Feller, "Basis set selection for molecular calculations." *Chem. Rev.* **86**, 681 (1986).

20. C. J. Cramer and D. G. Truhlar, "Density functional theory for transition metals and transition metal chemistry," *Phys. Chem. Chem. Phys.* **11**, 10757 (2009).

21. A. J. Cohen, P. Mori-Sánchez, and W. Yang, "Challenges for density functional theory," *J. Chem. Phys.* **112**, 289 (2012).

22. S. Shil, D. Bhattacharya, S. Sarker, and A. Misra, "Performance of the widely used Minnesota density functionals for the predictions of heats of formations, ionization potentials of some benchmarked first row transition metal complexes," *J. Phys. Chem. A* **117**, 4945 (2013).

23. A. D. Becke, "Fifty years of density-functional theory in chemical physics," *J. Chem. Phys.* **140**, 18A301 (2014).

24. A. P. Scott and L. Radom, "Harmonic vibrational frequencies: An evaluation of Hartree-Fock, Møller-Plesset, quadratic configuration interaction, density functional theory, and semi-empirical scale factors," *J. Phys. Chem.* **100**, 16502 (1996).

25. J. P. Merrick, D. Moran, and L. Radom, "An evaluation of harmonic vibrational frequency scale factors," *J. Phys. Chem. A* **111**, 11683 (2007).

Multiphoton Ionization and Third Harmonic Generation in Alkali Atoms

Introduction

The experiments described in this chapter introduce the student to the subject of multiphoton excitation using pulsed lasers as well as to the area of quantum defects in describing the energy levels in complex atoms in atomic physics. A complementary description of this experiment for undergraduate studies was presented earlier by Feigerle and Compton.[1] We expand considerably on this subject here. Studies of excitation of an atom before the advent of the laser involved the absorption of single photons. Maria Goeppert-Mayer [2] first introduced the concept of a simultaneous two-photon transition in 1931. Since that time, the theory of multiphoton excitation has been developed to a high degree of sophistication. The advent of the pulsed laser has ushered in an enormous interest in the study of excited states of atoms and molecules. Pulsed lasers can also be employed to produce a multiphoton ionization (MPI) event in which an electron is detached from the atom. Multiphoton ionization can occur via non-resonant excitation or be greatly enhanced by a resonant intermediate state, as illustrated in Figure 16.1.

For cases in which a multiple of the frequency of the laser is tuned in resonance with a real intermediate (i.e., long-lived stationary state), the ionization event is described as a resonantly enhanced multiphoton ionization or REMPI. It is common to describe this as an (n+m) REMPI process in which the number of photons to reach the real intermediate state is n and the total number of photons required to ionize from this excited state is m. Thus the resonant ionization process in Figure 16.1 is described as a (2+1) REMPI. A (1+1) REMPI process (sometimes known as resonant two-photon ionization or R2PI) can be a very sensitive detector of atoms and has been employed to detect essentially a single atom in the presence of Avogadro's number of other atoms (see for example [3]). Before describing the experiments on MPI of alkali atoms, we briefly discuss the basic physics of multiphoton ionization. Since we will be discussing (2+1) REMPI of alkali atoms, we restrict our discussion to the case in which an atom first absorbs two photons through a non-resonant intermediate state. The non-resonant state can be considered as a "virtual" state created by the interaction

Laser Experiments for Chemistry and Physics. First Edition. Robert N. Compton and Michael A. Duncan.
© Robert N. Compton and Michael A. Duncan 2016. Published in 2016 by Oxford University Press.

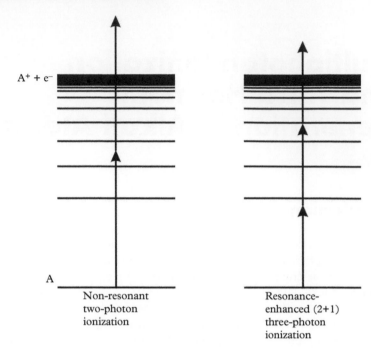

Figure 16.1 *Illustration of non-resonant two-photon ionization and resonantly enhanced three-photon ionization of an atom.*

of the incident light wave with the atom. This virtual state consists of contributions from all other possible bound and continuum states in accordance with their transition moment amplitudes, as well as their detuning (energy difference) from resonance. The two-photon transition moment for light of frequency ω can be written as

$$K_{f \leftarrow i} = \Sigma_j <f|\mu \cdot E> <j|\mu \cdot E>/(\omega_{ji} - \omega + i\Gamma/2) \qquad (16.1)$$

Where ω_{ji} represents the frequency difference between states i and j, and Γ is the combined width of states i and j. In principle, the summation is over all possible bound and continuum states j of the system. In practice only a few of the states nearest to the first photon contribute to the two-photon transition moment. Each term in the sum is a product of two single-photon transition moments and the states which these matrix elements couple are restricted by the usual one-photon dipole selection rules. The overall selection rule for two-photon absorption from i $\rightarrow$ f is the result of one-photon selection rules imposed on the individual transitions to and from the expansion states, j. For two-photon excitation of a ground-state alkali atom, with its valence electron in a ns orbital, the only expansion states allowable by the Laporte rule, $\Delta \ell = \pm 1$, are those formed from excitation of the ns electron to an np orbital. A second application of this parity restriction yields ns and nd excited states as the only possibilities for the final state, f, within the dipole excitation approximation. An equivalent Laporte rule for two-photon transitions is then found to be $\Delta\ell = 0, \pm2$. Thus, for example, the multiphoton ionization spectrum of the cesium atom is dominated by ns and nd Rydberg series resonances when using 600–700 nm excitation.

The discussion of the selection rules for two-photon excitation of ground-state alkali is based on treating the alkali as a "hydrogen-like" (one-electron) atom. This approximation can be used to understand the major features of the optical spectrum of Cs and the other alkalis because these atoms consist of a single electron outside of a closed shell in their ground electronic state configuration. However, the majority of atoms have more than one electron in their valence shell leading to an increased importance of electron–electron correlation and angular momentum coupling in determining their energy levels and optical spectra. The Laporte rule we invoked for analysis of the alkalis is a specific case of the more general requirement that parity must change in a dipole-allowed transition. Since parity remains a quantized property of multielectron atoms, both the one- and two-photon versions of this $\Delta\ell$ restriction still apply to multielectron atoms as long as the states involved are derived from predominantly one-electron configurations. Restrictions on other quantum numbers also exist. What quantum numbers are necessary to further specify the states involved in the transition depend on the strength of the fine and hyperfine interactions in the atom and the ability of the spectroscopic technique to resolve them. The selection rules for one- and two-photon dipole excitation,[4] in the limit that the states are described by Russell–Saunders (LS) coupling, are presented in Table 16.1.

The new feature produced for the two-photon case is the possibility of a change of angular momentum by two units. A good example of such a transition is the $6s\,^2S_{1/2} \rightarrow nd\,^2D_{5/2}$ or $nd\,^2D_{3/2}$ absorptions, which are prominent in both the two-photon and the (2+1) REMPI spectra of Cs.

The transition rate for an n-photon transition per atom, $R(\text{sec}^{-1})$ can be written as

$$R = \sigma_N\,F^N \tag{16.2}$$

and depends on the laser flux, F, to the Nth power times the generalized N-photon cross-section, σ_N ($\text{cm}^{2N}/\text{sec}^{N-1}$). To calculate the photon flux requires specification of the power output and focusing conditions of the laser. The nitrogen-pumped dye laser used in these experiments produces 400 psec pulses

Table 16.1 *Selection rules for Russell–Saunders (LS) coupled states of an atom.*

	ΔL	**ΔS**	**ΔJ**
one-photon	$0, \pm 1\ (0 \leftarrow/\rightarrow 0)$	0	$0, \pm1\ (0 \leftarrow/\rightarrow 0)$
two-photon	$0, \pm1, \pm 2\ (0 \leftarrow/\rightarrow 1)$	0	$0, \pm1, \pm 2\ (0 \leftarrow/\rightarrow 1)$

with greater than 100 μJ/pulse energy. For 650 nm light, this yields a rate of about 8×10^{23} photons/sec during the pulse. If this rate could be sustained, the laser would provide about a mole of photons per second! Unfortunately, the laser can only provide brief pulses at this intensity. To obtain the photon flux, we divide the number of photons/sec in the pulse by the cross-sectional area of the focused laser beam. The laser beam is focused to a smaller cross-sectional area than is directly available from the laser in order to increase the flux and subsequently the two-photon absorption rate. For a focused beam, the flux is highest at the beam waist, where the diameter, d, can be approximated as the product of the inherent divergence, θ, of the multimode laser beam times the focal length, f, of the lens

$$d = f\theta \qquad (16.3)$$

With a 0.5 milliradian divergence and f = 75 mm, the cross-sectional area of the beam at the waist is about 44×10^{-6} cm², yielding an instantaneous laser flux in the experiment of 1.9×10^{28} photons/sec/cm².

We can estimate the lifetime, τ, of the virtual intermediate state from the energy–time formulation of the Heisenberg uncertainty principle,

$$\Delta E \Delta t \geq h/2\pi \qquad (16.4)$$

The energy uncertainty, ΔE, is the detuning of the virtual state from some one-photon allowed stationary state and we take the uncertainty in time, Δt, to be the "lifetime" of the intermediate (virtual) state. If we take the energy uncertainty to be ~0.5 eV or $\Delta E = 4000$ cm⁻¹, then the lifetime of the intermediate resonance obtained from equation 16.4 is $\Delta t \sim 1.3 \times 10^{-15}$ sec. Finally, taking 1 Å as a typical radius of an atom, the cross-section can be estimated as $\sigma = 3 \times 10^{-16}$ cm². The probability of two-photon absorption, P, can be estimated from the product $P = F\tau\sigma$ to be $\sim 7 \times 10^{-3}$. This is a reasonably high probability and it can be concluded that the nitrogen-pumped dye laser has the ability to produce high enough photon flux for frequent "simultaneous" collisions of two photons with an atom or molecule.

Experimental details

The instrumentation for these MPI experiments includes a convenient source of alkali metal vapor provided by a "heat pipe," a nitrogen-pumped dye laser for the tunable electronic spectroscopy, and a detection scheme for the ions produced, provide by a collection wire situated within the heat pipe. A schematic diagram of the apparatus is presented in Figure 16.2.

Because the vapor pressure of Cs is only 1.3×10^{-6} Torr at room temperature,[5] a heat pipe is used to heat the metal sample and increase the vapor pressure to a few Torr. This same device can be used to obtain vapor from

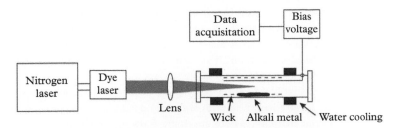

Figure 16.2 *Experimental setup of the heat pipe apparatus for the study of multiphoton ionization of Cs and Rb.*

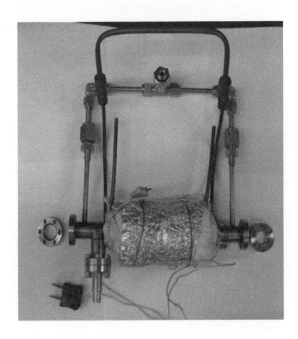

Figure 16.3 *Picture of a small heat pipe constructed of 1-in. conflat flanges with a stainless steel tube. The connection for the collector wire is shown at a right angle to the heat pipe.*

a variety of other metals. Figure 16.2 shows a simplified diagram of this device, and Figure 16.3 shows a photograph. A larger heat pipe arrangement is described in Chapter 27 on Stimulated Electronic Raman Scattering (SERS). The term heat pipe refers to any device which transfers heat by convection of an active or working medium. The working medium in this case is Cs or Rb metal and the latent heat of vaporization is transferred by evaporating liquid metal in the center of the heat pipe and condensing its vapor in the cooled region of the tube. The heated alkali metal must be isolated from air or moisture, and so the pipe is evacuated with a rough vacuum pump. A stainless steel mesh is used inside to "wick" the alkali metal to the center of the tube. The pipe is configured with 1-in. diameter windows (made of Pyrex) so that a laser beam can enter and exit, passing through the dense metal vapor in its center. A buffer gas such as argon is injected at the pipe ends and used to confine the metal vapor near the pipe center at the boundary of the heated zone and prevent its fogging of the windows. Metal vapor which is condensed at this boundary is recirculated back to the heated region by capillary action of the wick material. Definition of the heated region is further aided by external water cooling at the ends of the wick. The heat pipe is typically maintained at a fixed temperature between 200–300 °C with an Ar buffer gas pressure of ~3.5 Torr for MPI measurements.

Loading the alkali metal requires placing the heat pipe along with an ampule of Cs or Rb into a glove bag filled with argon. The student should use her or his ingenuity to fill the chamber. There are many ways to do this and each person should develop their own technique. However, one should be extremely cautious in handling alkali metals; face guards are required for safe usage of these materials.

Multiphoton ionization in the heat pipe is accomplished with a dye laser (PRA Laser LN107) pumped by a 2 MW pulsed nitrogen laser (PRA Laser LN1000). A Nd:YAG- or excimer-pumped dye laser would also work for these experiments. The nitrogen laser is extremely simple to operate and user friendly. On the other hand, wavelength tuning of this laser may require more expertise than is generally available from students attempting to do this for the first time. Dye changes and subsequent adjustments to the laser should therefore be done by the instructor or an assistant prior to the laboratory period. Different laser dyes, DCM and PBD, respectively, are required for the Cs and Rb experiments. Within the tuning curves of these dyes, the wavelength of the laser is controlled with a diffraction grating adjusted by a computer-controlled stepping motor (see discussion in Chapter 4). An iris is inserted into the beam path to spatially filter residual amplified spontaneous emission from the laser output prior to focusing it into the pipe with a 10 cm focal length plano-convex quartz lens.

The ions produced by multiphoton absorption are detected with a biased collection wire situated inside the heat pipe. This is mounted off-axis so as not to block the laser beam. The wire is connected to a BNC-type electrical feedthrough to provide easy monitoring of the ionization current. An external 9 V battery provides the bias to draw either electrons or positive ions formed from ionization to the collection wire. The MPI signals are large enough (~100 mV) to observe directly on a 1 MΩ impedance input of a 50-MHz oscilloscope without further amplification. Recording the intensity of this signal as a function of wavelength then yields a multiphoton ionization spectrum. The pulse height of the MPI signal varies from one laser shot to another by typically 25–50% due to smaller variations in the laser pulse intensity. The MPI signal is particularly sensitive to these variations because the ionization rate depends on the laser power raised to the power of the number of photons required for the process (see equation 16.2). The effect of these intensity variations is to introduce noise into the spectrum. This noise can be reduced significantly by integrating the signal within a narrow window in time just following the laser pulse and averaging the result over several shots of the laser using a gated integrator or boxcar averager. A Stanford Research Systems Model 250 boxcar averager was employed in these studies. The signal can also be recorded using an electrometer in order to measure the average current created by the pulsed laser or with a lock-in amplifier used at the laser pulse repetition rate. The output from any of these recording devices is then digitized (e.g., with an A/D converter) and transferred to a computer. Alternatively, the signal can be collected with a digital oscilloscope and saved directly within this device before transferral to the computer.

Results and discussion

Figure 16.4 presents a typical multiphoton ionization spectrum of the (2+1) REMPI of the cesium atom. The spectrum contains a series of Rydberg-type transitions whose spacing is smaller toward shorter wavelength (higher energy), converging near the ionization potential. In this spectrum the (2+1) signal due to the nd levels is more intense than that to the adjacent ns levels. The oscillator strength for two-photon excitation of the ns level is larger than that of the nd, however the cross-sections for photoionization of the nd levels are much larger, making the (2+1) REMPI signal greater for the nd levels, as seen in the figure. The MPI spectrum also exhibits an "envelope" like appearance, peaking at $\sim 11d$. This is due to the fact that the dye laser intensity is falling off rapidly for wavelengths longer than ~ 660 nm. In fact the true spectrum should exhibit the greatest intensity at the $9d$ level, and the intensity of the nd and ns levels should decrease with longer wavelengths scaling approximately as $1/n^3$. Figure 16.5 provides a diagram of the known energy levels of the cesium atom. These energy levels were taken from the compilation by Weber and Sansonetti.[6]

The spectroscopy of cesium illustrated here can be understood by comparing its energy levels to those of hydrogen, which were discussed in Chapter 14. The hydrogen energy levels rigorously fit the Rydberg or Ritz formula

$$E_n = E_\infty - R_\infty/n^2 \qquad (16.5)$$

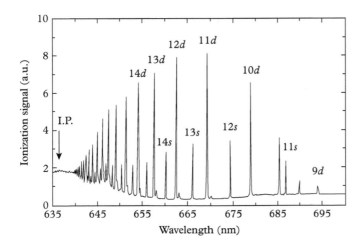

Figure 16.4 *(2+1) REMPI of cesium atom from 695 nm up to the ionization limit.*

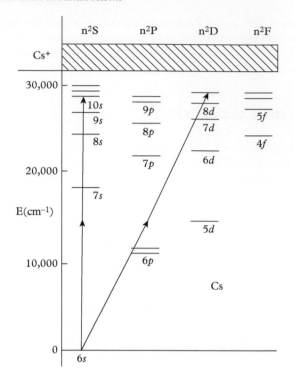

Figure 16.5 *Energy levels of* ns, np, nd, *and* nf *levels of neutral cesium. The hatched area indicates the ionization continuum.*

where R_∞ is the Rydberg constant (109,737.31568539(55) cm^{-1}; 13.60569253(30) eV) and E_∞ is the ionization potential for the H atom (for $n = 1$, $E_n = 0$, and $E_\infty = R_\infty$). The inverse n^2 dependence of the energy levels causes these to become more closely spaced at higher energies, similar to the behavior seen for cesium atoms in Figures 16.4 and 16.5. The simple inverse n^2 dependence causes the levels in hydrogen to be degenerate for different values of ℓ, the angular momentum quantum number. Thus, the *s, p, d*, etc., levels for a given value of n have the same energies. However, electron–electron repulsion causes these states to have different energies in multielectron atoms, and the spectral patterns become more complicated.

The energy levels for a multielectron atom such as cesium can be fit to a modified Ritz formula in which the principal quantum number n is replaced by $n - \mu_{n\ell}$ or

$$E_n = E_\infty - R/(n - \mu_{n\ell})^2 \tag{16.6}$$

where R is now the Rydberg constant for the $M^+ + e^-$ system. n is the principal quantum number, ℓ refers to the angular momentum for the state, and $\mu_{n\ell}$ is the "quantum defect," expressing the deviation from hydrogenic behavior. For a given value of n and ℓ, Weber and Sansonetti [6] write μ_n as an expansion

$$\mu_n = A + B/(n-A)^2 + C/(n-A)^4 \qquad (16.7)$$

Values of A, B, and C are tabulated [6] for the cesium atom in its *s, p, d, e,* and *f* excited states. Experimental values for the quantum defect of cesium can be determined from data such as that shown in Figure 16.4 and compared with that tabulated in the paper by Weber and Sansonetti.[6] The Rydberg constant for cesium is 109,736.86224 cm^{-1}. As described in [6], quantum defect values are about 4.0 for the excited *s* states, 3.6 for the excited *p* states, and 2.5 for the excited *d* states. The values for the quantum defect decrease with increasing ℓ, i.e., as the electron is further away from the atom and the angular momentum increases.

If the dye laser is tuned to higher energies, one-photon resonant transitions can be excited from the 6*s* ground state to the 6*p* excited states. These states lie at lower energies than those accessed by two photons above, and so two additional photons are required to ionize these states in a (1+2) REMPI process. The resulting spectrum is shown in Figure 16.6. The signal levels here are quite large because of the strong $s \rightarrow p$ transition in the first absorption step. The signal due to the fine-structure (spin-orbit) split ^{2}P states is much broader than

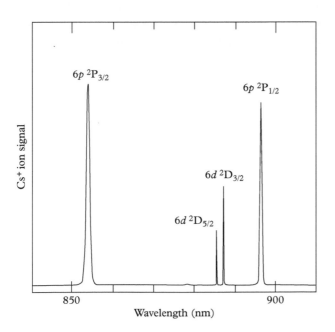

Figure 16.6 *Ionization signal in cesium due to (1+2) REMPI through the 6p ^{2}P$_{1/2}$ and 6p ^{2}P$_{3/2}$ levels and (2+1) REMPI through the 6d ^{2}D$_{1/2}$ and 6d ^{2}D$_{3/2}$ levels.*

that for the D states, indicating some saturation (power broadening) of the levels. In fact, the signal due to the $6p\,^2P_{3/2}$ state should be twice as large as that due to the $6p\,^2P_{1/2}$ state (the ratio of intensities is the ratio of degeneracies of the two levels $(2\cdot3/2 + 1)/(2\cdot1/2 + 1) = 2$). Saturation has the effect of depressing the relative intensities of the stronger transitions.

At greater metal atom densities and higher laser power, signal can also conceivably arise from secondary chemi-ionization processes such as

$$Cs^* + Cs^* \rightarrow Cs_2^+ + e^- \tag{16.8}$$

However, when the ionization signal is detected as a current in a wire, it is impossible to distinguish what the actual charge carriers are. More advanced experiments with mass spectrometry detection (see Chapter 13) could provide additional insight into this chemistry.

The same methods described here for cesium can also be applied to other alkali metals such as rubidium. The vapor pressure of rubidium is somewhat less than that of cesium, at about 2×10^{-7} Torr at room temperature,[5] and so slightly higher temperatures are required for its study. Figure 16.7 shows the resulting REMPI spectra for the Rydberg states of rubidium. As seen for cesium in Figure 16.4, the Rydberg levels get closer together as the ionization potential is approached. In this particular spectrum the *d* levels are saturated to a large extent so that the weaker *s* states are clearly visible.

These experiments on the REMPI of cesium or rubidium atoms have been performed routinely in the physics and chemistry laboratories at the University of Tennessee. Strangely, we have not been able to perform such experiments in

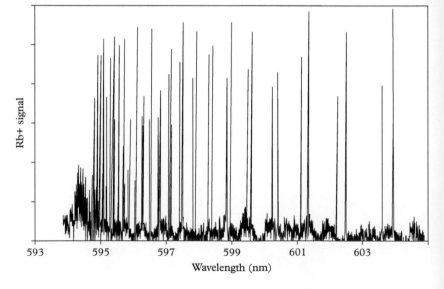

Figure 16.7 *(2+1) REMPI spectrum of rubidium in a heat pipe. The first peak on the right corresponds to the* n = 15 *level of the d state. The second peak from the right corresponds to the* n = 11 *level of the s state. Analysis of this data gives an average value of 1.35 for the quantum defect of the d states.*

the case of sodium. We do not understand this, since sodium is much easier to load into the heat pipe. This is unfortunate, since the atomic states of sodium are often studied by other methods (e.g., laser induced fluorescence) that would complement these REMPI studies.

Resonantly enhanced third harmonic generation and multiphoton ionization in cesium vapor

In addition to the REMPI processes described here, other multiphoton effects can arise in atomic systems. When a laser is in three-photon resonance with a gas, it is possible to generate third harmonic light at three times the pump frequency through a nonlinear laser interaction. Third harmonic generation (THG) is discussed in Chapter 26. In that chapter we discuss conversion of visible laser light into tunable VUV light with conversion efficiencies of $\sim 10^{-4}$. We describe here a more advanced experiment which could be performed as a senior project in which IR light from a laser is converted into the blue region with efficiencies of $\sim 1\%$.

In this experiment we describe measurements of THG and subsequent MPI using the same cesium vapor/heat pipe configuration described above. The basic setup is shown in Figure 16.8. To produce the IR light it is necessary to employ the hydrogen Raman shifter described in Chapter 27. Ions are detected as described above. A monochromator with a photomultiplier tube is added to detect the blue light generated by THG.

Because two stages of nonlinear optical conversion are required in this experiment, a higher pulse energy laser is required. Unfocused, tunable infrared light was generated by passing the light from a Nd:YAG-pumped dye laser (Quanta-Ray PDL-1) through a hydrogen Raman shifter. The resulting pulsed infrared light, with peak powers of $\sim 10^{6}$ W/cm^2, was passed through the cesium heat pipe at pressures ranging from 4 to 12 Torr. When tuning the laser near the three-photon resonance with the $7p\ ^2P_{3/2}$ and $7p\ ^2P_{1/2}$ levels, intense third harmonic

Figure 16.8 *Schematic of the apparatus to study THG and MPI in cesium in the region where two and three photons are in near resonance with the two-photon allowed 5d and three-photon allowed 7p levels, respectively (see Figure 16.5). The color changes in the laser beams indicate that blue light from the dye laser is shifted to the IR in the Raman shifter and the IR is shifted to blue light (THG) in the Cs vapor. A monochromator is shown to collect the THG light emitted from the interaction. However, the blue THG can also be separated from the infrared pump beam with a simple prism.*

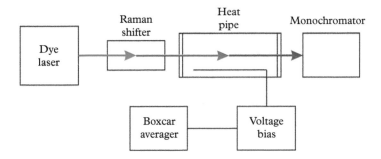

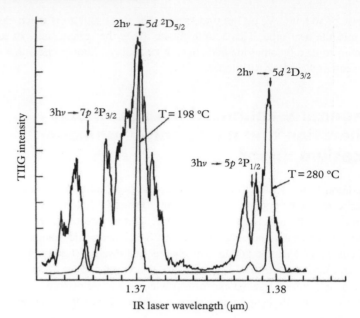

Figure 16.9 *Third harmonic light profiles as a function of the infrared laser wavelength for wall temperatures of 198 and 280 °C.*

radiation is observed. Figure 16.9 shows the THG intensity at two different temperatures (cesium densities). Inspection of Figure 16.5 reveals that when three photons are near resonance, two photons are in resonance with the $5d$ states. Thus the THG is most intense at this resonance.

The THG is in the visible blue region and can be detected with the monochromator, as shown in the figure, or it can also be separated by a prism and detected with a photodiode. The wavelength of the THG light varies from 454 to 460 nm as the IR pump light changes from 1.36 to 1.38 μm. The temperatures cited correspond to the wall of the heat pipe. For the T = 280 °C curve the conversion efficiency is ∼1% at the two-photon enhanced $5d$ resonance. One should consult Figure 16.5 to see how the two excitation schemes occur.

Notice that the THG occurs to the blue (higher energy) of the $7p$ resonances as expected from phase matching conditions. At low pressure in the absence of THG, the ionization due to the p and d states is prominent. At higher pressure the signal due to multiphoton ionization is particularly interesting. Figure 16.10 shows this ionization.

At the higher densities the ionization due to the $5d$ states remains high, whereas the ionization due to the three-photon excitation of the $7p$ states is absent. Ionization via both the $5d$ and $7p$ states requires 5 photons, $(2+3)$ and $(3+2)$, respectively. Chemi-ionization as discussed above would also be possible. The absence of ionization at the $7p$ levels may have an interesting explanation. It is well known [7,8] that in a dispersive medium three-photon excitation and third harmonic generation can be out of phase and cancel each other. This interference effect is discussed in the case of THG in rare gases in Chapter 26. In most cases

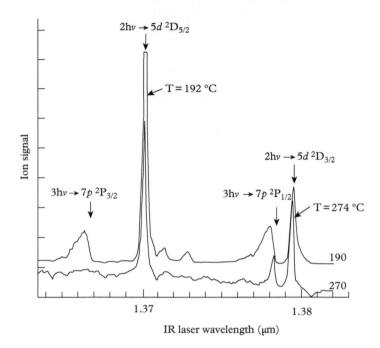

Figure 16.10 *Ionization profiles as a function of infrared laser wavelengths at wall temperatures of 192 and 274 °C.*

the ionization from a three-photon resonance can be restored by carefully reflecting the laser beam back upon itself. In this case the atom absorbs two photons traveling in one direction along with another traveling in the opposite direction. Under these conditions no THG is generated and ionization reappears. It is reasonable to attribute the disappearance of MPI at the $7p$ resonances as due to interference with the THG on resonance.

The above experiments involving nonlinear multiphoton processes in alkali metal vapors contained in a heat pipe are relatively easy to perform. Students are encouraged to explore the many other pathways to ionization and harmonic generation using alkali vapors.

....

REFERENCES

1. C. S. Feigerle and R. N. Compton, "Multiphoton ionization of cesium atoms," in *Physical Chemistry: Developing a Dynamic Curriculum*, R.W. Schwenz and R. J. Moore, eds., American Chemical Society Symposium Series, 1992, p. 178.
2. M. Goeppert-Mayer, "Elementary processes with two-quantum transitions," *Ann. Physik* **9**, 273 (1931).

3. G. S. Hurst, M. G. Payne, S. D. Kramer, and J. P. Young, "Resonance ionization spectroscopy and one-atom detection," *Rev. Mod. Phys.* **51**, 767 (1979).

4. K. D. Bonin and T. J. McIlrath, "Two-photon electric-dipole selection rules," *J. Opt. Soc. Am.* **B1**, 52 (1984).

5. A. N. Nesmeyanov, *Vapor Pressure of the Chemical Elements*, Elsevier, Amsterdam, 1963.

6. K.-H. Weber and C. J. Sansonetti, "Accurate energies of nS, nD, nF and nG levels of neutral cesium," *Phys. Rev. A* **35**, 4650 (1987).

7. J.C. Miller and R. N. Compton, "Third-harmonic generation and multiphoton ionization in rare gases," *Phys. Rev. A* **25**, 2056 (1982).

8. J. C. Miller, R. N. Compton, M. G. Payne, and W. R. Garrett, "Resonantly enhanced multiphoton ionization and third-harmonic generation in xenon gas," *Phys. Rev. Lett.* **45**, 114 (1980).

Electronic Absorption Spectroscopy of Molecular Iodine

17

Introduction

The rovibronic spectroscopy of molecular iodine has been extensively studied. In particular, the UV-vis spectroscopy of iodine has been a staple experiment in many physical chemistry laboratories for decades. The prominent $X\,^1\Sigma_g^+ \to B\,^3\Pi_{0u}^+$ transition covers ~550 to 500 nm, which corresponds to the yellow region of the electromagnetic spectrum (see Figure 17.1). Consequently, after passing through iodine vapor in a glass cell, white light appears purple in color. The many studies on the electronic spectra of the iodine molecule have been well summarized in the classic 1971 paper by Mulliken.[1] Numerous more recent studies have focused on the absorption spectrum of molecular iodine.[2–7]

Although the $X\,^1\Sigma_g^+ \to B\,^3\Pi_{0u}^+$ excitation is formally spin and angular momentum forbidden in one-photon absorption, the heavy iodine atom in the molecule relaxes these selection rules through spin–orbit coupling, giving rise to significant oscillator strength. In this experiment we consider the absorption spectrum of iodine from the v" = 0 and v" = 1 levels of the ground $X\,^1\Sigma_g^+$ state to the upper v' vibrational levels of the excited $B\,^3\Pi_{0u}^+$ state. The $X\,^1\Sigma_g^+$ and $B\,^3\Pi_{0u}^+$ potential energy curves are illustrated in Figure 17.1.

The Franck–Condon principle states that electronic/vibrational transitions are vertical (i.e., nuclei do not move during a photon absorption) and only high lying vibrational excitations (v') are produced in the $B\,^3\Pi_{0u}^+$ state, as inferred from Figure 17.1. The probability of finding molecules with a particular I–I internuclear separation (i.e., ψ^2) is shown for the two lowest vibrational levels, v" = 0 and v" = 1, represented by ψ_0 and ψ_1. Projecting the ψ_0^2 distribution onto the vibrational levels of the $B\,^3\Pi_{0u}^+$ state shows that only v' states from ~20 to 55 are expected to be strong, although using long path lengths and/or a heated cell and laser excitation it is possible to follow the vibrational levels closer to the dissociation limit of v' = 87. Many iodine absorption experiments in physical chemistry laboratories have been carried out using conventional UV-vis spectrometers.[8,9] In this experiment we employ tunable dye lasers operating in the visible region of the spectrum and use long path iodine cells. In some cases the laser allows multiple passes through the cell producing greater absorption.

Laser Experiments for Chemistry and Physics. First Edition. Robert N. Compton and Michael A. Duncan.
© Robert N. Compton and Michael A. Duncan 2016. Published in 2016 by Oxford University Press.

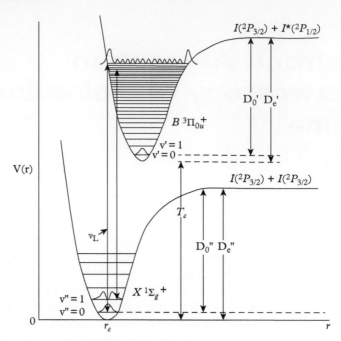

Figure 17.1 *Potential energy curves for the $X\,^1\Sigma_g^+$ and $B\,^3\Pi_{0u}^+$ states of molecular iodine.*

Experimental details

Iodine is a solid at room temperature but it has a significant vapor pressure (~0.131 mm) at 15 °C.[10] Figure 17.2 shows that slightly warming the tube to ~30 °C results in a vapor pressure of ~0.5 Torr, which is sufficient for absorption experiments.

Solid iodine crystals are placed in pyrex cells in such a way that the iodine can be frozen out with liquid nitrogen or dry ice. The cell is then pumped

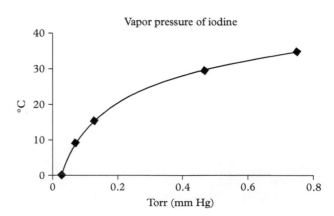

Figure 17.2 *Vapor pressure of iodine versus temperature (from [10]).*

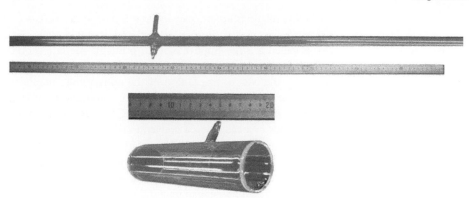

Figure 17.3 *Two glass cells containing flakes of iodine for measurements of the UV-vis absorption spectrum of I_2. The vapor pressure in the cell can be controlled with heating tapes. The top cell is ~2 m long and the lower cell is ~20 cm long. Viewing down the 2-m cell reveals a deep purple color.*

out and sealed. Two of the typical cells used in our experiments are shown in Figure 17.3. The iodine is initially placed in a side-arm off the tube which can be easily submerged in liquid nitrogen or an ice bath. Viewing down the 2-m cell produces a purple color while heating either cell to produce more vapor pressure produces a very deep purple color. When properly sealed, these tubes can last for years without attention.

Following the approximations to the energy levels derived using the solutions to the anharmonic Morse potential (see Chapter 14), the energy levels for the ground $X\,^1\Sigma_g{}^+$ state are given by

$$G(v'') = \omega_e'' \, (v'' + \tfrac{1}{2}) - \omega_e x_e'' \, (v'' + \tfrac{1}{2})^2 \qquad (17.1)$$

Similarly, the energy levels for the excited $B\,^3\Pi_{0u}{}^+$ state are

$$G(v') = T_e + \omega_e' (v' + \tfrac{1}{2}) - \omega_e x_e' (v' + \tfrac{1}{2})^2 \qquad (17.2)$$

We find the absorption spectrum for transitions from the ground vibrational state of the $X\,^1\Sigma_g{}^+$ state, $v'' = 0$, to the v' levels of the $B\,^3\Pi_{0u}$ state to be

$$G(v') - G(0) = T_e + \omega_e' (v' + \tfrac{1}{2}) - \omega_e x_e' (v' + \tfrac{1}{2})^2 - \tfrac{1}{2}\omega_e'' + \tfrac{1}{4}\omega_e x_e'' \qquad (17.3)$$

Thus the energy difference between any two vibrational levels of the $B\,^3\Pi_{0u}{}^+$ state is given by

$$\Delta v = \omega_e' - 2\omega_e x_e' (v' + 1) \qquad (17.4)$$

From this relationship one can see that a plot of Δv versus $(v'+1)$ should yield a straight line with a slope of $-2\omega_e x'_e$ and an intercept at $(v'+1)=\omega'_e$. Such a plot is called a Birge–Sponer plot. In principle the line should go through zero at the $(v'+1)$ value corresponding to the dissociation energy of the $B\,^3\Pi_{0u}$ state. However, the Morse description of the anharmonic potential energy curves is inadequate here at large internuclear separation and a different extrapolation is needed. It is traditional to record an $X\,^1\Sigma_g^+ \rightarrow B\,^3\Pi_{0u}{}^+$ absorption spectrum for iodine vapor in a cell using a conventional UV-vis absorption spectrometer and apply the linear Birge–Sponer extrapolation to determine the molecular constants for the $B\,^3\Pi_{0u}{}^+$ state. Using a nitrogen-pumped dye laser (or other appropriate laser), it is possible to measure longer path lengths and multiple passes through the cell. This makes it possible to obtain the absorption spectrum up to higher vibrational levels and with greater precision. Figure 17.4 shows typical laser absorption spectra recorded by students in an undergraduate physical chemistry laboratory in one afternoon (~3 h).

These spectra were recorded using multiple passes (one to three) through the short cell shown in Figure 17.3. Multiple passes through either a heated cell or a heated long pass cell in Figure 17.3 should allow one to approach the dissociation limit ($v' \approx 87$). Included in the figure is a typical UV-vis spectrum recorded by the students showing $v''=0$ to v' as well as $v''=1$ to v' levels (so-called hot bands). In order to construct the Birge–Sponer plot, it is first necessary to assign the quantum numbers to the vibrational features ("dips") in the signal. The vibrational quantum number, v', for a few of the vibrational levels are taken from the literature [11] (see Table 17.1).

Figure 17.5 shows a typical Birge–Sponer plot for the $X\,^1\Sigma_g^+ \rightarrow B\,^3\Pi_{0u}{}^+$ transitions for molecular iodine based upon some of the data in Figure 17.4 and using the vibrational assignments given in Table 17.1.

From equation 17.4 we have $\Delta v = \omega'_e - 2\omega_e x'_e\,(v'+1)$, thus the intercept in Figure 17.4 gives the vibrational frequency of the $B\,^3\Pi_{0u}{}^+$ state to be $\omega'_e = 131\ \mathrm{cm}^{-1}$. From the slope of the Birge–Sponer plot we obtain $-2\omega_e x'_e = -1.9848\ \mathrm{cm}^{-1}$ and $\omega_e x'_e = 0.9924\ \mathrm{cm}^{-1}$. Literature values are $\omega'_e = 125.68_7\ \mathrm{cm}^{-1}$ and $\omega_e x'_e = 0.7642$.[12] Figure 17.5 shows a typical set of data and analysis

Figure 17.4 *Visible absorption spectrum of molecular iodine originating from its ground and first vibrational state. The complete line shows the visible absorption spectrum using a conventional UV-vis spectrometer recorded for a single pass through the 20-cm cell.*

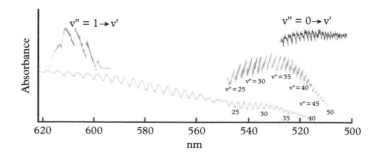

Table 17.1 *Vibrational assignments for the $X\,^1\Sigma_g^+ \rightarrow B\,^3\Pi_{0u}^+$ transition in I_2 (taken from [11]).*

v'	v"	λ, nm	v'	v"	λ, nm	v'	v"	λ, nm
27	0	541.2	18	1	571.6	13	2	595.7
28	0	539.0	19	1	568.6	14	2	592.0
29	0	536.9	20	1	565.6	15	2	588.5

selected from many experiments performed over the years in the physical chemistry laboratory at the University of Tennessee. Other experiments gave ω_e' and $\omega_e x_e'$ much closer to the presently accepted values.

Estimates of the dissociation energies for the ground $^1\Sigma_g^+$ and excited $^3\Pi_{0u}^+$ states can be obtained by inspection of Figure 17.1. The bond energy, D_0', of the excited $^3\Pi_{0u}^+$ state is the sum of all of the vibrational energies from $v' = 0$ to the maximum value (i.e., at the dissociation limit). If we consider the data in the Birge–Sponer plot to be a straight line, the dissociation energy, D_0', would be the area under the Birge–Sponer plot. Also, D_e' is just D_0' plus the zero-point energy or $D_e' = D_0' + \omega_e'/2 - \omega_e x_e'/4$.

The dissociation energy of the ground $^1\Sigma_g^+$ state can be estimated by measuring the total energy to one of the known vibrational energy levels of the $^3\Pi_{0u}^+$ state from the ground state and then adding to that the energy to reach the $^3\Pi_{0u}^+$ state dissociative continuum. This energy is the difference between $v'' = 0$ and the energy to produce $I\,(^2P_{3/2})$ and $I^*\,(^2P_{1/2})$. The energy difference between $I\,(^2P_{3/2})$ and $I^*\,(^2P_{1/2})$ is accurately known [13] to be 7603.15 cm^{-1}.

From the Birge–Sponer plot in Figure 17.5 one might estimate that the vibrational level, v'_{max}, required to dissociate the $^3\Pi_{0u}^+$ state is about $v'_{max} = 66$.

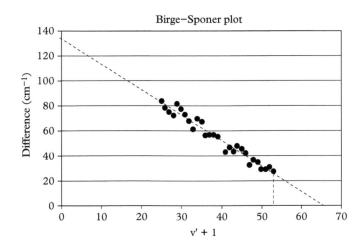

Figure 17.5 *Birge–Sponer plot of the $X\,^1\Sigma_g^+ \rightarrow B\,^3\Pi_{0u}$ transitions for molecular iodine.*

However, acquiring more data points on this curve shows an upturn. The plot is no longer linear and extends out much further. This is a result of the anharmonicity of this potential ($\omega_e x_e'$ is a sizable fraction of ω_e'). Thus the Birge–Sponer treatment of the vibrations for high vibrational levels fails and higher corrections beyond the simple Morse potential are necessary.

The Birge–Sponer plot of ΔG vs. $v' + 1$ assumes that the interaction between the two iodine atoms in the molecule are described by a Morse potential. However, at large internuclear distances, the two atoms interact more through an electrostatic potential of the form

$$V(r) = D - C_1/r + C_2/r^2 - C_3/r^3 + C_4/r^4 - C_5/r^5 + \cdots \qquad (17.5)$$

The higher vibrational energies of the iodine $X\ ^1\Sigma_g^+ \rightarrow B\ ^3\Pi_{0u}^+$ transition can be analyzed using a LeRoy–Bernstein [14] plot in order to reach the dissociation limit. A value of $v'_{max} = 87$ has been proposed. Thus this value far exceeds that seen in the Birge–Sponer plot in Figure 17.5. Steinfeld [15] has a complete discussion of the Birge–Sponer and LeRoy–Bernstein treatment of the dissociation energy of iodine. In Chapter 14 we derived a simple expression for v_{max} in terms of ω_e' and $\omega_e x_e'$, i.e., $v_{max} = \omega_e'/(2\omega_e x_e') - 1$. Applying this to the literature values cited above for ω_e' and $\omega_e x_e'$ gives $v_{max} = 81$. Using the values from the Birge–Sponer plot in Figure 17.5 gives $v_{max} = 65$ which is very close to the extrapolated value. Clearly the $B\ ^3\Pi_{0u}^+$ I_2 potential energy curve deviates markedly from the Morse potential at large internuclear distances. The use of multiple passes through the 2-m iodine tube in Figure 17.3 using the laser method described in this chapter might be able to reach close to $v'_{max} = 87$.

Finally, tunable lasers with higher resolution can also be used to study the closely spaced rotational levels in each of the vibrational levels for the I_2 transitions. Close inspection of the vibrational levels in Figure 17.4 reveals partially resolved structures between the vibrational states. A careful scan of these levels produces a rotational structure which can be used to give further molecular information.

....

REFERENCES

1. R. S. Mulliken, "Iodine revisited," *J. Phys. Chem.* **55**, 288 (1971).
2. E. L. Lewis, C. W. P. Palmer, and J. L. Cruickshank, "Iodine molecular constants from adsorption and laser fluorescence," *Am. J. Phys.* **62**, 350 (1994).
3. R. B. Snadden, "The iodine spectrum revisited," *J. Chem. Educ.* **64**, 919 (1987).
4. R. D'Alterio, R. Mattson, and R. Harris, "Potential curves for the I_2 molecule: An undergraduate Physical Chemistry experiment," *J. Chem. Educ.* **51**, 282 (1974).

5. L. Lessinger, "Morse oscillators, Birge-Sponer extrapolation, and the electronic absorption spectrum of I_2," *J. Chem. Educ.* **71**, 388 (1994).

6. I. J. McNaught, "The electronic spectrum of iodine revisited," *J. Chem. Educ.* **57**, 101 (1980).

7. R. D. Verma, "Ultraviolet spectrum of the iodine molecule," *J. Chem. Phys.* **32**, 738 (1960).

8. See e.g. R. J. Sime, *Physical Chemistry, Methods, Techniques, and Experiments*, Saunders College Publishing, Philadelphia, 1990, p. 660.

9. C. W. Garland, J. W. Nibler, and D. P. Shoemaker, *Experiments in Physical Chemistry*, eighth edition, McGraw-Hill, Boston, 2009.

10. *Handbook of Chemistry and Physics*, seventy-eighth edition, D. R. Lide, ed., CRC Press, Boca Raton, FL, 1997.

11. J. I. Steinfeld, R. N. Zare, I. Jones, M. Lesk, and W. Klemperer, "Spectroscopic constants and vibrational assignment for the $B\,^3\Pi_{0u}{}^+$ state of iodine," *J. Phys. Chem.* **42**, 25 (1965).

12. K. P. Huber and G. Herzberg, *Molecular Spectra and Molecular Structure: IV. Constants of Diatomic Molecules*, Van Nostrand Reinhold Company, New York, 1979.

13. A. Kramida, Y. Ralchenko, J. Reader, and NIST ASD Team (2014). *NIST Atomic Spectra Database* (version 5.2), [Online]. Available:http://physics.nist.gov/asd. National Institute of Standards and Technology, Gaithersburg, MD.

14. R. J. LeRoy and R. B. Bernstein, "Dissociation energy and long-range potential of diatomic molecules from vibrational spacings of higher levels," *J. Chem. Phys.* **52**, 3869 (1970); R. J. LeRoy, "Spectroscopic reassignment and ground-state dissociation energy of molecular iodine," *J. Chem. Phys.* **52**, 2678 (1970); R. J. LeRoy and R. B. Bernstein, "Dissociation energies and long-range potentials of diatomic molecules from vibrational spacings: The halogens," *J. Mol. Spec.* **37**, 109 (1971).

15. J. I. Steinfeld, *An Introduction to Modern Molecular Spectroscopy*, second edition MIT Press, Cambridge, MA, 2014, p. 131.

18

Electronic Spectroscopy of Iodine Using REMPI

Introduction

Resonance-enhanced multiphoton ionization (REMPI) of an atomic vapor was described in Chapter 16, and we saw how one-photon forbidden electronic states can be easily studied via two-photon excitation using a laser. Multiphoton excitation has also been effectively employed to access new one-photon forbidden excited states in many symmetric molecules. To illustrate this, consider a homonuclear diatomic molecule such as O_2 or N_2. A homonuclear diatomic molecule possesses inversion symmetry and the electronic states of that molecule are either symmetric (g = gerade) or antisymmetric (u = ungerade) with respect to inversion about the center of symmetry. Since the dipole moment operator is antisymmetric with respect to inversion, dipole excitation can only occur between states of opposite symmetry, i.e., $g \leftrightarrow u$. This is expressed as the dipole moment transition matrix element

$$<f|\mu \cdot E|i> \tag{18.1}$$

(equation 18.1) which connects the initial state i with the final state, f. Here μ and E are the dipole moment operator and electric field vector for linearly polarized light, respectively. Since $\mu = eE$ is an odd function of r, the transition moment integrated over all space is zero if the initial and final states are of the same symmetry with respect to r (even or odd). As discussed before for atoms, we refer to this g or u symmetry as the parity. Therefore the excitation of homonuclear diatomic molecules between states of the same parity is dipole forbidden. This is often referred to as the Laporte rule in spectroscopy. Now we consider the case of two-photon excitation from an initial state i to a final state f through an intermediate state j as depicted in Figure 18.1.

The intermediate state j does not need to be a bound stationary state. We speak of this state as a "virtual" level, which represents a non-stationary state created by the interaction of the light with the molecule and consists of an admixture of all possible bound and continuum states. The two-photon transition moment, K, can be calculated from the expansion state description of the level, j, from second-order perturbation theory,

$$K_{f \leftarrow i} = \sum_j \left[(<f|\mu \cdot E|j><j|\mu \cdot E|i>) / (h\nu_{ji} - h\nu + i\Gamma/2) \right] \tag{18.2}$$

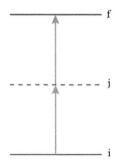

Figure 18.1 *Illustration of two-photon absorption from state i to f using a non-resonant intermediate step* j.

Laser Experiments for Chemistry and Physics. First Edition. Robert N. Compton and Michael A. Duncan.
© Robert N. Compton and Michael A. Duncan 2016. Published in 2016 by Oxford University Press.

where $h\nu_{ji}$ is the energy difference between quantum states i and j, $h\nu$ is the photon energy, and Γ is the combined energy width of the states i and j. The summation is over all bound and continuum states of the system.

In this experiment, we describe the measurement of multiphoton ionization of molecular iodine. I_2 is one of the first molecules studied in multiphoton ionization and demonstrates the great potential for observing transitions which were previously forbidden in one-photon absorption experiments. In this case, the two photons are in resonance with rovibrational levels of the 1_g electronic excited state of the I_2 molecule, followed by photoionization of this excited state by one photon. This is referred to as a $(2 + 1)$ REMPI process.

Experimental details

Figure 18.2 presents a scan of the multiphoton ionization of iodine vapor from the UV (3700 Å) to the visible (6000 Å) region of the spectrum. Multiphoton ionization studies in an undergraduate laboratory can be performed in any of the regions shown in Figure 18.2. Of particular interest is the region labeled $B\,^3\Pi_{0u}{}^+$ leading up to the dissociation limit of this state at 4990.8 Å. Chapter 17 describes measurements of the absorption spectrum and the derivation of the dissociation energy of this state. Note that energies above 4990.8 Å produce excited iodine atoms which undergo further resonance enhanced ionization. In this chapter we

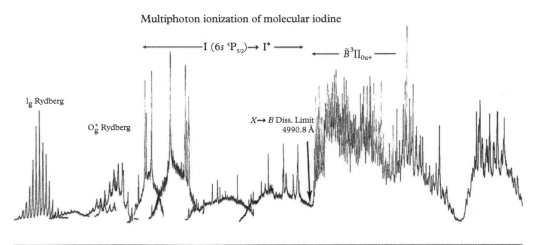

Multiphoton ionization of molecular iodine

Figure 18.2 *Survey of the REMPI signal obtained by tuning a dye laser focused into iodine vapor from the UV (3700 Å) to the visible (6000 Å) region of the spectrum. The "humps" in the background of the nine distinct regions are due to the changing intensity of the laser over the dye range covered.*

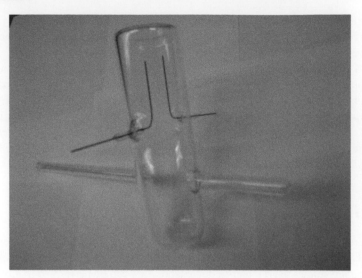

Figure 18.3 *Glass cell used to contain iodine vapor at room temperature. The focused laser is sent through the end of the glass cell between the two internal metal electrodes.*

focus on the so-called Dalby Band 1_g Rydberg state in the region from 3700 to 3800 Å which is easily accessible using a nitrogen-pumped dye laser.[1]

The nitrogen laser is tightly focused (~4 cm lens) between wires inside a glass sealed ionization cell and the ionization signal can be electrons or I_2^+ ions. A glass cell is fabricated using two glass-to-metal seals as shown in Figure 18.3.

Iodine pellets are placed in the tube shown projecting toward the right. The cell is then pumped out through the tube extending to the left with a mechanical pump having a liquid nitrogen trap. This trap is used to reduce the background pressure and to minimize the amount of corrosive iodine that might get into the pump. The finger containing the iodine is also kept initially at liquid nitrogen temperature to allow the glass tube to be pumped down prior to closing it off. Upon closing off the pump and removing the liquid nitrogen from the iodine finger, a faint purple color can be observed inside the tube. The vapor pressure of iodine as a function of temperature was shown in Figure 17.2.

The first multiphoton ionization study of a g → g two-photon transition was the (2+1) REMPI of I_2 reported by Dalby and coworkers.[1] The overall process consists of a $X\ ^1\Sigma_g^+ \rightarrow 1_g$ two-photon excitation followed by one-photon ionization leading to $I_2^+ + e^-$ continuum. Absorption of a fourth photon leading to dissociation of I_2^+ to $I^+ + I$ can also occur.[2] The branching ratio between the formation of I_2^+ and I^+ depends upon the laser power. However, unless the excitation step is saturated, the rate limiting step in the formation of both ions is still the two-photon excitation. As a result, the ion signal is observed clearly only when the laser wavelength is in resonance with this step. A spectrum of the Dalby bands, recorded in our undergraduate physical chemistry laboratory (UT), is shown in Figure 18.4 along with a schematic illustration of the potential curves for the states involved.

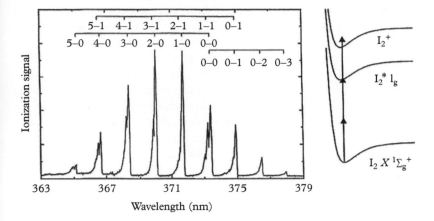

Figure 18.4 *(2+1) REMPI spectrum of the Dalby bands of the iodine molecule. The individual peaks are due to vibrational fine structure in the two-photon excitation part of the REMPI process.*

The various peaks observed in the spectrum are due to transitions originating from and terminating on different vibrational levels of the $X\ ^1\Sigma_g^+ \to 1_g$ transition. Since iodine is a heavy molecule, the rotational levels are not resolved and only vibrational levels are observed. These are indicated in Figure 18.4 with the labels v' – v" where v" and v' refer to vibrational quantum numbers for the ground and excited state, respectively. Because of the small frequency for the vibrations in I_2, even room temperature is sufficient to significantly populate excited vibrational levels in the $I_2\ X\ ^1\Sigma_g^+$ state. Transitions from these excited vibrational levels both enrich the spectrum and complicate the assignment of the vibrational origin, the 0–0 band. The origin is, however, readily identified from inspection of the energy differences between the peaks in the spectrum, since the vibrational frequencies for the ground and excited electronic states are $\omega_e'' = 214.50$ cm^{-1} and $\omega_e' = 241.41$ cm^{-1}, respectively.[1,3] In particular, the peaks which are identified as transitions from v" = 0 to v' = 0, 1, 2, 3, 4, and 5 are observed in the spectrum separated (in two photons) by around 240 cm^{-1}, the upper state vibrational spacing. In contrast, the peaks which are identified as transitions from v" = 0, 1, 2, and 3 to v" = 0 are separated (again in two photons) by around 214 cm^{-1}, the ground state vibrational spacing. These two sequences of peaks are labeled in the spectrum and can be seen to share one peak in common, the vibrational origin. Additional sequences can also be identified. In particular, a sequence of transitions which originate from v" = 1 is also labeled in Figure 18.4. As mentioned before, rotational fine structure is not resolved in the peaks but is evident from the blue (higher energy side) "shading" of the bands.

Spectral analysis

The spectrum in Figure 18.4 can be analyzed to produce a measurement of the vibrational frequencies, ω_e'' and ω_e', for the $X\ ^1\Sigma_g^+$ and 1_g states of I_2.

The anharmonicities are quite small for both states ($\omega_e x_e'' = 0.6147$ cm^{-1} and $\omega_e x_e' = 0.66$ cm^{-1}) and are difficult to extract from the spectrum at this resolution. While this limits the extent over which the spectrum can be analyzed, it also indicates that the simple harmonic oscillator approximation should provide a good representation of the potentials in the regions sampled by the spectrum. If this were a one-photon absorption spectrum, a normal procedure at this point would be to perform a Franck–Condon analysis based on harmonic potentials to extract the bond length and rotational constant for the excited electronic state (see Chapter 17). However, the spectrum in Figure 18.4 is a result of a two-photon excitation of I_2 followed by subsequent ionization to I_2^+. To determine whether or not a Franck–Condon analysis is a valid approach for this system, two issues need to be addressed: (i) the influence of the ionization step on the relative intensities of the vibrationally resolved peaks, and (ii) the appropriate Franck–Condon factor for a two-photon transition.

In general, the probability of photoionization is energy dependent. What is necessary for our purposes is that the probability must vary slowly enough over the wavelengths scanned so that the ionization step only produces a minor perturbation on the intensities of excitation. This is frequently the case except in the region of autoionization resonances, shape resonances (though these are typically broad), or a "Cooper" minimum.[4,5] Unfortunately, data on the energy dependence of the ionization cross-section from excited states are rare. One of the present applications of REMPI is to investigate the energy dependence and dynamics of ionization from excited states.[1] The spectrum shown in Figure 18.4 shows a smooth progression in the intensities indicating that there are no sharp (autoionizing) resonances in the ionization cross-section in this region.

The kind of Franck–Condon factor needed for a two-photon transition has been determined by Fiegerle [6] using an analysis of the two-photon transition moment embodied in equation 18.2. A key element in this analysis is the recognition that when the laser frequency is sufficiently off resonance in one photon, inclusion of vibrations does little to change the denominator of equation 18.2, and one can average over the vibrational components of the expansion states, j. The effect of this averaging is to produce a vibrationally resolved transition moment, $K_{f \leftarrow i}$ (v_f, v_i), which separates in the Condon approximation into an electronic two-photon transition moment $K_{f \leftarrow i}$, times the vibrational overlap integral, $<v_f | v_i>$,

$$K_{f \leftarrow i} (v_f, v_i) \approx <v_f | v_i> K_{f \leftarrow i} \tag{18.3}$$

Multiplying equation (18.3) by its complex conjugate yields a probability for the vibrationally resolved two-photon transition.

$$P(\alpha_f v_f \leftarrow \alpha_i v_i) \approx |K_{f \leftarrow i}|^2 \, q(v_f, v_i) \tag{18.4}$$

Here, the Franck–Condon factor for the two-photon transition, $q(v_f, v_i)$, is seen to be identical to that for the one-photon case.

$$q(v_f, v_i) = |<v_f | v_i>|^2 \qquad (18.5)$$

We conclude that as long as inclusion of vibrations does not introduce any resonance effects in either the excitation to the virtual level or the ionization from the bound excited state, then we should be able to analyze the vibrational intensities of a multiphoton ionization spectrum just as if it were due to a one-photon absorption.

The equation for a one-photon Franck–Condon factor is analytically solvable for simple harmonic oscillator wavefunctions.[7] Programs exist (or alternatively the solutions can be readily programmed) for the evaluation of $q(v_f, v_i)$ for specific cases. The Franck–Condon factors for an electronic transition in I_2 where $\omega_e' = 241.41$ cm^{-1} and $\omega_e'' = 214.50$ cm^{-1} are shown in Figure 18.5 as a function of the difference between the upper and lower state bond lengths.

A general characteristic of the vibrational intensities is that for increasing $\Delta(R_e'' - R_e')$, the maximum overlap of the wavefunction for $v'' = 0$ occurs at a correspondingly larger value of v'. As shown in the figure, the relative intensities are extremely sensitive to changes in $\Delta(R''-R')$, with changes of 0.001 Å giving rise to easily discernable differences in the Franck–Condon factors. The relative intensities of the $v'' = 0$ to $v' = 0, 1, 2, 3, 4,$ and 5 peaks from the spectrum shown in Figure 18.4 are also plotted in Figure 18.5. The intensities have been normalized to meet the vibrational sum rule

$$\sum_{vf} q(v_f, v_i) = \sum_{vfi} q(v_f, v_i) = 1 \qquad (18.6)$$

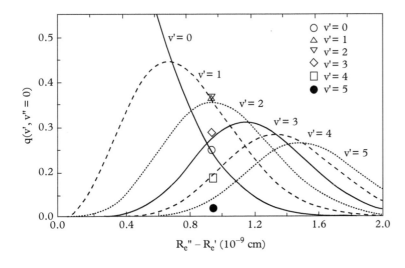

Figure 18.5 *Franck–Condon factors calculated as a function of the difference in the equilibrium bond lengths for the ground X $^1\Sigma_g^+$ and 1_g excited state of I_2. The experimentally observed intensities for the Dalby bands which originate from $v'' = 0$ are shown compared with those predicted by the calculation at $R_e'' - R_e' = 0.095$ Å.*

and then positioned at the $\Delta(R''–R')$ value which provides the optimum agreement with calculated values. From this analysis we determine that the bond length for the 1_g excited state differs from that of the ground state by 0.095 Å. This is in good agreement with the literature value of $\Delta(R_e''–R_e') = 0.099$ Å.[1] The difference here is most likely due to our lack of normalizing the intensities to the laser power.

In general, a Franck–Condon factor analysis based on harmonic potentials is not capable of distinguishing the sign of $\Delta(R_e''–R_e')$. This must be inferred from other considerations. In the case of I_2, the transition involves the excitation of an antibonding electron to a Rydberg $n\sigma_g$ non-bonding orbital and therefore the equilibrium bond length should be smaller in the excited state. Using $R_e'' = 2.6663$ Å, we obtain a bond length for the excited state of 2.571 Å. This bond length can also be used to obtain the rotational constant, B_e'. For completeness, the results of Dalby et al. [1] gave a Rydberg state of symmetry 1_g with $T_e' = 53,562.75$ cm^{-1}, $B_e' = 0.04029$ cm^{-1}, and $\omega_e' = 241.41$ cm^{-1}.

We have described two experiments employing a low-power tunable dye laser which illustrate the utility of multiphoton ionization atomic and molecular spectroscopy. These experiments have been performed by students in an undergraduate physical chemistry laboratory and the graduate course on experimental methods in the physics laboratory at the University of Tennessee. As seen in Chapter 16, students are attracted by the precision with which one is able to determine the ionization potential of an atom using the Rydberg series. The analysis goes beyond the hydrogen atom energy level treatment and introduces the concept of a quantum defect. In the present chapter, the analysis of the $g \rightarrow g$ forbidden transition for diatomic iodine allows the student to observe a Franck–Condon progression as well as hot-band transitions. Such analysis gives the vibration frequencies, energy levels, moment of inertia, and internuclear separation of an excited state. The student should also be impressed by the fact that multiphoton ionization of iodine reveals electronic states which were not seen previously! This experiment certainly conveys the potential applications of nonlinear optics in spectroscopy.

REFERENCES

1. G. Petty, C. Tai, and F. W. Dalby, "Nonlinear resonant photoionization spectra of molecular iodine," *Phys. Rev. Lett.* **34**, 1027 (1975); F. W. Dalby, G. Petty-Sil, M. H. L. Pryce, and C. Tai, "Nonlinear resonant photoionization spectra of molecular iodine," *Can. J. Phys.* **55**, 1033 (1977).
2. J. C. Miller and R. N. Compton, "Multiphoton ionization of iodine and benzene: Photoelectron and ion kinetic energy distributions," *J. Chem. Phys.* **75**, 2020 (1981).

3. K. P. Huber and G. Herzberg, *Molecular Spectra and Molecular Structure I. Constants of Diatomic Molecules*, Van Nostrand Reinhold, New York, 1979.

4. J. L. Dehmer, A. C. Parr, and S. H. Southworth, *Handbook on Synchrotron Radiation*, Vol. II, G. V. Marr, ed., North-Holland, Amsterdam, 1986.

5. L. E. Cuellar, R. N. Compton, H. S. Carman, Jr., and C. S. Feigerle, "Photoelectron angular distributions for ns ($n = 8$–12) subshells of cesium: Relativistic effects," *Phys. Rev. Lett.* **65**, 163 (1990).

6. C. S. Feigerle, "The relative intensities of vibrational fine structure in two-photon transitions," *J. Mol. Spec.* **146**, 1 (1991).

7. E. Hutchinsson, "Band spectra intensities for symmetrical diatomic molecules," *Phys. Rev.* **36**, 410 (1930).

19

Raman Spectroscopy Under Liquid Nitrogen

Introduction

Raman spectroscopy has been a powerful tool for the characterization of the structure of matter in the gaseous, liquid, and solid phases since the discovery of the Raman effect in 1928 by Sir Chandrasekhara Venkata Raman and Kariamanickam Srinivasa Krishnan. Raman spectroscopy is also one of the main methods in the analytical chemist's tool box. As described in Chapter 14, Raman and IR spectroscopy provide complementary information on molecular vibrations and in cases of high molecular symmetry (e.g., I_h, O_h, etc.), the two methods are "mutually exclusive," i.e., vibrational modes are either Raman active or IR active, but not both. The selection rules for the interaction of light and matter predict that asymmetric vibrations are excited in IR absorption and symmetric vibrations are active in Raman scattering. Together these two complementary methods of molecular spectroscopy have been essential techniques for understanding the structure of organic and inorganic molecules since the 1930s. In recent years, Raman spectroscopy has also proven to be a useful tool for the analysis of biological systems. Conformational changes of peptides and proteins in aqueous solutions, as well as nucleic acids and polynucleotides have been studied. Another recent application has been in the study of fullerene, carbon nanotube, and graphene materials. A general discussion of the comparisons of Raman and IR spectroscopy can be found in the book by Ingle and Crouch [1] and that by Nakamoto.[2] An excellent introduction to Raman spectroscopy can be found in the general education article by Tobias.[3] Other sources for elementary discussions of Raman spectroscopy are the books by Engel,[4] Struve,[5] Steinfeld,[6] and Harris and Bertolucci.[7] Before introducing Raman experiments, we provide a brief introduction to this technique in the following elementary treatment.

When electromagnetic radiation is incident upon an atom or molecule, the fundamental charges present experience forces exerted on them whose direction depends upon the orientation of the nuclei relative to the direction of the light's polarization. The electric field of the light, E, shifts the positive nuclei and negative electrons in opposite directions, giving rise to an induced polarization, P. This polarization produces an induced dipole moment, μ, given by

$$P = \mu = \alpha E \qquad (19.1)$$

where α is the polarizability. For highly symmetric molecules such as CCl_4 (T_d symmetry) or SF_6 (O_h symmetry), the induced polarizability, P, is parallel to the electric field, E. In general, however, P and E are not parallel but related through a polarizability tensor, α:

$$\begin{pmatrix} P_x \\ P_y \\ P_z \end{pmatrix} = \begin{pmatrix} \alpha_{xx} & \alpha_{xy} & \alpha_{xz} \\ \alpha_{yx} & \alpha_{yy} & \alpha_{yz} \\ \alpha_{zx} & \alpha_{zy} & \alpha_{zz} \end{pmatrix} \begin{pmatrix} E_x \\ E_y \\ E_z \end{pmatrix} \tag{19.2}$$

Linearly polarized electromagnetic radiation with frequency ν can be represented by a time dependent electric field given by

$$E(t) = E_0 \cos(2\pi\nu t) \tag{19.3}$$

This electric field induces the polarization, P, within the molecule which we write as a time-dependent dipole moment, $\mu(t)$.

$$P = \mu(t) = \alpha E_0 \cos(2\pi\nu t) \tag{19.4}$$

The change in the polarizability α is proportional to the change in the bond length about the equilibrium position of the nuclei, r_e, or $r_e + r(t)$. Expanding the polarizability $\alpha(r)$ about the equilibrium position in a Taylor–McLaurin series, and keeping only the first term, we can write

$$\alpha(r) = \alpha(r_e) + d\alpha/dr|_{r=re}(r-r_e) + \cdots \tag{19.5}$$

Vibration of the molecule about r_e to an extreme position r_{max} produces a time variation given by

$$r(t) = r_{max} \cos(2\pi\nu_{vib} t) \tag{19.6}$$

where ν_{vib} represents the natural vibrational frequency of the molecule.

Equation 19.4 can be rewritten as

$$\mu(t) = \alpha(r_e) E_0 \cos(2\pi\nu t) + [d\alpha/dr|_{r=re}]r_{max} \cos(2\pi\nu_{vib}t) E_0 \cos(2\pi\nu t) \tag{19.7}$$

Using the trigonometric identity $2\cos\theta\cos\varphi = \cos(\theta+\varphi) + \cos(\theta-\varphi)$, we can write

$$\mu(t) = \alpha(r_e) E_0 \cos 2\pi\nu t + (1/2)[d\alpha/dr|_{r=re}]r_{max} \\ E_0[\cos(2\pi t(\nu+\nu_{vib})) - \cos(2\pi t(\nu-\nu_{vib}))] \tag{19.8}$$

Introduction of the trigonometric identity produces an expression for the induced polarizability (dipole moment) containing a $\cos(2\pi\nu t)$ term that gives rise to a scattered wave at the frequency of the initial field, which we call elastic scattering (no change in energy) or Rayleigh scattering. The second term involves $\cos(2\pi(\nu - \nu_{vib}))$ which represents scattering in which the initial wave has lost an energy corresponding to the vibrational quantum ($h\nu_{vib}$); this is called Stokes Raman scattering. Finally the last term is attributed to the anti-Stokes Raman scattering term involving $\cos(2\pi(\nu + \nu_{vib}))$, in which the initial wave has gained a vibrational quantum of energy. Thus scattering occurs at the frequency of the incident light (Rayleigh scattering), and at frequencies decreased (Stokes Raman) or increased (anti-Stokes Raman) by the vibrational frequency of the molecule. These processes are illustrated in Figure 14.1 of Chapter 14. The Raman intensities in equation 19.8 contain the term $(\frac{1}{2})[d\alpha/dr|_{r=re}]r_{max}$ which turns out to be about 10^{-4} times that of the leading term $\alpha(r_e)$. Thus the Rayleigh line is approximately 10^4 times larger than the Raman Stokes lines. The intensities of the anti-Stokes lines vary with the temperature of the sample. The presence of the more intense Rayleigh scattering represents a major problem in recording Raman spectra, especially for low energy vibrations close to the Rayleigh line. Stray laser light scattering inside the spectrometer can also be a significant problem. The scattering at the laser line can be reduced using a double monochromator and an interference or notch filter at the laser line frequency. It is also possible to use a tunable laser for the excitation adjusted to the frequency of an allowed $s \to p$ transition in an alkali vapor that can be totally attenuated with a metal vapor cell (see for example [8]). Typically a diode or Ti:sapphire laser can be tuned to the $5s$ $^2S_{1/2} \to 5p\ ^2P_{3/2}$ transition of potassium at 2780.24 nm. Potassium vapor at room temperature in a glass cell, along with a moderate pressure of argon to broaden the atomic line to match the laser bandwidth, can work well to deplete the Rayleigh line and scattered light. This method allows measurements to be made to within a few wavenumbers of the excitation wavelength. The K line filter cell is simple to make and employ. However, this method requires a tunable high-resolution laser.

The observation of an anti-Stokes line in the spectrum requires the molecule to be in an excited vibrational state. When molecules are in thermal equilibrium, and assuming harmonic oscillator behavior, the lowest energy is that of the zero point level ($h\nu/2$) and the first vibrational state has an energy of $3h\nu/2$. Assuming that the degeneracy of each level is the same (or each are non-degenerate) one can write the ratio of scattered intensities of the anti-Stokes to Stokes lines as

$$\frac{I(\text{anti-Stokes})}{I(\text{Stokes})} = \left[\frac{(\nu_L - \nu_v)^4}{(\nu_L + \nu_v)^4}\right]\frac{e^{-3h\nu/kT}}{e^{-h\nu/2kT}} \tag{19.9}$$

Where ν_L is the laser frequency and ν_v is the vibrational frequency. The factor $((\nu_L + \nu_v)^4/(\nu_L - \nu_v)^4)$ takes into account the λ^4 factor in the scattering intensities. Neglecting this factor at a temperature of 300 K with k = 0.695035 cm^{-1}/K, the ratios of the anti-Stokes to Stokes intensities for vibrational energies of 100 and 1000 cm^{-1} are ~0.6190 and 0.00827, respectively.

Experimental details

Molecules at room temperature can populate higher vibrational and rotational states, as well as often existing in a number of different molecular conformations. Cooling the sample to a low temperature can freeze out higher rovibrational states, as well as significantly reduce the presence of higher energy conformers. Performing Raman spectroscopy with the sample submerged *under* liquid nitrogen (T = 77 K) provides a simple yet effective way to perform high-resolution Raman spectroscopy of liquid or solid samples. The sample mounting scheme is shown in Figure 19.1. The sample holder is an aluminum block with indentations to hold the liquid or solid samples, which is submerged under liquid nitrogen in the double-Styrofoam-reservoir configuration shown in the figure. A few holes are placed in the coffee cup for the liquid nitrogen to flow between the two Styrofoam containers. Sometimes it is necessary to place a screen grid over the solid sample to prevent it from floating in the liquid nitrogen. In most cases the vapor of the boiling liquid nitrogen prevents fogging of the lens in the Raman apparatus by purging the water vapor in the air away from the lens.

It is interesting to look at the Raman spectrum of liquid nitrogen itself. Figure 19.2 shows a spectrum in which liquid nitrogen has been sitting out for some time in the room, thereby picking up oxygen. The spectrum shows two narrow features at 2328.4 cm^{-1} and 2290.2 cm^{-1} which are due to $^{14}N_2$ and $^{14}N^{15}N$, respectively. Previously measured values for the N_2 Raman lines are 2327.0 cm^{-1} and 2288.8 cm^{-1}.[9] The liquid nitrogen vibration is red shifted by only ~3.4 cm^{-1} from the gas-phase value.[10] These peaks are always present and serve to provide a convenient wavelength calibration for the spectrum. The width of the

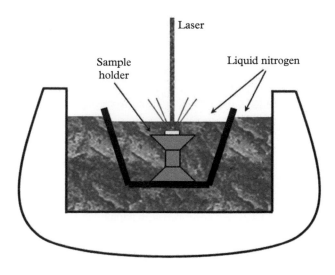

Figure 19.1 *Setup of the Raman Under Nitrogen (RUN) spectroscopy experiment. The outer reservoir of liquid nitrogen is held in a Styrofoam packing container. The inner container is a Styrofoam coffee cup supported by three toothpicks.*

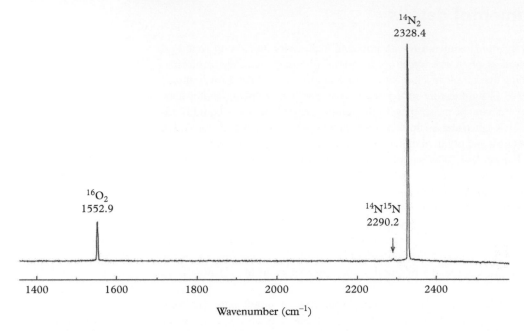

Figure 19.2 *Raman spectrum of liquid nitrogen with some dissolved oxygen. Oxygen is generally not present in fresh samples.*

$^{14}N_2$ peak can also be used to provide an estimate of the resolution of the apparatus. These lines are very narrow and the positions of the peaks are sensitive to atmospheric pressure. The presence of liquid oxygen is also seen as a peak at 1552.9 cm^{-1}.

The first RUN spectrum was recorded in an undergraduate physical chemistry laboratory at the University of Tennessee. Carbon tetrachloride was selected as the first sample since it was one of the liquids studied by Chandrasekhara Raman. Its spectrum is shown in Figure 19.3.

The upper part of Figure 19.3 shows all of the allowed Raman bands of CCl$_4$. The structure in the region around 800 cm^{-1} is attributed to what is called a Fermi resonance (see [11], page 272). A Fermi resonance occurs whenever two states of the same symmetry are nearly degenerate and are coupled by anharmonic interactions. In this case the $\nu_1 + \nu_4$ combination is close to the ν_3 fundamental vibration, as shown in the figure. Fermi resonances are discussed further in Chapter 22.

The structure in the $\nu_1(a_1)$ band (symmetric stretch) shown in the bottom of Figure 19.3 is attributed to the natural abundances of the ^{25}Cl and ^{37}Cl isotopic components of CCl$_4$. As described in Chapter 12, there are many online programs that calculate the distribution of probabilities for a complex molecule

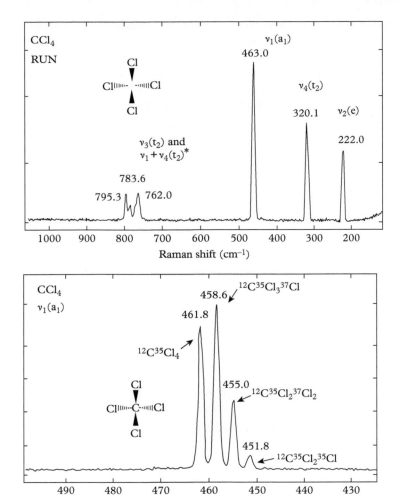

Figure 19.3 *Raman Under Nitrogen of carbon tetrachloride. The top frame shows a broad spectrum of all of the allowed Raman vibrations. The bottom frame shows an expanded view of the $v_1(a_1)$ band.*

made of many atoms with varying isotopic abundances. The intensities of the isotope peaks in CCl_4 shown in the bottom of Figure 19.3 can be calculated from the probability mass function (binomial distribution) in equation 19.10.

$$F(k, n, p) = [n! / (k!(n-k)!] \, p^k (1-p)^{n-k} \qquad (19.10)$$

where n represents the number of trials, k the number of successes, and p the probability of a success. One should apply the procedure above for the CCl_4 molecule to check the intensities in Figure 19.3 (^{35}Cl and ^{37}Cl are 75.77% and 24.23% abundant, respectively), neglecting the isotopes of carbon. To calculate the mass distribution for a multi-atom molecule with each atom having a number of naturally occurring isotopes, the binomial distribution method described above must be generalized. To do this, we take into account multiple kinds of atoms as well as those which have more than two isotopes. The former is done by utilizing a product of distributions, because the isotopic distribution of each atom is independent from the others while the latter extends the binomial distribution to the multinomial distribution:

$$\Pr\left(k; \vec{n}, p\right) = \prod_i \Pr_i\left(\vec{k}_i; n_i, \vec{p}_i\right) = \prod_i n_i \left(\prod_j \frac{p_{ij}}{k_{ij}}\right) \tag{19.11}$$

Table 19.1 *Isotope abundances of carbon and oxygen in the CO_2 molecule.*

Isotope	Abundance
carbon-12 (^{12}C)	0.9890
carbon-13 (^{13}C)	0.0110
oxygen-16 (^{16}O)	0.9976
oxygen-17 (^{17}O)	0.0004
oxygen-18 (^{18}O)	0.0020

Each k_{ij} is the number of isotopes j, of atom i, p_{ij} is the probability of that isotope, and n_i is the total number of atoms of type i. As an example of this method, consider the carbon dioxide (CO_2) molecule. Its isotopes and abundances are listed in Table 19.1.

Suppose we want to calculate the probability of observing $^{12}C^{17}O^{18}O$. Beginning with each type of atom individually, this gives

$$\Pr\left(^{12}C\right) = \frac{1!}{1!\,0!}(0.9890)^1(0.0110)^0 = 0.9890$$

$$\Pr\left(^{17}O^{18}O\right) = \frac{2!}{1!\,1!\,0!}(0.9976)^0(0.0004)^1(0.0020)^1 = 8 \times 10^{-7}$$

$$\overset{\text{yields}}{\rightarrow} \Pr\left(^{12}C^{17}O^{18}O\right) = (0.9890)\left(8 \times 10^{-7}\right) = 7.91 \times 10^{-7}$$

In another example of RUN considering the isotopic contributions, we show a Raman spectrum for the carbon disulfide molecule submerged under liquid nitrogen in Figure 19.4. This further illustrates the high resolution spectra possible using this method. The isotopes of sulfur and their calculated and measured abundances are shown in the inset of the figure.

The RUN method was first introduced for a series of air sensitive metallocene molecules. The utility of recording Raman spectra under liquid nitrogen was demonstrated for ferrocene, uranocene, and thorocene.[12] Using RUN, low-temperature, liquid nitrogen-cooled Raman spectra for these compounds exhibited higher resolution than that in previous studies and new vibrational features were reported.

As a typical example of the advantages of RUN spectroscopy, a spectrum for the ferrocene molecule is shown in Figure 19.5. At room temperature the sample can be damaged from heating by the laser and, more importantly, the spectral features are broad. The effect of rovibrational cooling in the RUN spectra is further illustrated in Figure 19.6 and compared to the room temperature spectrum for the region around 1100 cm^{-1}.

It is possible to calculate the vibrational frequencies and identify the various Raman and infrared modes for relatively large molecules using the Gaussian

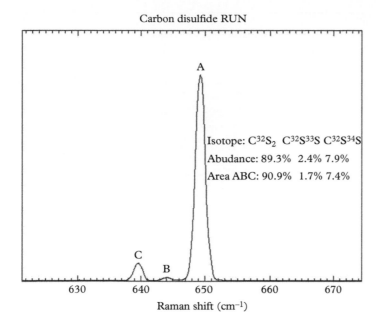

Carbon disulfide RUN

Isotope: $C^{32}S_2$ $C^{32}S^{33}S$ $C^{32}S^{34}S$
Abudance: 89.3% 2.4% 7.9%
Area ABC: 90.9% 1.7% 7.4%

Raman shift (cm^{-1})

Figure 19.4 *RUN spectrum of carbon disulfide, CS$_2$, showing the Raman peaks due to the sulfur isotopes.*

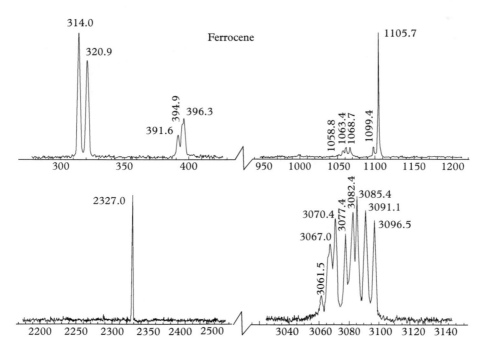

Figure 19.5 *Raman spectrum of the ferrocene molecule showing the vibrational structure in all regions of the spectra, especially in the C–H stretching region (3060 to 3100 cm^{-1}). The peak at 2327 cm^{-1} is from the liquid nitrogen. (Reproduced with permission from J. Chem. Phys. 120, 2708 (2004). Copyright 2004, AIP Publishing LLC.)*

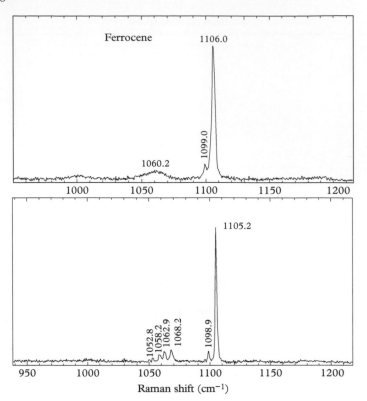

Figure 19.6 *Comparison of the room temperature and RUN spectrum for ferrocene in the region of 1100 cm⁻¹. The spectrum is dominated by a signal at 1105.2 cm⁻¹, assigned to the ring-breathing mode of the cyclopentadienyl ligands. (Reproduced with permission from J. Chem. Phys. 120, 2708 (2004). Copyright 2004, AIP Publishing LLC.)*

program [13] or other methods such as HyperChem, etc. (see Chapter 15). We have computed the frequencies for all 57 vibrations of ferrocene at the B3LYP/Def2/TZVPD level. The results of these calculations for the allowed Raman transitions are shown in Figure 19.7. The three transitions labeled represent the symmetric breathing modes in which the volume of the molecules changes the most upon vibration. The data in Figure 19.5 and the calculations in Figure 19.7 show many lines in the C–H stretching region around 3100 cm⁻¹. The large number of vibrational peaks here suggests that RUN may represent a new method for structure analysis similar to NMR spectroscopy.

A further example of the application of Raman spectroscopy for the determination of molecular structure is found for the 1,2-dichloroethane molecule. In Chapter 29 (Raman Spectroscopy Applied to Molecular Conformational Analysis) the RUN spectrum for the *anti-* and *gauche-* forms of 1,2-dichloroethane are presented along with Gaussian calculations of the vibrational modes and their energies. There are many other potential applications of Raman spectroscopy for molecular structure determination. A motivated student should choose a molecule of interest and perform a RUN spectrum followed by a comparison with calculated vibrational spectra.

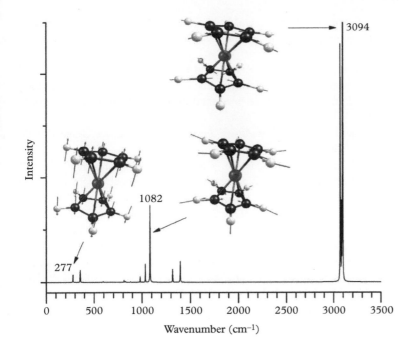

Figure 19.7 *Raman transitions predicted by theory for the ferrocene molecule. The three vibrations labeled are symmetric breathing modes. Vibrations from the calculations were scaled by a factor of 0.96 for comparison to the experiment.*

REFERENCES

1. J. D. Ingle, Jr. and S. R. Crouch, *Spectrochemical Analysis*, Prentice Hall, Englewood Cliffs, NJ, 1988, pages 494–524.
2. K. Nakamoto, *Infrared and Raman Spectra of Inorganic and Coordination Compounds*, John Wiley and Sons, Inc., New York, 1997.
3. R. S. Tobias, "Raman spectroscopy in inorganic chemistry," *J. Chem. Ed.* **44**, 2 (1967).
4. T. Engel, *Quantum Chemistry and Spectroscopy*, Pearson-Benjamin Cummings, San Francisco, 2006.
5. W. S. Struve, *Fundamentals of Molecular Spectroscopy*, John Wiley & Sons, New York, 1989.
6. J. I. Steinfeld, *Molecules and Radiation: An Introduction to Modern Molecular Spectroscopy*, MIT Press, Cambridge, MA, 1985.
7. D. C. Harris and M. D. Bertolucci, *Symmetry and Spectroscopy: An Introduction to Vibrational and Electronic Spectroscopy*, Dover Publications, Inc., New York, 1978.
8. W. Lee and W. R. Lempert, "Spectrally filtered Raman/Thomson scattering using a rubidium vapor filter," *AIAA Journal* **40**, 2504 (2002).

9. K. F. Everitt, J. L. Skinner, and B. M. Ladanyi, "Vibrational energy relaxation in liquid oxygen (revisited) and in liquid nitrogen," *J. Chem. Phys.* **116**, 179 (2002).

10. M. Musso, F. Matthai, D. Keutel, and K.-L. Oehme, "Isotropic Raman line shapes near gas-liquid critical points: The shift, width, and asymmetry of coupled and uncoupled states of fluid nitrogen," *J. Chem. Phys.* **116**, 8015 (2002).

11. P. F. Bernath, *Spectra of Atoms and Molecules*, Oxford University Press, Oxford, 1995.

12. J. S. Hager, J. Zahardis, R. M. Pagni, R. N. Compton, and J. Li, "Raman under nitrogen. The high-resolution Raman spectroscopy of crystalline uranocene, thorocene, and ferrocene," *J. Chem. Phys.* **120**, 2708 (2004).

13. M. J. Frisch, G. W. Trucks, H. B. Schlegel, G. E. Scuseria, M. A. Robb, J. R. Cheeseman, J. A. Montgomery, Jr., T. Vreven, K. N. Kudin, J. C. Burant, J. M. Millam, S. S. Iyengar, J. Tomasi, V. Barone, B. Mennucci, M. Cossi, G. Scalmani, N. Rega, G. A. Petersson, H. Nakatsuji, M. Hada, M. Ehara, K. Toyota, R. Fukuda, J. Hasegawa, M. Ishida, T. Nakajima, Y. Honda, O. Kitao, H. Nakai, M. Klene, X. Li, J. E. Knox, H. P. Hratchian, J. B. Cross, C. Adamo, J. Jaramillo, R. Gomperts, R. E. Stratmann, O. Yazyev, A. J. Austin, R. Cammi, C. Pomelli, J. W. Ochterski, P. Y. Ayala, K. Morokuma, G. A. Voth, P. Salvador, J. J. Dannenberg, V. G. Zakrzewski, S. Dapprich, A. D. Daniels, M. C. Strain, O. Farkas, D. K. Malick, A. D. Rabuck, K. Raghavachari, J. B. Foresman, J. V. Ortiz, Q. Cui, A. G. Baboul, S. Clifford, J. Cioslowski, B. B. Stefanov, G. Liu, A. Liashenko, P. Piskorz, I. Komaromi, R. L. Martin, D. J. Fox, T. Keith, M. A. Al-Laham, C. Y. Peng, A. Nanayakkara, M. Challacombe, P. M. W. Gill, B. Johnson, W. Chen, M. W. Wong, C. Gonzalez, and J. A. Pople, Gaussian 03 (Revision D.02), Gaussian, Inc., Pittsburgh PA, 2003.

Optical Rotary Dispersion of a Chiral Liquid (α-pinene)

20

Introduction

The mirror image of the left hand of a human being is the right hand. Since a hand has a top and a bottom (i.e., it is three dimensional), the right and left hands cannot be superimposed upon each other. Like the hand, certain molecules whose mirror images are non-superimposable are designated as chiral (Greek for *handed*). Likewise any spinning object moving along its axis of spin cannot be superimposed upon its mirror image and is defined as chiral. This is true even if the object is flat but moving in a direction out of the plane. Thus twining plants, sea shell conchs, spiral galaxies, and moving electrons are chiral.[1] Because photons have unit spin angular momentum and are always moving, they are also chiral. Chiral molecules exhibit readily observable effects termed optical activity. Optical activity is observed as: (i) *circular dichroism*, the difference in extinction for right- and left-circularly polarized light passing through the medium, and (ii) *optical rotation*, the rotation of the plane of polarization of linearly polarized light passing through the medium. A measurement of the optical rotation as a function of the wavelength of the light is called optical rotary dispersion (ORD). The circular dichroism as a function of light frequency shows peaks in the electronic region (electronic circular dichroism, ECD) or in regions of vibrational excitation (vibrational circular dichroism, VCD). Because linearly polarized light can be viewed as a coherent superposition of right- and left-circularly polarized light, the two phenomena are related. Formally, circular dichroism and optical rotation are related through the Kramers–Kronig relations;[2] knowledge of the complete spectrum of one allows the calculation of the other. In practice, however, the exercise is not so tractable because of the experimental difficulty of recording complete spectra.

The optical rotation angle, φ (in radians), of the plane of polarization of light through a chiral medium is often written as [3]

$$\varphi = \pi(n_L - n_R)\,\ell/\lambda \qquad (20.1)$$

where $n_L - n_R$ is the difference in the indices of refraction for left- and right-circularly polarized light of the optically active medium at a wavelength λ, and ℓ is the path length. The precise angle through which the linearly polarized light vector is rotated for a given chiral medium is determined by the density or concentration of the optically active component, the path length, the wavelength, and the temperature. The specific rotation has been adopted as a useful quantitative measure of these effects. The specific rotation for a chiral solution is a function of wavelength and temperature and is defined as

$$[\alpha]_{\lambda}^{T} = \frac{\theta}{c\ell} \qquad (20.2)$$

where θ is the measured angle of rotation (in deg), ℓ is the distance (in dm) of the optically active solution through which the light travels, and c is the concentration (in g/ml). Often the specific rotation will be dependent on the concentration in a mixed solution. To standardize the reporting of specific rotation, values are typically given at room temperature using the standard wavelength of the sodium D line (589 nm). A plot of the specific rotation of a chiral medium as a function of wavelength is called an *optical rotatory dispersion* curve and it can be useful for gaining insight into the electronic structure of the medium of interest. For wavelengths that are well outside a single resonance absorption region of the material, the optical rotatory dispersion can be described by the Drude expression: [4]

$$[\alpha]_{\lambda}^{T} = \frac{A}{\lambda^2 - \lambda_0^2} \qquad (20.3)$$

where A is the rotation constant and λ_0 is the dispersion constant. If several optically allowed transitions contribute in the region of wavelength λ, the optical rotation is a sum over their contributions. When the specific rotation of a compound decreases with increasing wavelength as in equation 20.3, and there are no local extrema in the curve, the optical rotatory dispersion (ORD) curve is called a *plain curve*. Many articles have appeared describing demonstrations of optical rotation.[5–16] Mahurin et al. [17] described a simple but pedagogical demonstration of optical rotation, which can be easily instituted in the undergraduate chemistry or physics laboratory. In this demonstration, linearly polarized light from a He-Ne, argon ion, or other laser is directed down a 1-m cylindrical glass tube filled with a solution containing chiral molecules (e.g., sucrose). Polarized light is scattered at right angles to the direction of propagation as a result of a combination of Tyndall and Rayleigh scattering. Maximum scattering intensity occurs at right angles to the plane of polarization. The rotation of the beam is clearly seen as alternating bright and dark regions as the light polarization vector "spirals" along the tube. In this study, various wavelengths (456.9, 476.5, 488.0, 514.5 nm) supplied by an argon ion laser, the 632.8 nm line from a He-Ne laser, and the 785 nm line from a diode laser were used to create the ORD and extinction curves for sucrose in water (see [17]). The dark and bright scattering at a right angle to the propagation of light down the sugar solution is shown in Figure 20.1.

The ORD curve was used to fit the Drude expression (equation 20.3) to obtain values for the rotation and dispersion constants. If the tube is mounted in

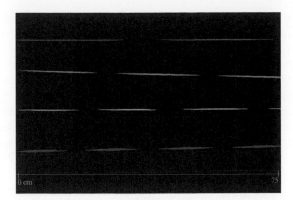

Figure 20.1 *Laser light scattering in a tube containing a sucrose (chiral) solution. The red beam from a He-Ne laser (632.8 nm) at the top is followed by that from a multi-line argon ion laser (reading top to bottom after the red He-Ne laser, 514, 488, and 457 nm). It is interesting to pass a bright linearly polarized white light source through the tube and observe a "barber pole" of spiraling colors of the rainbow (this was first performed by Michael Faraday).*

an upright position, students can walk around the tube and watch the dark and bright bands move up or down the tube depending upon the sign of the ORD. Provided that Kayro syrup is introduced into the tube, the setup can be kept for years for students and visitors to observe. Unfortunately, sucrose will turn cloudy after a period of time due to the action of bacteria.

Experimental details

A more conventional experimental arrangement for the measurement of optical rotation is shown in Figure 20.2. This experiment investigates the ORD of the two enantiomers of α-pinene, although any chiral molecule solution can be employed. The natural product α-pinene is a bicyclic monoterpene which can be rendered from the resins of conifers and other plants. The α-pinene molecule consists only of carbon and hydrogen atoms ($C_{10}H_{16}$) and the molecule is rigid with no isomers aside from the β-pinene structural isomer. The two enantiomers of α-pinene are shown in Figure 20.3.

The experimental arrangement can be configured in many different ways with varying degrees of sophistication. In essence, linearly polarized light is passed through the sample and the angle of rotation is determined with a second polarizer. Figure 20.4 shows the generic experimental geometry for an ORD experiment.

Figure 20.2 *Experimental arrangement for determining optical rotation in a chiral gas or liquid. In the figure the electric vector of the light is rotating clockwise as seen by an observer looking into the light source or (+) rotation. A (+) sign indicates that the compound is dextrorotatory and rotates light clockwise while a (−) sign indicates that the compound is levorotary and rotates the E-vector of the light counterclockwise, as seen by an observer looking at the light source.*

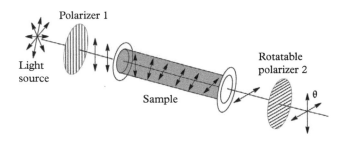

Figure 20.3 *Structure of (–) and (+) α-pinene. The student should identify the R- and S-enantiomers and the chiral carbon(s).*

If a light source and monochromator are used, as shown in Figure 20.4, the Glan polarizing prisms can be replaced by polarizing sheets and the detector can be the eye. It is important to note that in using either the eye or other detectors it is much easier and more accurate to detect a minimum in the light intensity as opposed to a maximum to determine the optical rotation. The ORD can be determined using a number of different lasers including laser pointers, a nitrogen-pumped dye laser, or the 337.1 nm light from a nitrogen laser. Measurements of optical rotation do not depend upon the light intensity, only the rotation angle. Figure 20.5 shows the data from one student using the monochromator and

Figure 20.4 *The labels in the upper part of the figure depict the generic experimental arrangement for ORD measurements. The lower part of the picture shows a typical arrangement. In some experiments a flashlight such as the one on the table or a laser pointer can serve as the light source. In the picture shown, the light source was a xenon lamp and wavelength analysis was accomplished with a Jobin Yvon monochromator.*

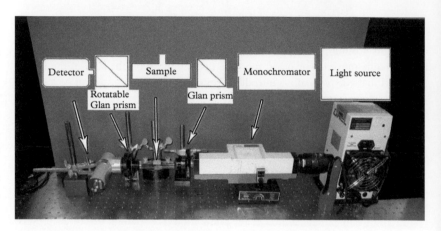

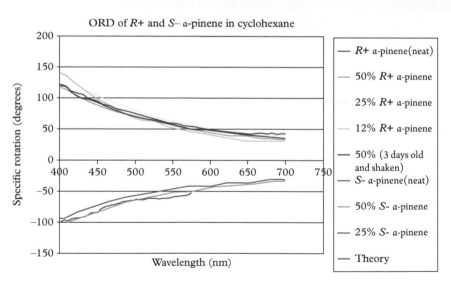

Figure 20.5 *ORD of α-pinene from 700 to 400 nm for various concentrations in cyclohexane. Theoretical calculations for the ORD are also given for the R+ α-pinene.*

xenon lamp light source for both *R*-(+) and *S*-(-) enantiomers of α-pinene, for neat α-pinene and different concentrations in cyclohexane. There is no detectable dependence on concentration at the levels studied. This is not always the case. (e.g., students have performed ORD experiments with sucrose and observed a significant dependence on concentration, a long-known fact.) Two high school students working in the UT laboratory compared the ORD for sucrose in solution with that for sucrose crystals grown from evaporation of supersaturated sugar solutions. Crystals approaching 0.5 cm could be prepared. The optical rotation "per molecule" can be calculated from the ORD measurements and compared. Another student measured the ORD of α-pinene as a function of sample temperature and found very little dependence upon temperature in accordance with the fact that α-pinene has only one conformer. In a previous study, a slight temperature dependence of the ORD was detected.[18] The small change detected over a large temperature range was most likely due to the contribution of vibrationally excited α-pinene molecules at higher temperature.

Some molecules have multiple conformers for a given enantiomer and these can have optical rotations with opposite sign. For, example the equatorial and axial forms of the 3-methylcyclopentanone molecule possess almost equal circular dichroism magnitudes but of opposite sign![19] The equatorial form is lower in energy by only by ~4 kJ/mol, so both equatorial and axial conformers are present at room temperature and above. An interesting experiment for students is to measure the specific rotation of *R*-3-methylcyclopentanone at 337.1 nm as a function of temperature. The observed change in specific rotation can be used to make a van't Hoff plot to determine the enthalpy difference between the equatorial and axial forms of 3-methylcyclopentanone. The specific rotation for 3-methylcyclopentanone at 337.1 nm is very large (~1000 deg) so only a short path length is necessary. (Question: Given the information above, will the optical rotation increase or decrease with increasing temperature?)

The Gaussian computational program can be employed to calculate the ECD, VCD, and ORD for many chiral molecules. In Chapter 15 we describe the directions for performing such calculations. The calculations can also be performed for molecules "embedded" in a solvent. The student is encouraged to perform these calculations for various basis sets and compare with the measured values. Figure 20.5 shows the excellent agreement between experiment and theory for this molecule. We also note that in this case measurements and calculations allow one to determine the absolute configuration (*R* or *S*) for α-pinene!

Over many forests, large amounts of monoterpenes such as (+) and (−) α-pinene are emitted into the atmosphere by conifers and other plants. They are emitted partly to attract pollinators and to repel herbivores. These terpenes are believed to be responsible for the "blue haze" seen over many forests such as those in the Smoky Mountains, from which they derive their name. Many groups have studied the effects of these molecules and their photoreaction products on air pollution, etc. Interestingly, some of these studies have shown that for the dominant α-pinene, the (–)-form was measured in large excess over the (+)-form over tropical rainforests, whereas the reverse was observed over boreal forests [20] (see also [21]). Much effort has been devoted to detecting pinene in the forest environment (see [20,21] and others cited therein). Perhaps ORD measurements could be of some importance in this effort. An interested student may wish to employ a green laser pointer to attempt to measure the presence of (+) or (–) α-pinene using a long optical path over a forest canopy (e.g., between mountains) at night. Performing the experiment at night is believed to be necessary because α-pinene is destroyed photochemically by atmospheric reactions with ozone, OH, or NO_3 during the day [22] (as your nose can readily verify). Also, it is easier to search for the green laser pointer at night. A green laser is better than a He-Ne laser because the optical rotation is larger at the shorter wavelength. One might also use a sodium street lamp as the light source and two sheets of polarizing plastic, one to define the polarization of the light and the second as the polarization rotation detector. A plumb bob can be employed to define the direction of polarization at the source and a second plumb bob at a distance with a rotatable polarizer to determine the rotation.

REFERENCES

1. R. N. Compton and R. M Pagni, "The chirality of biomolecules," *Adv. Atom., Mol., Opt. Phys.* **48**, 219 (2002).
2. R. de L. Kronig, "On the theory of the dispersion of X-rays," *J. Opt. Soc. Am.* **12**, 547 (1926).
3. A. Fresnel, "Extrait d'un mémoire sur la double refraction particuliere que presente le cristal de roche dans la direction de son axe," *Ann. Chim. Phys.* **28**, 147 (1825).
4. P. Drude, *Lehrbuch der Optik*, Hirzel, Leipzig, 1900, p 379. English translation, Dover, New York, 1959.
5. C. R. Noller, "Apparatus for lecture demonstration of optical activity," *J. Chem. Educ.* **26**, 269 (1949).
6. D. J. Kolb, "Some ideas from the past," *J. Chem. Educ.* **64**, 805 (1987).
7. H. C. Dorn, H. Bell, and T. Birkett, "A simple polarimeter and experiments utilizing an overhead projector," *J. Chem. Educ.* **61**, 1106 (1984).
8. G. Freier and B. G. Eaton, "Optical activity demonstration," *Am. J. Phys.* **43**, 939 (1975).
9. G. F. Hambly, "Optical activity: An improved demonstration," *J. Chem. Educ.* **65**, 623 (1988).
10. J. W. Hill, "An overhead projection demonstration of optical activity," *J. Chem. Educ.* **50**, 574 (1973).
11. J. E. Fernandez, "A simple demonstration of optical activity," *J. Chem. Educ.* **53**, 508 (1976).

12. J. B. Kinney and J. F. Skinner, "A device for easy demonstration of optical activity and optical rotatory dispersion," *J. Chem. Educ.* **54**, 494 (1977).

13. W. K. Dean, addendum to "Simple demonstration of optical activity," *J. Chem. Educ.* **54**, 494 (1977).

14. R. J. Shavitz, "Easily constructed student polarizer," *J. Chem. Educ.* **55**, 682 (1978).

15. B. Knauer, "A demonstration of the optical activity of a pair of enantiomers," *J. Chem. Educ.* **66**, 1033 (1989).

16. P. Crabbe, *ORD and CD in Chemistry and Biochemistry: An Introduction*, Academic Press, New York, 1972.

17. S. M. Mahurin, R. N. Compton, and R. N. Zare, "Demonstration of optical rotatory dispersion of sucrose," *J. Chem. Educ.* **76**, 1234 (1999).

18. K. Wiberg, Y.-Q. Wang, M. J. Murphy, and P. H. Vaccaro, "Temperature dependence of optical rotation: α-pinene, β-pinene pinane, camphene, camphor and fenchone," *J. Phys. Chem. A* **108**, 5559 (2004).

19. W. Al-Basheer, R. M. Pagni, and R. N. Compton, "Spectroscopic and theoretical investigation of (R)-3-methylcyclopentanone. The effect of solvent and temperature on the distribution of conformers," *J. Phys. Chem. A* **111**, 2293 (2007).

20. J. Williams, N. Yassaa, S. Bartenbach, and J. Lelieveld, "Mirror image hydrocarbons from tropical and boreal forests," *Atmos. Chem. Phys. Discuss.* **6**, 9583 (2006).

21. H. W. Wilson, "Infrared absorption bands of α- and β-pinenes in the 8–14 μm atmospheric window region," *Appl. Opt.* **18**, 3434 (1979).

22. B. J. Finlayson-Pitts and J. N. Pitts, *Chemistry of the Upper and Lower Atmosphere*, Academic Press, San Diego, CA, 2000.

21

Faraday Rotation

Introduction

In 1845 Michael Faraday (1791–1867) made the first experimental observation of a relationship between light and magnetism.[1] This discovery followed earlier studies by Oersted (1819), which showed that an electrical current can produce a magnetic field. Faraday demonstrated that linearly polarized light, upon passing through a transparent dielectric medium, will be rotated by a magnetic field which is collinear with the propagation of the light. This is now referred to as the Faraday Effect or Magneto-Optic Effect. Later, in 1853, Émile Verdet (1824–66) made the observation that "the ratio between the strength of the magnet and the amount of rotation is constant," i.e. the rotation, θ, is linearly proportional to the magnetic field, B, and path length, L. This constant of proportionality is known as the Verdet constant, V:

$$\theta = V\,L\,B \tag{21.1}$$

At the suggestion of James Clerk Maxwell, J. E. H. Gordon made the first determination of the Verdet constant in absolute units in 1876.[2] An interesting history of the Faraday Effect, together with the notebook account of the actual experiment, is presented in an article by E. Scott Barr.[3]

Among his many scientific contributions, Faraday also discovered paramagnetism.[4] For example, he noted that bubbles containing oxygen were attracted by a magnetic field (i.e., the oxygen molecule was paramagnetic [4]). We will see later that if the transparent material is diamagnetic the Verdet constant is positive, and the light vector will rotate clockwise as viewed toward the light source when the direction of propagation is in the same direction as that of the magnetic field. Likewise, for a paramagnetic material [4] the Verdet constant is negative and light will rotate its polarization counterclockwise when the direction of propagation is parallel to and in the same direction as the magnetic field. An illustration of the counterclockwise rotation (to the observer) for a positive Verdet constant is shown in Figure 21.1.

In Chapter 20 we described the rotation of polarized light by chiral molecules. This is called optical activity or optical rotation. Although both optical rotatory dispersion (ORD) and the Faraday Effect are characterized by the rotation of the polarization of light, the two phenomena are very different. In the case of natural optical activity, the rotation is independent of the direction of propagation of light for a given enantiomer, whereas the Faraday Effect depends on the direction of

Laser Experiments for Chemistry and Physics. First Edition. Robert N. Compton and Michael A. Duncan.
© Robert N. Compton and Michael A. Duncan 2016. Published in 2016 by Oxford University Press.

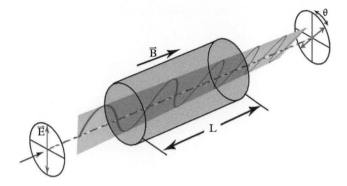

Figure 21.1 *Illustration of the rotation of the plane of polarization of the E-vector for linearly polarized light passing through a diamagnetic material. If a wire coil is used to produce the magnetic field, the plane of polarization is rotated in the same sense as the electrical current.*

propagation relative to the magnetic field direction. This means that reflection of light back through a chiral material rotates the electric vector back to its original polarization direction (i.e., no change in rotation). However, in the case of the Faraday Effect the rotation will double if light is passed through a material and reflected back upon itself. In principle, as shown in this experiment, the angle of rotation can be amplified n times upon observing n passes of polarized light through a material in a magnetic field.

Before describing the experiment performed in the modern optics and modern physics laboratories at the University of Tennessee, we introduce the underlying physics behind the Faraday Effect. The Faraday Effect has been treated in many optical physics textbooks.[5–8] In addition, many Faraday rotation experiments have been described in the literature.[9–13] In 1930 Rosenfeld first presented a detailed theory of the Faraday Effect.[14]

As an introduction we briefly refer to a classical description of the Faraday Effect which is related to the normal Zeeman Effect (see the treatment in the book by Guenther [6]). The Zeeman Effect refers to the splitting of spectral lines when an atom is subject to a magnetic field. For example, if a magnetic field is applied to a one-electron atom which has a natural absorption frequency of ω_0, the energy of the atom will be split into three levels: an unshifted level at ω_0 as well as levels shifted up, ω_+, or down, ω_-, in frequency relative to ω_0. Quantum mechanics is required to properly describe the absorption of right- or left-circularly polarized light for the ω_+ or ω_- levels, respectively. The shifted levels are proportional to the applied field given by the Larmor precession frequency

$$\omega_L = \left(\frac{e}{2m}\right) B \qquad (21.2)$$

where e and m are the charge and mass of the electron, respectively. The signs, + or −, refer to the fact that ω_+ and ω_- are excited by right- and left-circularly polarized light, respectively.

Writing down the equations of motion for the system and making the approximation that the Larmor frequency, ω_L, is much less than the natural frequency, ω_0, one can derive an expression for the index of refraction for right- and left-circularly polarized light of frequency ω to be

$$n_{\pm} = 1 + \frac{\omega_p^2}{((\omega_0 \pm \omega_L)^2 - \omega^2)} \tag{21.3}$$

ω_p represents the plasma frequency, $\omega_p^2 = Nq^2/\varepsilon_0 m$, where m and q represent the mass and charge of the electron, respectively and N is the charge density.

Inspection of equation 21.3 shows that the speed of light is different for left- and right-circularly polarized light, which nicely illustrates the physical origin of the Faraday rotation effect. The angle of rotation, θ, can be calculated from the general equation

$$\theta = (\omega L/c)(n_- - n_+) \tag{21.4}$$

where c is the speed of light. Since $\omega_L \ll \omega_0$ and $(n_- + n_+) \approx 2n_0$ we can write ω

$$(n_- - n_+) = \frac{(n_-^2 - n_+^2)}{2n_0} \tag{21.5}$$

Following the collection of terms (see [5]) we have

$$\theta = \left(\frac{L}{cn_0}\right)\left[\frac{2\omega\omega_0\omega_L\omega_p^2}{(\omega_0^2 + \omega_L^2 - \omega^2)^2 - 4\omega_0^2\omega_L^2}\right] \tag{21.6}$$

Note that the magnetic field dependence arises through the Larmor frequency in equation 21.2. An approximate plot of θ versus ω illustrating the dispersion nature of the Faraday Effect for an alkali atom is shown in Figure 14.10b in the book by Gunther.[6] Inspection of equation 21.6 shows that θ becomes negative when $\omega = \omega_0$ and becomes positive for ω values above and below ω_0, passing through zero at ω_+ and ω_-.

Van Baak [15] has described an excellent combined experimental and theoretical analysis of Faraday rotation in an atomic vapor of rubidium. In this experiment a diode laser is tuned in the vicinity of the D_2 line at 780 nm in a room temperature vapor of Rb, and the optical rotation is observed as a probe of atomic dispersion. Van Baak's theoretical treatment also includes the Doppler motion of the atoms in fitting the absorption and dispersion interaction of the light. An analysis similar to that above gives the Faraday rotation as

$$\theta(v) = \left(\frac{N\lambda^2 L}{32\pi^2\tau}\right)\left[\frac{(v_{0+} - v)}{\left((v_{0+} - v)^2 + \left(\frac{\Delta v}{2}\right)^2\right)} - \frac{(v_{0-} - v)}{\left((v_{0+} - v)^2 + \left(\frac{\Delta v}{2}\right)^2\right)}\right] \tag{21.7}$$

Where $2\pi v = \omega$, τ is the spontaneous lifetime of the upper state, and Δv is the corresponding "natural linewidth," i.e., $\Delta v = 1/(2\pi\tau)$. Using the equation for the index of refraction,

$$n(v) = 1 + \left(\frac{N\lambda^3}{32\pi^3\tau}\right)\left[\frac{(v_0 - v)}{\left((v - v_0)^2 + \left(\frac{\Delta v}{2}\right)^2\right)}\right] \qquad (21.8)$$

and combining equations 21.7 and 21.8 under the assumption of small magnetic field strengths, Van Baak [15] arrives at a familiar expression for the Faraday rotation

$$\theta(v) = -\left(\frac{e}{2mc}\right)v\frac{dn}{dv} = V L B \qquad (21.9)$$

In the literature, $\theta(v) = -(e/2mc)v\, dn/dv$ is referred to as the Becquerel equation and is sometimes alternatively written as $\theta(\lambda) = (e/2mc)\lambda\, (dn/d\lambda)$.

Van Baak points out that the frequency dependence of the rotation is equal to the dimensionless quantity $v(dn/dv)$ times $e/2mc$, which is equal to 293.34 rad/Tm. Note also that for normal dispersion dn/dv is negative and θ is positive. The change in the index of refraction, n, around a resonance can be large giving rise to a measurable rotation for an atomic gas.

The experiment described by Van Baak [15] represents an excellent example that can be performed in an undergraduate physics laboratory due to the wide availability of diode lasers and the ease of making glass cells containing rubidium, as well as commercial assistance with the detection apparatus.[16]

The Magneto-Optic Effect (Faraday Effect) for molecules was treated rigorously by Robert Serber in 1932.[17] This article also discussed the early experiments on Faraday rotation by R. W. Wood involving the $X\,^1\Sigma_g^+$ to $B\,^3\Pi_{0u}^+$ transition in molecular iodine [18] (see Chapter 17) as well as molecular transitions in sodium dimer.[19] Serber considered the inner electrons of a molecule as perturbations to the one-electron atomic problem discussed above and introduced a factor γ to account for this in equation 21.9 through

$$\theta(v) = -\gamma\left(\frac{e}{2mc}\right)v\frac{dn}{dv} = \gamma\left(\frac{e}{2mc}\right)\lambda\frac{dn}{d\lambda} \qquad (21.10)$$

Serber points out that experiments for atoms tend to have "normal" Verdet constants ($\gamma = 1$) whereas γ for H_2, N_2, CO_2, and N_2O are 1.00, 0.63, 0.56, 0.34, respectively. He further pointed out that the Verdet constant for oxygen at high pressure was anomalous and attributed this to the presence of O_2 dimers. As shown below, the Verdet constant for liquids also deviates from equation 21.10. Today the factor γ is referred to as the magneto-optical anomaly.

Measurement of the Verdet constant

The Verdet constant which quantifies the Faraday Effect can be determined by observing the rotation in a strong constant magnetic field or by using an alternating magnetic field and using lock-in detection. The use of lock-in detection allows for accurate measurement of small Faraday rotations with modest magnetic field strengths. Turvey [13] first used both AC and DC magnetic fields to determine V. Jain et al. [9] presented a particularly lucid treatment of the measurement of a Verdet constant as a function of wavelength using many lines from different fixed frequency lasers. We follow this treatment and first cast the general theoretical treatment of the Faraday Effect in a manner which relates well to the measurement of the Verdet constant using the AC detection method.

Consider linearly polarized light propagating in the z-direction in which the electric vector is in the x-direction. The electric vector can be expressed in the Jones matrix form as

$$E = \begin{pmatrix} 1 \\ 0 \end{pmatrix} E_0 \exp(-i\omega t + ikz) \tag{21.11}$$

The magnetic field is also directed along the z-direction. Upon passing through the material in the presence of a magnetic field, the E-vector rotates through an angle θ and the resulting E becomes

$$E = \begin{pmatrix} \cos\theta \\ \sin\theta \end{pmatrix} E_0 \exp(-i\omega t + ikz) \tag{21.12}$$

We now place a polarization analyzer in line with the beam at an angle ϕ and the E-vector of the light passing through the analyzer becomes

$$E = \begin{pmatrix} \cos(\phi - \theta)\cos\theta \\ \cos(\phi - \theta)\sin\phi \end{pmatrix} E_0 \exp(-i\omega t + ikz) \tag{21.13}$$

Since the light intensity, I, is proportional to E^2 we have

$$I = E_0^2 \cos(\phi - \theta)^2 \tag{21.14}$$

Light is detected after the analyzer and we set the angle ϕ in order to obtain the largest difference in signal, $\Delta I/I_0$. Therefore we take the derivative of I with respect to θ and set it equal to zero, i.e.,

$$\frac{\partial I}{\partial \theta} = \sin 2(\phi - \theta)E_0^2 = 0 \tag{21.15}$$

Thus $(\phi - \theta) = 45°$ and since θ is very small $\phi \approx 45°$. Therefore using $\cos(\phi - \theta) = \cos(\phi)\cos(\theta) + \sin(\phi)\sin(\theta) \approx 1/\sqrt{2} + \theta/\sqrt{2}$. Plugging this result into equation 20.14 and dropping the θ^2 term results in

$$I = \frac{1}{2}(1 + 2\theta)E_0^2 \tag{21.16}$$

If the magnetic field is modulated with a sinusoidal current with a frequency Ω then the rotation angle also follows a form $\theta = \theta_0 \sin(\Omega t)$. Equation (20.16) can be rewritten as

$$I = (E_0^2/2)[1 + 2\theta_0 \sin(\Omega t)] = I_0 + \Delta I \sin(\Omega t) \qquad (21.17)$$

Noting equation 21.14, and using the small angle equation at $\phi = 45°$, one can show that $E_0^2 = 2I_0$. Inserting $2I_0$ for E_0^2 in Equation 21.16 results in $\theta_0 = \Delta I/2I_0$. Thus the amplitude of the angle of rotation at 45° can be determined by measuring $\Delta I/2I_0$. After determination of θ_0 (rad), B_0 (Tesla) and the path length L (meter) the Verdet constant can be found from

$$V = \theta_0/B_0L \quad \text{rad/Tm} \qquad (21.18)$$

The interested student should refer to the many undergraduate experiments on measurements of the Faraday rotation and the determination of the Verdet constant using static or oscillating magnetic fields discussed above. The experiment performed at the University of Tennessee using the AC method is described briefly below.

A polarized diode laser ($\lambda = 650$ nm) is passed through a liquid or glass sample that is situated in a magnetic field provided by two Helmholz coils. The alternating magnetic field is produced using an AC current, I, from a laboratory power supply. The magnetic field was calculated from the Helmholz equation

$$B = (4/5)^{3/2}(\mu_0 NI)/R \qquad (21.19)$$

where N (200 turn) and R (0.105 m) designate the number of turns and radius of each coil and μ_0 is the permeability of free space ($4\pi \times 10^{-7}$ Tm/A). A polarizer is placed at 45° (see equation 21.15) with respect to the polarization of the incident laser and a photodiode is used to detect the light passing through the polarizer. The AC signal detected by the photodiode is processed by a lock-in amplifier which displays the RMS voltage amplitude of the AC signal. The DC output light level (I_0) from the lock-in amplifier was recorded on an oscilloscope. The entire experimental setup is shown in Figure 21.2. The data were acquired at various magnitudes of the current through the Helmholtz coil.

Figure 21.3 shows the measured θ/L for a 5 cm path length sample of CS_2 as a function of the magnetic field. The rotation corresponding to the thin glass cuvette holding the CS_2 is very small and has been subtracted from the rotation of the combined CS_2 plus glass.

The experimental arrangement for measuring the Faraday rotation corresponding to three passes is shown in Figure 21.4, while Faraday rotation measurements for the three passes are shown in Figure 21.5.

The measured Verdet constants for one and three passes through CS_2 are shown in Figure 21.6 in comparison to the published results of Villaverde and

Figure 21.2 *General experimental arrangement used to determine the Faraday rotation from which one can determine the Verdet constant. The AC signal from the alternating magnetic field is processed by a lock-in amplifier and displayed on an oscilloscope. The DC signal from the detector (I_0) is also read from the oscilloscope. The sample shown is a piece of glass.*

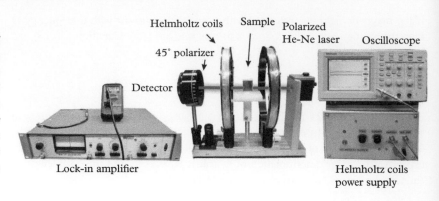

Figure 21.3 *A plot of the experimental data obtained from the single pass measurement of the Verdet constant. The slope of the plot is the measured Verdet constant for CS_2.*

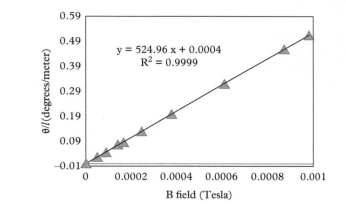

Figure 21.4 *Experimental arrangement for measurement of the Faraday rotation using three passes through the CS_2 sample. The laser used in this experiment is a diode laser (650 nm). It is possible to make many passes through the sample with proper adjustment of the light angles between the two mirrors.*

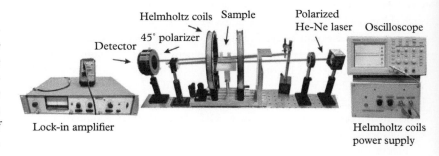

Donatti et al. [20] over a range of wavelengths. These values also correlate well with earlier measurements at 632.8 nm.[21,22]

The Faraday rotation results for liquid carbon disulfide can be compared to the classical Becquerel equation above (see equations 21.9 and 21.10):

$$\theta(\lambda) = -\gamma \left(\frac{e}{2mc} \right) \lambda \frac{dn}{d\lambda} \tag{21.20}$$

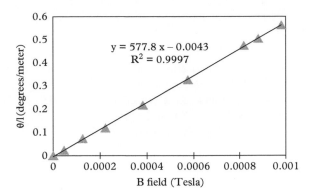

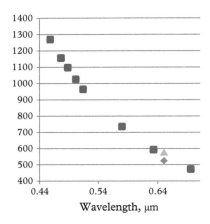

Figure 21.5 *A plot of the experimental data obtained from a three-pass measurement of the Verdet constant. The slope of the plot is the measured Verdet constant for CS_2.*

To determine $dn/d\lambda$ we use the Cauchy (Louis Cauchy, 1789–1857) equation relating the index of refraction with wavelength.[4]

$$n = C_1 + C_2/\lambda^2 + C_3/\lambda^4 - C_4/\lambda^6 + C_5/\lambda^8 \ldots \quad (21.21)$$

Taking only the first two terms we can write $dn/d\lambda = -2C_2/\lambda^3$, and equation 21.20 becomes

$$\theta(\lambda) = \gamma \left(\frac{e}{mc}\right) C_2/\lambda^2 \quad (21.22)$$

Samoc [23] has analyzed a vast amount of data on the index of refraction for CS_2 and arrives at a best fit to the Cauchy relation as

$$n = 1.580826 + 15238.9/\lambda^2 + 4.8578 \times 10^8/\lambda^4$$
$$-8.2863 \times 10^{13}/\lambda^6 + 1.4619 \times 10^{19}/\lambda^8 \quad (21.23)$$

where λ is in nm. One can easily verify that terms beyond $1/\lambda^2$ can be neglected for our purposes. Keeping only the first two terms we find that $dn/d\lambda = -2C_2/\lambda^3 = -1.11 \times 10^{-4}$/nm at 650 nm. Also, all of the constants can be consolidated into a simple expression for the Verdet constant. From Van Baak [15] and others we can write $(e/2mc) = 293.34$ rad/Tm $= 16,807$ deg/Tm, and therefore we can calculate the Faraday rotation at 650 nm.

$$\theta(650 \text{ nm}) = \gamma(e/2mc)\lambda 2C_2/\lambda^3 = \gamma(16,807)(650)1.11 \times 10^{-4} = 1213\gamma \quad (21.24)$$

Since the measured value for the Faraday rotation at 650 nm is 550 deg/Tm, we arrive at a value for the magneto-optical anomaly of $\gamma = 0.45$. Three other values at 632 nm give $\gamma = 0.399$ (see [20]).

Finally we comment that optical rotation in alkali atoms using the metal vapor heat pipes described in Chapter 16 could be easily studied using a dye laser tuned close to an atomic resonance.

Figure 21.6 *Verdet constant measurements (ordinate) in units of deg/Tm for one (triangle) and three (diamond) passes through CS_2 in comparison to the data of Villaverde and Donatti et al.[20] Notice that the ordinate is displaced by 400 deg/Tm.*

REFERENCES

1. M. Faraday, "On the magnetization of light and the illumination of magnetic lines of force," *Phil. Trans. R. Soc. London* **136**, 1 (1846).

2. J. E. H. Gordon, "Determination of Verdet's constant in absolute units," *Proc. Roy. Soc. London* **25**, 144 (1876).

3. E. S. Barr, "Men and milestones in optics. V: Michael Faraday," *Appl. Opt.* **6**, 632 (1967).

4. Diamagnetism refers to the property of any material to create a magnetic field in opposition to an externally applied magnetic field. Some materials also exhibit paramagnetism (attraction) and in some cases the paramagnetism can be much greater than its inherent diamagnetism.

5. E. Hecht, *Optics*, fourth edition, Addison-Wesley, San Francisco, 2002, pages 366–8.

6. R. Guenther, *Modern Optics*, John Wiley & Sons, New York, 1990, pages 590–6.

7. F. A. Jenkins and H. E. White, *Fundamentals of Optics*, third edition, McGraw-Hill, New York, 1957, pp. 596–8.

8. G. R. Fowles, *Introduction to Modern Optics*, Dover Publications, New York, 1968.

9. A. Jain, J. Kumar, F. Zhou, L. Li, and S. Tripathy, "A simple experiment for determining Verdet constants using alternating current magnetic fields," *Am. J. Phys.* **67**, 714 (1999).

10. S. Steingiser, G. Rosenblit, R. Custer, and C. E. Waring, "A precision Faraday effect apparatus," *Rev. Sci. Instrum.* **21**, 109 (1950).

11. F. J. Loeffler, "A Faraday rotation experiment for the undergraduate physics laboratory," *Am. J. Phys.* **51**, 661 (1983).

12. F. L. Pedrotti and P. Bandettini, "Faraday rotation in the undergraduate advanced laboratory," *Am. J. Phys.* **58**, 542 (1990).

13. K. Turvey, "Determination of Verdet constant from combined AC and DC measurements," *Rev. Sci. Instrum.* **64**, 1561 (1993).

14. L. Rosenfeld, "Zur theorie des Faradayeffekts," *Zeits. f. Phys.* **57**, 835 (1929).

15. D. A. Van Baak, "Resonant Faraday rotation as a probe of atomic dispersion." *Am. J. Phys.* **64**, 724 (1996).

16. http://www.teachspin.com/instruments/faraday/index.shtml.

17. R. Serber, "The theory of the Faraday effect in molecules," *Phys. Rev.* **41**, 489 (1932).

18. R. W. Wood and G. Ribaud, CVII, "The magneto-optics of iodine vapour," *Phil. Mag.* **27**, 1009 (1914).

19. F. W. Loomis, "Vibrational levels and heat of dissociation of Na_2," *Phys. Rev.* **31**, 323 (1928).

20. A. B. Villaverde and D. A. Donatti, "Verdet constant of liquids; measurements with a pulsed magnetic field," *J. Chem. Phys.* **71**, 4021 (1979).

21. N. George, R. W. Waniek, and S. W. Lee, "Faraday effect at optical frequencies in strong magnetic fields," *Appl. Opt.* **4**, 253 (1965).

22. K. Dismukes, S. H. Lott, Jr., and J. P. Barach, "Faraday effect measurements with pulsed magnetic fields," *Appl. Opt.* **5**, 1246 (1966).

23. A. Samoc, "Dispersion of refractive properties of solvents: Chloroform, toluene, benzene, and carbon disulfide in ultraviolet, visible, and near-infrared," *J. Am. Appl. Phys.* **94**, 6167 (2003).

22 Fermi Resonance in CO$_2$

Introduction

As discussed elsewhere in this book, infrared and Raman spectroscopy are critical tools throughout chemical analysis and they provide fundamental information about the structures of molecules. The vibrations usually seen in these measurements are *fundamentals*, in which one quanta of excitation is active in a normal mode of vibration. However, in addition to the allowed fundamental transitions, *overtone* and *combination* bands are also sometimes observed. Overtones represent vibrations which are approximately a multiple of a fundamental frequency (e.g., $n\nu_i$, where $n = 2$ is the first overtone, $n = 3$ second overtone, etc.). Combination bands are those excitations which involve mixed fundamental vibrations, in which two or more modes are active (e.g., $\nu_i + \nu_j$). The energies of overtone or combination transitions are not exactly equal to multiples or sums of the fundamental frequencies because of anharmonicities in the potential functions. As discussed in Chapter 14, diatomic molecules with only one vibrational degree of freedom have potentials best described as anharmonic (Morse) oscillators. Polyatomic molecules also have such anharmonicities in each individual vibrational coordinate, but in addition to this they have so-called "off-diagonal" anharmonicities that shift vibrations when more than one coordinate is involved. Because of these effects, the frequencies of overtone bands are usually slightly smaller than $n\nu_i$ and the frequencies of combination bands are usually slightly lower than $\nu_i + \nu_j$.

In addition to the observation of overtones and combination bands, *perturbations* to the expected vibrational spectra may occur. One of the most interesting perturbations in vibrational spectroscopy is the *Fermi resonance*. In this effect, two or more vibrations in a molecule interact in such a way as to shift the vibrational resonances away from their expected frequencies. The Raman spectrum of the carbon dioxide molecule provides the classic demonstration of this effect. From the discussion in Chapter 14, recall that the symmetric and linear CO$_2$ molecule is predicted to exhibit one Raman active (ν_1) and three IR active (ν_3 plus a degenerate ν_2 vibration) bands. A calculation of these bands using the B3LYP level of DFT and the 6–311++G(d,p) basis set gives the result shown in Figure 22.1. The scaling factor recommended here is 0.964, and so the frequencies calculated are multiplied by this factor before comparing them to the experimental values. As discussed in Chapter 15, such scaling can account for the anharmonic shift to lower frequencies in an individual mode, but it does not allow the

Laser Experiments for Chemistry and Physics. First Edition. Robert N. Compton and Michael A. Duncan.
© Robert N. Compton and Michael A. Duncan 2016. Published in 2016 by Oxford University Press.

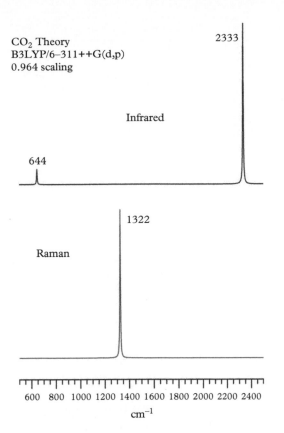

Figure 22.1 *Calculations of the infra-red (top) and Raman (bottom) bands of CO_2 predicted by DFT at the B3LYP level of theory. Notice the three predicted lines and the mutual exclusion between the IR and Raman excitations. Using the standard vibrational scaling factor of 0.964 to account for anharmonicity,[1] B3LYP calculations give the symmetric stretch ν_1 band at 1322 cm⁻¹, the asymmetric stretch ν_3 band at 2333 cm⁻¹, and the degenerate out-of-plane bending ν_2 vibration at 644 cm⁻¹.*

calculations to reproduce other anharmonic effects such as overtones, combination bands, or Fermi resonances.

The vibrational bands in Figure 22.1 are therefore those *expected* in the Raman or IR spectrum, assuming that there are no overtones, combinations, or perturbations. However, the molecule is not limited by attempts of theory to describe its motion! Although only one Raman active frequency is predicted, early measurements of the gas-phase Raman spectrum of CO_2 exhibited two strong peaks, one at 1355 cm⁻¹ and another at 1285 cm⁻¹.[2] The Raman spectrum of dry ice shows very similar features. Figure 22.2 shows a Raman spectrum of dry ice recorded under liquid nitrogen, i.e. a RUN spectrum of solid CO_2, using the experimental arrangement described in Chapter 19.

The early Raman spectroscopy observations of CO_2 raised the question: Why are there two strong Raman transitions where only one is expected? This question was answered by Enrico Fermi in 1931.[3] Fermi noted that the frequency of the first overtone of the doubly-degenerate bending vibration (ν_2) was approximately equal to that of the predicted symmetric stretch (ν_1) vibration. By symmetry,

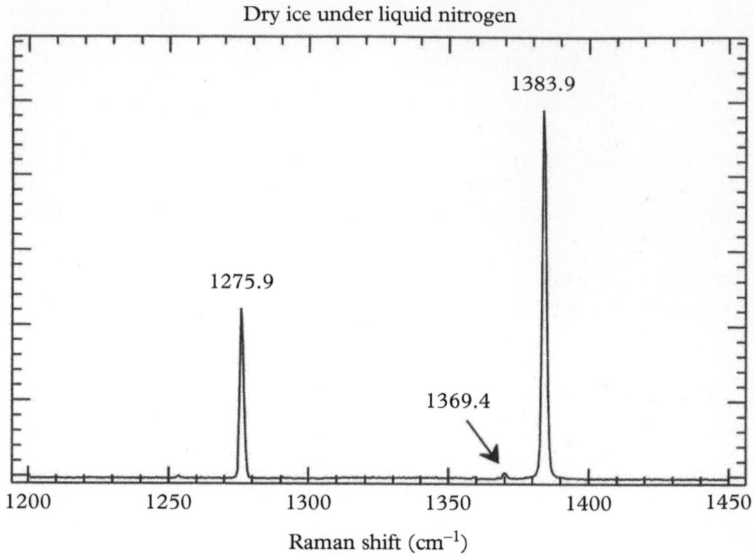

Dry ice under liquid nitrogen

1383.9

1275.9

1369.4

1200   1250   1300   1350   1400   1450

Raman shift (cm^{-1})

Figure 22.2 *Raman spectrum of dry ice under liquid nitrogen (77 K). The peak at 1369.4 cm^{-1} is attributed to the $^{13}C^{16}O_2$ isotopic species. Previous studies of the Raman spectrum of solid CO_2 at dry ice temperature (88 K) gave peaks at 1278 cm^{-1} and 1385 cm^{-1}.[2]*

excitation of the v_2 vibration is a forbidden Raman transition but the first overtone $2v_2$ vibration is Raman allowed. Fermi noted that both v_1 and $2v_2$ have the same symmetry, Σ_g^+, and postulated that when two levels are sufficiently close together in energy and are of the same symmetry they will "repel" each other, one going up and the other going down in energy, and the weaker transition would "borrow" intensity from the one-photon allowed transition. This is illustrated in Figure 22.3.

We repeat the Fermi resonance condition stated above in a more formal manner: The first excited state of v_2 is designated as 01^10 (i.e., no vibrational quanta in v_1 or v_3 and one quanta in v_2). The v_2 bending mode has ±1 unit of angular momentum about the molecular axis and thus the v_2 bending vibration of CO_2 has symmetry Π_u. Placing two quanta in the v_2 mode leads to states with symmetry $(\Pi_u \times \Pi_u)^+ = \Sigma_g^+ + \Delta_g$ or, 02^00 and 02^20, respectively. We calculated v_1 to be 1322 cm^{-1} whereas v_2 is calculated to be 644 cm^{-1} (see Figure 22.1).

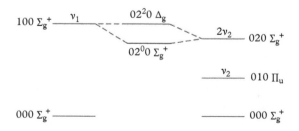

$100\ \Sigma_g^+$ ——— v_1 $02^20\ \Delta_g$

$2v_2$ ——— $020\ \Sigma_g^+$

$02^00\ \Sigma_g^+$

v_2 ——— $010\ \Pi_u$

Figure 22.3 *Illustration of the energy levels of the v_1, v_2 and first overtone $2v_2$ of CO_2 giving rise to the Fermi resonance states $02^20\ \Delta_g$ and $02^00\ \Sigma_g^+$.*

$000\ \Sigma_g^+$ ———   ——— $000\ \Sigma_g^+$

Thus, $2\nu_2$ and ν_1 are predicted to be within 78 cm^{-1} of each other. However, the two Fermi resonance states are actually measured to be separated by ~108 cm^{-1}.

The Fermi resonance can be understood by considering first-order perturbation theory, in which the Hamiltonian giving the energy levels of CO_2 can be written as an unperturbed zeroth order Hamiltonian, H_0, plus the perturbation term, H′,

$$H = H_0 + H' \qquad (22.1)$$

where the energies of the unperturbed states Ψ_{100} and Ψ_{020} are determined from

$$H_0\Psi_{100} = E_{100}\Psi_{100} \text{ and } H_0\Psi_{020} = E_{020}\Psi_{020} \qquad (22.2)$$

Also E_{000} is set identical to zero and thus $E_{100} = h\nu_1$ and $E_{020} = 2h\nu_2$ (neglecting the anharmonicity). We define the Fermi mixed wavefunctions as

$$\Psi_+ = a\Psi_{100} + b\Psi_{020} \qquad (22.3)$$

where a^2 and b^2 represent the contribution of Ψ_{100} and Ψ_{020}, respectively, and $a^2 + b^2 = 1$. The mixed states lead to two energy levels which we label E_+ and E_- and the energies of the split states are determined from

$$(H_0 + H') \Psi_\pm = E_\pm \Psi_\pm \qquad (22.4)$$

The secular equation can be written as

$$\begin{pmatrix} (E_{100} - E) & E' \\ E' & E_{020} - E \end{pmatrix} \begin{pmatrix} a \\ b \end{pmatrix} = 0 \qquad (22.5)$$

where the matrix elements are dictated by the eigenvalue problem

$$M_{ij} = \int \Psi_i H \Psi_j d\tau - E\delta_{ij} \qquad (22.6)$$

Solution of the secular determinant requires solving the simple quadratic equation

$$(E_{100} - E)(E_{020} - E) - E'^2 = 0 \qquad (22.7)$$

This results in two solutions for the energy levels, $E_\pm$

$$E_\pm = \frac{(E_{100} + E_{020}) \pm \sqrt{(E_{100} - E_{020})^2 + 4(E'^2 - E_{100}E_{020})}}{2} \qquad (22.8)$$

From rearrangement of equation 22.8 it is easily seen that

$$E_+ + E_- = E_{100} + E_{020} \quad \text{and} \quad E_+ - E_- = \sqrt{(E_{100} - E_{020})^2 + 4\left(E'^2 - E_{100}E_{020}\right)} \qquad (22.9)$$

Also from equation 22.5 we can write

$$(E_{100} - E)a + E'b = 0 \quad \text{and} \quad E'a + (E_{020} - E)b = 0 \tag{22.10}$$

Using the normalization condition $a^2 + b^2 = 1$, together with the system of equations (equations 22.8, 22.9 and 22.10), one can solve for two sets of coefficients corresponding to each eigenvalue $E_{\pm}$. The square of the coefficients, a^2 and b^2, are interpreted as the relative contribution of the symmetric stretch, Ψ_{100}, and the bending mode, Ψ_{020}, respectively. One should use the data given above and in Figure 22.2 to obtain the constants a and b along with the perturbation energy E'.

As stated above, the Fermi resonance concept was first postulated to explain the two Raman lines in the carbon dioxide spectrum. However, this topic has been the subject of numerous theoretical studies over the years. In 2007, Rodriguez-Garcia et al. presented a first-principles calculation of the energies of the eight lowest-lying anharmonic vibrational states of CO_2 including the Fermi resonance states.[4] They obtained excellent agreement between experiment and theory: (theoretical) and experimental anharmonic frequencies of the Fermi doublet are (1288.9) 1285.4 and (1389.3)1388.2 cm^{-1}.

One notes from Figure 22.2 that the energy difference between the two Raman bands is 108 cm^{-1} and the intensity ratio is ~2:1. Experimentally it has been found that both the frequency and intensity ratio of these bands change upon application of pressure to CO_2 in the condensed phase. This behavior has been used as a spectroscopic *geobarometer* for minerals with CO_2 inclusions. For an interesting discussion of this field, see [5].

The same analysis outlined above for CO_2 can be applied to the carbon disulfide molecule, CS_2. CS_2 is a volatile liquid and it is easy to record Raman spectra for both liquid and under liquid nitrogen (RUN) samples. Like CO_2, CS_2 freezes at 77 K. However, CS_2 is toxic and the student should avoid breathing its vapor or allowing it to contact her/his skin.

Fermi resonances are found in infrared and Raman spectra for many molecules. Their identification comes from examining the regions of overtone bands of allowed transitions in the region of other single-photon allowed transitions. In fact, many organic molecules exhibit such Fermi resonances, as the overtones of their CH_n bending modes fall at about the same frequencies as their C–H stretches. It is partly because of this that the C–H stretching region of organic molecules is not regarded as definitive in determining their structures. Instead, the lower frequency region known as the "fingerprint region," where fewer overtones, combinations, and Fermi resonances are possible, is better for this.

....

REFERENCES

1. J. P. Merrick, D. Moran, and L. Radom, "An evaluation of harmonic vibrational frequency scale factors," *J. Phys. Chem. A* **111**, 11683 (2007).

2. J. E. Cahill and G. E. Leroi, "Raman spectra of solid CO_2, N_2O, N_2 and CO," *J. Chem. Phys.* **51**, 1324 (1969).

3. E. Fermi, "Uber den Ramaneffekt den Kohlendioxyds," *Z. Phys.* **71**, 250 (1931).

4. V. Rodriguez-Garcia, S. Hirata, K. Yagi, K. Hirao, T. Taketsugu, I. Schweigert, and M. Tasumi, "Fermi resonance in CO_2: A combined electronic coupled-cluster and vibrational configuration-interaction prediction," *J. Chem. Phys.* **126**, 124303 (2007).

5. O. Sode, M. Keceli, K. Yagi, and S. Hirata, "Fermi resonance in solid CO_2 under pressure," *J. Chem. Phys.* **138**, 074501 (2013).

23

Photoacoustic Spectroscopy of Methane

Introduction

The photoacoustic effect is the production of sound in a material induced by the absorption of light. Absorption generates heat, followed by sudden expansion of the material and generation of pressure waves emanating outward from the absorption point, i.e., sound waves. The effect was discovered by Alexander Graham Bell,[1,2] in an experiment in which he focused a beam of sunlight into air and turned it on and off rapidly with a mechanical spinning blade chopper. The effect can in principle be observed in liquids or solids, but it is studied primarily in gases. Modern lasers, particularly those which are pulsed, provide a convenient way to concentrate the generation of heat in space and time, which leads to effective generation of sound. This method is employed for various forms of ultrasensitive spectroscopic analysis,[3] but it is also used widely for the routine calibration of lasers. Because the experiment detects a *consequence* of light absorption rather than absorption itself, it is regarded to be a form of so-called "action" spectroscopy. The present experiment demonstrates this effect for the infrared spectrum of methane.

Because of its high symmetry, methane (aka, natural gas) has only four unique vibrations.[4] The symmetric C–H stretch is not IR active, but is detected in Raman experiments at 2917.0 cm^{-1}. A bending mode at 1533.6 cm^{-1} is also Raman active. The infrared-active modes include a bending vibration at 1306.2 and a single asymmetric C–H stretch at 3019.5 cm^{-1}.[4] Because of its high symmetry, methane is classified as a "spherical top" with respect to its rotational structure, and its rotational constants are all equal ($A = B = C$), giving rise to a relatively simple pattern of rovibrational levels. Because the carbon atom lies at the center of mass, and only the light hydrogen atoms contribute to its moment of inertia, the single rotational constant is relatively large ($B = 5.24$ cm^{-1}) producing rotational fine structure on the vibrational bands that can be resolved with spectrometers or lasers having only modest resolution. This spectrum and its rotational structure are well known.[5–7] This methane spectrum is employed widely for diagnostics of this gas in the Earth's atmosphere or in planetary environments such as demonstrated in the recent missions to Mars.[8] Additional details about the infrared spectrum of methane are presented in Chapter 27.

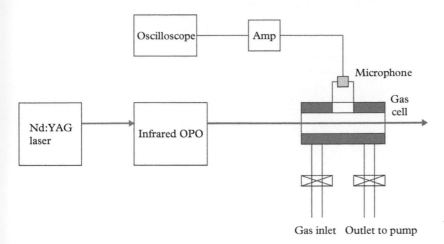

Figure 23.1 *The experimental configuration used for photoacoustic spectroscopy.*

Experimental procedure

Detection of spectroscopy via the photoacoustic effect requires a tunable light source, a sample holder, a microphone to detect the sound, and recording electronics such as an oscilloscope connected to a PC. Figure 23.1 shows the experimental setup employed here. Methane gas, which is available from bench-top outlets in many chemistry labs, is introduced into a cylindrical metal tube fitted with appropriate IR windows (e.g., CaF_2). An inexpensive microphone (e.g., Radio Shack "PC condenser Mic Element," model 270–0090) is mounted in a hole drilled in the side wall of this tube. The pressure of methane can be measured directly with a suitable pressure gauge, or controlled in a crude fashion using the input and exhaust valves connecting gas and vacuum lines to the tube. The pressure is not critical; pressures at or just below atmosphere work well. The tunable laser employed here is the infrared optical parametric oscillator (OPO) made by LaserVision,[9] pumped by a pulsed Nd:YAG laser, which is described in Chapter 4. This laser provides pulsed tunable IR radiation with 1–10 mJ/pulse energy in the 2100–4500 cm^{-1} region, with a linewidth of about 1.0 cm^{-1}. The output of the microphone is amplified and detected with a digital oscilloscope as a function of the wavelength as the laser scans through the spectrum.

Results and discussion

Figure 23.2 shows the infrared spectrum of methane recorded in the 3000 cm^{-1} region using photoacoustic detection. The vibrational band is centered just above 3000 cm^{-1}, and has a central Q-branch with P- and R-branches to either side. Table 23.1 presents the literature values for the line positions in this spectrum.[5]

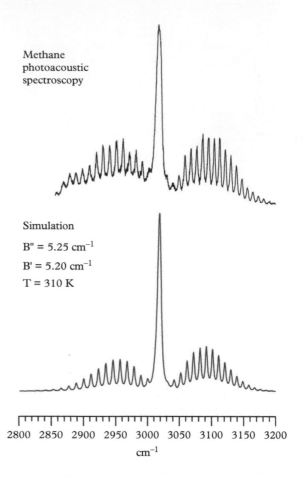

Methane
photoacoustic
spectroscopy

Simulation

B" = 5.25 cm^{-1}

B' = 5.20 cm^{-1}

T = 310 K

2800 2850 2900 2950 3000 3050 3100 3150 3200

cm^{-1}

Figure 23.2 *The infrared photoacoustic spectrum measured for methane in the region of the asymmetric C–H stretch compared to a simulation of the rotational structure in this vibrational band done with the program PGOPHER.*

This spectrum is deceptively simple in appearance, resembling the rovibronic patterns seen for diatomic molecules because of the high symmetry. A full discussion of the spectroscopic analysis is beyond the scope of the present chapter; the interested reader is referred to Chapter 14 and to the presentation by Herzberg [6] or that by Childs [7] for more detail. Because there is only one rotational constant, B, the rotational energy level patterns are given by the simple expression

$$E(J) = BJ(J + 1) \tag{23.1}$$

This is the rigid rotor version of the energy level equation, which should be corrected for terms describing centrifugal distortion and vibration–rotation coupling. At an experimental resolution of about 1 cm^{-1}, the former is insignificant and can be ignored, but the latter has a noticeable effect on the spectrum even at low resolution. Vibration–rotation interaction is usually treated so that the molecule has rotational constants that depend on the vibrational state; here we need $B(v = 0)$ and $B(v = 1)$ values, indicated by convention as B_0 and B_1. Because

Table 23.1 *Line positions in the rovibrational structure of the infra-red spectrum of CH_4.[5] "0" is the Q-branch and the other lines are numbered relative to that.*

Line number	Frequency (cm⁻¹)	Wavelength (μm)
−13	2886.1	3.4649
−12	2896.7	3.4522
−11	2907.0	3.4400
−10	2917.7	3.4274
−9	2927.8	3.4155
−8	2938.1	3.4036
−7	2948.3	3.3918
−6	2958.6	3.3800
−5	2968.8	3.3684
−4	2979.1	3.3567
−3	2989.2	3.3454
−2	2999.7	3.3357
0	3019.6	3.3117
+1	3029.1	3.3013
+2	3039.0	3.2906
+3	3048.6	3.2802
+4	3058.1	3.2700
+5	3067.5	3.2600
+6	3077.0	3.2499
+7	3086.3	3.2401
+8	3095.4	3.2306
+9	3104.7	3.2209
+10	3113.6	3.2117
+11	3122.6	3.2025
+12	3131.6	3.1933
+13	3140.3	3.1844
+14	3149.2	3.1754
+15	3157.8	3.1668

the B_1 value corresponds to a molecule whose bonds are stretched with respect to those in the ground state, the moment of inertia for the $v = 1$ state is usually larger than that for the ground state and the B value is smaller. We thus expect $B_1 < B_0$. The original assignment of this spectrum obtained values of $B_0 = 5.252$ and $B_1 = 5.201$ cm⁻¹.[7] More refined constants have of course been determined in subsequent higher resolution studies of this system.[4] With these constants, the stacks of rotational energy levels can be constructed for the $v = 0$ and $v = 1$ vibrational levels. The rotational selection rule for such a system is $\Delta J = 0, \pm 1$.

This gives rise to the P-, Q-, and R-branch structure. The other main variable is the temperature, which determines the thermal population in rotational states. This, together with an appropriate degeneracy factor [6] for each J of $(2J+1)^2$, determines the relative intensities for lines originating in different J states. It should be noted that methane has four equivalent hydrogens and this nuclear spin statistical consideration may be expected to cause some sort of line intensity alternation, as discussed in Chapter 14 for the H_2 molecule. These effects have been analyzed in detail by E. B. Wilson, one of the pioneers of molecular spectroscopy.[10] It turns out that due to the special symmetry in this system, there is no J-dependent alternation in degeneracy or line intensity in this system. All the levels seen have the same weighting factors.

The rotational structure within a vibrational band such as this can be simulated with appropriate values for the rotational constants, temperature, linewidth, etc., as indicated above. Although several programs are available for this, and many practicing spectroscopists write their own personal versions, one convenient open source program is PGOPHER.[11] PGOPHER calculates the transition energies between the ground and excited J states (J″ and J′, respectively) and intensities for each transition for those systems whose energy levels and transitions follow a standard form. These transitions can be printed as a list or plotted as a simulated spectrum. A linewidth can be applied to each transition to make the simulated spectrum appear more like the measured one. The lower frame of Figure 23.2 presents such a simulation of the methane spectrum generated using this program. As shown, the simulation reproduces the qualitative appearance of the spectrum nicely. However, as noted by early observers, there is a systematic deviation in the line positions predicted in this way from those observed.[7]

To improve the assignments, it was recognized that degenerate vibrations such as the methane asymmetric stretch can have angular momentum which couples with the rotational angular momentum.[7] An additional coupling parameter ζ was introduced to account for this Coriolis effect, which splits individual J states for the excited vibration and shifts the observed transitions slightly.[7] Instead of the energies $E(J') = B_1 J'(J' + 1)$, the energies are instead split into three levels having energies $B_1(J' + \zeta)(J' + \zeta + 1)$, $B_1 J'(J' + 1)$ and $B_1(J' - \zeta)(J' - \zeta + 1)$. The allowed transitions are:

$$R\text{-branch} : J'' \rightarrow J' - \zeta + 1$$
$$Q\text{-branch} : J'' \rightarrow J'$$
$$P\text{-branch} : J'' \rightarrow J' + \zeta - 1$$

The ζ parameter was found to have the value of 0.0503 for the asymmetric stretch band. Including these new terms into a manual or modified computer calculation of the transitions, which can be done with Excel or other database software, predicts the rotational line positions more accurately.

It should be noted that this rotational structure, and its comparison to simulations such as that provided by PGOPHER, is an ideal way to determine the *temperature* of the sample. The temperature determines the thermal populations in J" states via the Boltzmann relationship, and this in turn determines the line intensities for each J" $\rightarrow$ J' transition. This is particularly useful when well-known molecules are identified in extreme environments such as flames, the upper levels of the Earth's atmosphere, comet tails, or planetary atmospheres. The rotational patterns seen provide a measurement of the temperature in those environments. As a variation on the present experiment, the cell containing the methane could be heated (e.g., with a heating tape) or cooled (e.g., with ice-water or dry-ice/acetone bath) to see how temperature changes the appearance of the spectrum. These effects can of course also be seen easily in the PGOPHER simulations.

The experiment described here for methane is easily extended to any number of common gases having convenient infrared spectra. Examples include carbon dioxide, acetylene, water vapor, etc. In these systems, the rotational structure is somewhat more complex and requires a more advanced knowledge of spectroscopy for its analysis, but the experimental data collection is straightforward. In all of these cases, the same spectra can of course be measured with an ordinary FT-IR spectrometer equipped with a gas cell. A standard experiment often found in undergraduate laboratories is the infrared absorption spectrum of HCl, which exhibits a similar rotational structure to that seen here and has the additional features of isotopic bands from the naturally occurring chlorine isotopes, ^{35}Cl and ^{37}Cl.[12] This same experiment could also be done with the present infrared laser and photoacoustic detection. Likewise, isotopic effects can be investigated for the spectrum of methane by using CD_4 or even $^{13}CH_4$ gases in the present experiment.

In modern chemical physics research, the infrared OPO system employed here has become an essential tool, e.g., in the spectroscopy of ions and their clusters produced in molecular beams and size-selected with mass spectrometers.[13–16] In laboratories doing those experiments and others, the methane photoacoustic measurement described here is employed for wavelength calibration of the infrared OPO lasers.

Many other experiments based on the general concept of photoacoustic spectroscopy have also been developed in modern research. For example, when molecules in surface films absorb infrared laser radiation, the induced heating causes the local material to expand and contract. This motion can be detected with an atomic force microscope (AFM) to obtain sub-diffraction-limited IR spectra of molecules on surfaces.[17]

....

REFERENCES

1. A. G. Bell, "On the production and reproduction of sound by light," *Am. J. Sci.* **20**, 305 (1880).
2. A. G. Bell, "The production of sound by radiant energy," *Science* **28 May**, 242 (1881).

3. A. C. Tam, in *Ultrasensitive Laser Spectroscopy*, D. S. Kliger, ed., Academic Press, New York, 1983, p. 2.

4. T. Shimanouchi, "Molecular Vibrational Frequencies" in *NIST Chemistry WebBook*, NIST Standard Reference Database Number 69, P. J. Linstrom and W. G. Mallard, eds. National Institute of Standards and Technology, Gaithersburg MD, 20899 (http://webbook.nist.gov).

5. A. H. Nielsen and H. H. Nielsen, "The infrared absorption bands of methane," *Phys. Rev.* **48**, 864 (1935).

6. G. Herzberg, *Molecular Spectra and Molecular Structure II. Infrared and Raman Spectra*, D. Van Nostrand Co., New York, 1945, pp. 453–8.

7. W. H. J. Childs, "The structure of the near infrared bands of methane I. General survey and a new band at 11050 Å," *Proc. Royal Soc. London Series A* **153**, 555 (1936).

8. C. R. Webster, P. R. Mahaffy, S. K. Atreya, G. J. Flesch, M. A. Mischna, P.-Y. Meslin, K. A. Farley, P. G. Conrad, L. E. Christensen, A. A. Pavlov, J. Martín-Torres, M.-P. Zorzano, T. H. McConnochie, T. Owen, J. L. Eigenbrode, D. P. Glavin, A. Steele, C. A. Malespin, P. D. Archer, Jr., B. Sutter, P. Coll, C. Freissinet, C. P. McKay, J. E. Moores, S. P. Schwenzer, J. C. Bridges, R. Navarro-Gonzalez, R. Gellert, M. T. Lemmon, and the MSL Science Team, "Mars methane detection and variability at Gale crater," *Science* **347**, 415 (2015).

9. W. R. Bosenberg and D. R. Guyer, "Broadly tunable, single-frequency optical parametric frequency-conversion system," *J. Opt. Soc. Am. B* **10**, 1716 (1993).

10. E. B. Wilson, "The statistical weights of the rotational levels of polyatomic molecules, including methane, ammonia, benzene, cyclopropane and ethylene," *J. Chem. Phys.* **3**, 276 (1935).

11. C. M. Western, PGOPHER, a program for simulating rotational structure, University of Bristol, http://pgopher.chm.bris.ac.uk.

12. C. W. Garland, J. W. Nibler, and D. P. Shoemaker, *Experiments in Physical Chemistry*, eighth edition, McGraw-Hill, Boston, 2003, p. 416.

13. T. Ebata, A. Fujii, and N. Mikami, "Vibrational spectroscopy of small-sized hydrogen-bonded clusters and their ions," *Int. Rev. Phys. Chem.* **17**, 331 (1998).

14. E. J. Bieske and O. Dopfer, "High resolution spectroscopy of cluster ions," *Chem. Rev.* **100**, 3963 (2000).

15. M. A. Duncan, "Infrared spectroscopy to probe structure and dynamics in metal ion-molecule complexes," *Int. Rev. Phys. Chem.* **22**, 407 (2003).

16. W. H. Robertson and M. A. Johnson, "Molecular aspects of halide ion hydration: The cluster approach," *Annu. Rev. Phys. Chem.* **54**, 173 (2003).

17. A. Dazzi, R. Prazeres, F. Glotin, and J. M. Ortega, "Local infrared microspectroscopy with subwavelength spatial resolution with an atomic force microscope tip used as a photothermal sensor," *Opt. Lett.* **30**, 2388 (2005).

Optogalvanic Spectroscopy

24

Introduction

The optogalvanic effect occurs when light absorption by species in an electrical discharge causes the discharge current to change.[1–7] Electrical discharges contain both positive and negative ions, electrons, neutrals, etc., all in a complex distribution of ground and excited states. Current flow through the discharge is established by applying a field across the electrodes. Energetic electron collisions cause excitation and ionization of neutral species. Ion–electron recombination also populates neutral excited states. Emission from excited states to lower levels causes the characteristic spectra known for many gas discharge lamps, leading to relaxation of excited states. Some of the many physical processes involved were discussed in Chapter 1. At any given time in the operation of a discharge, there is a complex equilibrium of excitation and relaxation processes contributing to the species present, the states populated, and the net flow of current through the system. Anything which disrupts this equilibrium can cause a momentary change in the charge flow. In the optogalvanic effect, light absorption moves population from one state to another in this complex mixture, thus perturbing the equilibrium. This can be detected as a change in the current flow through the discharge (either increasing or decreasing) occurring during or immediately following the absorption. The wavelength dependence of this effect provides optogalvanic spectroscopy. Like photoacoustic spectroscopy (Chapter 23), the signal detected is a consequence of light absorption rather than absorption itself. Therefore, optogalvanic spectroscopy is another form of "action" spectroscopy.

The optogalvanic effect was discovered originally by Penning when he observed the effect of one neon discharge lamp on another one operating next to it.[7] Others also noted the effect, but it became recognized as a technique for spectroscopy when Green and coworkers first employed a tunable dye laser for the optical excitation.[1] It is now recognized that discharges containing almost any kind of atomic species,[1,2,7] small molecules,[3–5] or even ions,[6] can exhibit this effect. A general purpose discharge tube, fitted with electrodes and allowing flow-through or static-fill samples of many different gases, can be constructed as shown in Figure 24.1. In this design, the added pressure of gas or the discharge conditions can be optimized for the signal under study. However, another popular option is the use of sealed hollow cathode discharge lamps like those in Atomic Absorption (AA) spectrometers, common in analytical chemistry labs. AA experiments measure the emission spectra of samples added to a flame. The spectrum

Laser Experiments for Chemistry and Physics. First Edition. Robert N. Compton and Michael A. Duncan.
© Robert N. Compton and Michael A. Duncan 2016. Published in 2016 by Oxford University Press.

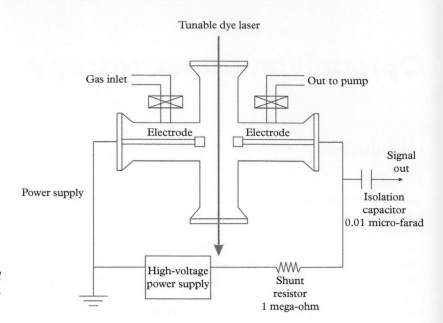

Figure 24.1 *The discharge cell and wiring configuration used for optogalvanic spectroscopy.*

is compared to the patterns for standard emission lamps containing different elements to obtain elemental analysis. Standard lamps are available for many atomic or molecular gases (e.g., hydrogen, helium, neon, argon, mercury, various metals, etc.) from companies such as Perkin-Elmer (http://www.perkinelmer.com/), Agilent (http://www.chem.agilent.com/), or Photron (http://www.photron.com/). Optogalvanic spectroscopy at visible wavelengths using these standard lamps is commonly used in research labs together with etalon transmission fringes (see Chapter 2; Figures 2.18 and 2.19) for the calibration of tunable dye lasers or OPOs.

Experimental details

To begin, the kind of discharge lamp to be used must be selected. A general purpose discharge cell like that shown in Figure 24.1 can be constructed with the help of a good glassblower. This design is desirable if many experiments using different gases will be carried out. The vacuum requires a mechanical pump capable of achieving mTorr pressures. Gases can be supplied from standard cylinders or from lecture bottles. *Safety Note: A glass vacuum line cannot withstand pressures above atmospheric. Use special care with a needle valve or fine pressure control valve when adding gases from high pressure cylinders!* The operating pressures are typically a few Torr. This can be measured with an appropriate vacuum gauge or determined empirically by gradually increasing the pressure up from vacuum with high voltage applied until a discharge forms. The high-voltage power supply must be able to provide approximately 0–2000 V; typical voltages are in the 500–1000 V range. For experiments using only a few selected gases, it is convenient to buy sealed hollow cathode discharge lamps such as those mentioned earlier. The external power supply and circuit must be constructed, but the glass vacuum line, pumps, and gas cylinders can be eliminated in this way. A picture of a hollow cathode discharge used for optogalvanic spectroscopy is shown in Figure 1.13.

The discharge properties of the gas to be studied must be matched to the available tunable laser. As an example, a neon lamp exhibits many atomic emission lines in the red–yellow region of the spectrum. These lines can be seen with a handheld prism- or grating-based spectroscope, as shown in Figure 24.2. To study these transitions in optogalvanic spectroscopy, a dye laser or OPO is needed that is tunable in this same wavelength region. A pulsed dye laser or OPO is recommended, such as those pumped by pulsed Nd:YAG lasers. A collection of discharge lamps and spectroscopes is convenient for the introduction to atomic line spectra, and these complement the optogalvanic spectroscopy measurements. Discharge lamps and spectroscopes for this purpose were provided previously by CENCO, but are now available through Ward's Science (https://wardsci.com).

Using either a multi-gas cell or an individual discharge lamp, extra circuit elements must be added to measure the discharge current and how it changes following optical excitation. The high voltage must be applied to drive the lamp, as shown in Figure 24.1. A shunt resistor reduces the current requirement for the power supply and damps out discharge voltage fluctuations. The main issue is how to sample the discharge current, since high voltages are present that can damage an oscilloscope. This is accomplished using an isolation capacitor, as shown in the figure. Because of its RC time constant, this acts as a high-pass filter, transmitting only the fast-time variations in the current induced by a pulsed laser, but holding off the DC high voltage.

Figure 24.2 *The emission spectrum of a neon discharge lamp viewed through a handheld spectroscope and photographed with a digital camera.*

Results and discussion

Figure 24.3 shows an example of an optogalvanic spectrum measured with a neon hollow cathode discharge lamp (Jarrell-Ash Type SI; model 45479) using a Nd:YAG-pumped dye laser (Lumonics Hyperdye) in the DCM dye region. The dye laser pulse energy was 3 mJ/pulse. The atomic transitions of neutral neon atoms seen are labeled in the figure. It is important to note that in the data collection, signals may go in either positive or negative direction, depending on how the transition affects the current in the circuit. It makes sense that most signals would produce an increase in signal, since excitation of atomic transitions helps atoms along the way to ionization, thus producing more ions and a greater current. However, some transitions work against the production of more current and negative-going signals are often seen.

Many of the systems that are easiest to study with optogalvanic spectroscopy involve electronic transitions of atoms. These kind of transitions have of course been known for many years and the literature values for these are tabulated in the NIST Atomic Spectra Database.[8] However, as shown by many investigators, molecular species and small ions also can be used, providing a rich variety of spectroscopy experiments.[1–7]

Johnston has described an intriguing variation of the optogalvanic effect, using the discharge inside a He-Ne laser.[9] When an external dye laser beam is introduced into this discharge, ionization rates of different atomic states are affected, but some of these are connected to the levels that emit laser radiation. The laser output can then be increased or decreased depending on the states excited. As shown in Figure 24.4, the red laser output can be measured with a photodiode rather than detecting the current through the discharge. A number of neon atomic

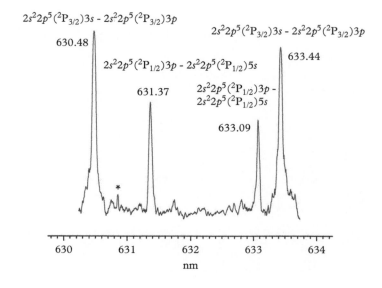

Figure 24.3 *Optogalvanic spectroscopy of a neon lamp measured with a hollow cathode discharge lamp. The atomic transitions are indicated. The ∗ indicates a weak transition of neon cation.*

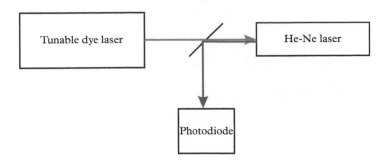

Figure 24.4 *The experimental config-uration for detecting the optogalvanic ef-fect through the light emitted by a He-Ne laser.*

state resonances were detected in this way.[9] However, care must be taken to de-tect the red He-Ne beam at 632.8 nm, and to distinguish this from the dye laser tuning in the same wavelength region.

This experiment is attractive because of its simplicity and because the applica-tion of high voltages and the extraction of current are not necessary. This method could also be implemented with other gas discharge lasers (e.g., argon ion). Also, other excitation sources, such as a CW tunable diode laser, could be substituted for the pulsed dye laser.

....

REFERENCES

1. R. B. Green, R. A. Keller, G. G. Luther, P. K. Schenck, and J. C. Travis, "Galvanic detection of optical absorptions in a gas discharge," *Appl. Phys. Lett.* **29**, 727 (1976).
2. D. S. King and P. K. Shenck, "Optogalvanic spectroscopy," *Laser Focus* **14**, 50 (1978).
3. C. R. Webster and C. T. Rettner, "Laser optogalvanic spectroscopy of molecules," *Laser Focus* **19**, 1 (1983).
4. C. Demuynck and J. L. Destombes, "Optogalvanic spectrum of molecular iodine," *IEEE J. Quant. Elec.* **17**, 575 (1981).
5. C. R. Webster and R. T. Menzies, "Infrared laser optogalvanic spectroscopy of molecules," *J. Chem. Phys.* **78**, 2121 (1983).
6. R. Walkup, R. W. Dreyfus, and Ph. Avouris, "Laser optogalvanic detection of molecular ions," *Phys. Rev. Lett.* **50**, 1846 (1983).
7. B. Barbieri and N. Beverini, "Optogalvanic spectroscopy," *Rev. Mod. Phys.* **62**, 603 (1990).
8. A. Kramida, Yu Ralchenko, J. Reader, and NIST ASD Team (2014). *NIST Atomic Spectra Database* (ver. 5.2). Available: http://physics.nist.gov/asd. National Institute of Standards and Technology, Gaithersburg, MD.
9. T. F. Johnston, Jr., "Measuring He-Ne line profiles with the optogalvanic effect," *Laser Focus* **14**, 58 (1978).

25

Diode Laser Atomic Spectroscopy

Introduction

Tunable diode lasers represent relatively inexpensive and reliable sources of narrow-band (<1 MHz) tunable laser light which can be effectively used for performing high-resolution spectroscopy over selected absorption features in atoms and molecules. A description of the extended cavity diode laser (ECDL) is found in Chapter 4 and in [1] and [2]. The applications of diode lasers in performing (i) single-photon absorption spectroscopy, (ii) saturated absorption spectroscopy, and (iii) Doppler-free two-photon spectroscopy for undergraduate experiments involving alkali vapors have been described in many publications (see [3–6] and those described below). Most of these experiments involve rubidium or cesium atoms because of the relatively high vapor pressure of these elements at nominal temperatures and the availability of diode lasers having outputs in the spectral region of the first ns-np transition for these atoms. Below we describe three such experiments.

Diode laser absorption experiments in cesium and rubidium vapor

The random motion of atoms or molecules in a low-pressure gas, where collisions are rare, results in a Doppler-broadened absorption line represented by a Gaussian function centered at the resonance frequency (ν_0) or wavelength (λ_0). Atoms or molecules of mass m at temperature T will exhibit a Gaussian absorption profile given by:

$$P_\lambda(\lambda) = \sqrt{\frac{mc^2}{2\pi kT\lambda_0^2}} \, \exp-\left[\frac{mc^2}{2kT\lambda_0^2}(\lambda-\lambda_0)^2\right] d\lambda \qquad (25.1)$$

where c is the speed of light in vacuum and k is the Boltzmann constant, T is in K and m in kg. The wavelength λ represents the wavelength of the absorbing light

Laser Experiments for Chemistry and Physics. First Edition. Robert N. Compton and Michael A. Duncan.
© Robert N. Compton and Michael A. Duncan 2016. Published in 2016 by Oxford University Press.

and λ_0 is the wavelength corresponding to the resonant transition. The linewidth (full-width-half-maximum, FWHM) of this Doppler-broadened transition is

$$\Delta\lambda_{\text{FWHM}} = \sqrt{\frac{8kT\ln2}{mc^2}}\,\lambda_0 \qquad (25.2)$$

Using frequency instead of wavelength, the linewidth is

$$\Delta\nu_{1/2} = \sqrt{\frac{8kT\ln2}{mc^2}}\,\nu_0 \qquad (25.3)$$

The Doppler width (FWHM) for ^{87}Rb at 780 nm (3.85×10^{14} Hz) at room temperature (300 K) is 1026 MHz (0.034 cm^{-1}).

In addition to the inhomogeneous Doppler broadening described above there is also homogeneous broadening due to the natural lifetime of the state, as well as that from collisions if the pressure is higher. Inhomogeneous broadening refers to cases in which every atom experiences a different broadening, whereas homogeneous refers to cases in which each atom experiences an identical broadening. Both of these effects essentially shorten the lifetime of the excited state leading to homogeneous broadening and a Lorentzian line shape. The lifetime and the energy width of the state are related through the uncertainty principle, $\Delta E\Delta t \geq \hbar$. The experimentally observed line shape is therefore a convolution of the Gaussian and Lorentzian line shapes resulting in what is called a Voigt profile.

Leahy, Hastings, and Wilt [7] have described an undergraduate laboratory experiment designed to demonstrate the temperature dependence of Doppler broadening in rubidium vapor using a diode laser. Doppler broadening is demonstrated by simply detecting the absorption profiles for tunable diode laser light passing through a heated cell containing rubidium atoms. Diode laser absorption of rubidium vapor is reported for several resolved hyperfine levels in the ^{85}Rb ($5s$ $^2S_{1/2}$ F = 2 to ^{85}Rb ($5p$ $^2P_{3/2}$ F = 1, 2, 3) transitions near 795 nm and the hyperfine resolved ^{87}Rb ($5s$ $^2S_{1/2}$ F = 1 to ^{87}Rb ($5p$ $^2P_{3/2}$ F = 1 and 2) transitions near 780 nm. Their measurements were carried out between 296 and 500 K. The Doppler broadening observed in these experiments was in the range of 500 to 650 MHz. The combined linewidth attributed to collision broadening [8] and the finite lifetime [9] is only ~10 MHz and is small enough to easily extract the Doppler width from the Voigt profiles due to Doppler and lifetime broadening. Excellent agreement between theory and experiment is obtained for the Doppler broadening. This experiment is well described in [7] and requires no further discussion. As pointed out by the authors, the experimental setup is similar to that described in the book by Brandenberger.[10]

Doppler broadening can be greatly reduced by performing saturated absorption spectroscopy.[11] We first describe the physics behind Doppler-free saturated absorption diode laser spectroscopy, then review a number of previous reports of undergraduate laboratory experiments, followed by a description of the experiment employed in the modern physics laboratories at the University of Tennessee.

Saturated absorption spectroscopy

In saturated absorption spectroscopy the beam from a tunable diode laser is split into two beams and these are counter-propagated through an atomic vapor. When the laser frequency v_1 is within the Doppler profile of the absorption feature but different from the resonant frequency for the atom at rest, v_0, one of the beams will excite atoms with some velocity $+V_z$ and the counter-propagating beam will interact with only those atoms moving in the $-V_z$ direction. However, when the laser is tuned to the resonant frequency, v_0, the two beams interact with the same group of atoms, i.e., those with velocities equal to zero along either direction of propagation. Since some of the atoms with $V_z = 0$ are moving perpendicular to the laser beam (V_x and V_y) there is a finite number of these interactions. Now if one of the beams (pump beam) is strong enough to deplete the population of the atoms in the ground state, the profile of the weaker beam (probe beam) will show a much reduced absorption only over a very narrow region. The "hole burning" of the ground-state population appears as a pronounced dip (Lamb dip) in the Doppler profile of the probe beam transmission. With proper adjustment of the parameters available (laser power, etc.) the width of the saturated absorption feature can approach that of the Doppler-free line width of the transition.

Saturated absorption spectroscopy experiments designed for undergraduate laboratories have been described in detail by MacAdam et al., [4] Rao et al. [5], and Preston.[12] A quick search on the Internet will reveal many other laboratory descriptions of undergraduate experiments in this area. The treatment of Preston [12] describes a clever method of using two probe beams of nearly equal intensity created from the front and back of a beam splitter which allows the intense pump beam through. Reflection of the diode laser from the front and back of the thick beam splitter allows spatial separation of the two probe beams, but they still overlap the volume of the pump beam. Subtracting the signal from the two probe beams provides direct detection of the allowed hyperfine transitions for $F = 2 \rightarrow F' = 1, 2, 3$ as well as the crossover transitions. The crossover transitions occur when two transitions come from the same ground state but differ in frequency by less than the Doppler linewidth (see Preston [12] for details).

Rubidium has two naturally occurring isotopes, ^{87}Rb (28%) and ^{85}Rb (72%). The ^{85}Rb isotope has nuclear spin $I = 5/2$ whereas the ^{87}Rb has $I = 3/2$. For the ground state $L = 0$ and $J = 1/2$ giving the hyperfine levels $F = 3$ and 2 for ^{85}Rb and $F = 1$ and 2 for ^{87}Rb. For the $5p\ ^2P_{3/2}$ (so-called D state) the F' levels for ^{85}Rb are 1, 2, 3, 4 and for ^{87}Rb are 0, 1, 2, 3. The hyperfine energy levels for both of the isotopes of rubidium are shown in Figure 25.1.

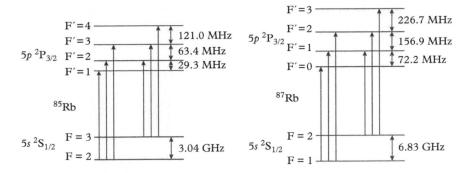

Figure 25.1 *Hyperfine energy levels of the two naturally occurring isotopes of rubidium,* ^{87}Rb *(28% abundant, I = 3/2) and* ^{85}Rb *(72% abundant, I = 5/2).*

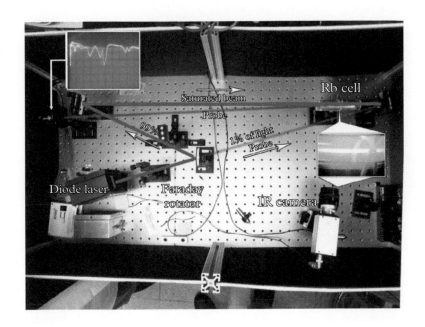

Figure 25.2 *Experimental setup and signal display of the saturated absorption spectroscopy experiment at the University of Tennessee. The inset on the right shows the IR image of the fluorescence from the Rb due to the presence of the laser beam.*

Saturated absorption spectroscopy experiments have been performed in the modern physics laboratory at the University of Tennessee directed by Dr. J. E. Parks for a number of years. The experimental geometry used in this laboratory for these experiments is given in Figure 25.2. The diode laser has an extended cavity design. A Faraday rotator (isolator) is used to block unwanted feedback into the diode laser and is not employed in many experiments described in the literature. The IR camera is handy to monitor the wavelength when the laser is in resonance with an allowed transition. A handheld IR viewer can also be employed. The output of the probe beam detector is fed into an oscilloscope whose x-axis

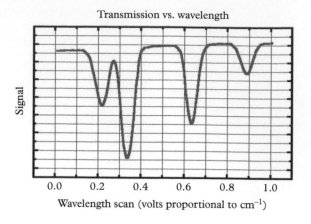

Figure 25.3 *A typical scan of the Doppler-broadened absorption spectrum of rubidium vapor.*

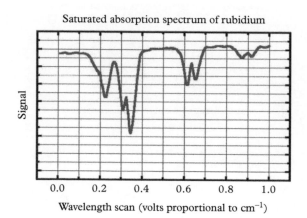

Figure 25.4 *Saturated absorption spectrum of rubidium.*

is synchronized with the laser frequency. Figure 25.3 shows a typical diode laser absorption spectrum for rubidium vapor.

Introduction of the saturated (strong) diode laser beam which is counter-propagated with the weak beam (see Figure 25.2) produces "dips" in the Doppler-broadened profiles, as shown in Figure 25.4.

One can readily observe the Lamb dips due to the saturated absorption. The Doppler broadening seen in Figure 25.3 is absent in the narrow, Doppler-free, features evident in Figure 25.4. A more dramatic Doppler-free spectroscopy can be seen in the following discussion of two-photon absorption spectroscopy. Also, the interested student should refer to Chapters 16 and 18 that describe multiphoton ionization spectroscopy of alkali atoms and molecular iodine, respectively.

Doppler-free two-photon spectroscopy

With the addition of a photomultiplier detector, a heated cell, and a blue filter to the above experimental geometry, it is possible to perform Doppler-free two-photon spectroscopy. Doppler-free multiphoton spectroscopy is discussed at great length in the book by Demtröder.[12] We consider here the absorption of two photons when their energies add up to the excited state of the atom. An atom at rest (E_0) can absorb two photons through an allowed transition when their energies add up to the excited state level (E^*), i.e.,

$$E^* - E_0 = h(\omega_1 + \omega_2) \tag{25.4}$$

where $\omega = 2\pi\nu$. If the atom is moving with a velocity v in the laboratory frame the atom will experience a Doppler shift in the radiation given by

$$\omega' = \omega - k \cdot v \tag{25.5}$$

where k is the wavevector $2\pi/\lambda$ of the laser. Thus the resonance condition above can be rewritten

$$E^* - E_0 = h\left[(\omega_1 + \omega_2) - v \cdot (k_1 + k_2)\right] \tag{25.6}$$

Now, if the two photons have the same frequency and are counter-propagating (i.e., $k_1 = -k_2$) the Doppler shift becomes *equal to zero* for all molecules regardless of their energies or direction of motion. Thus all molecules absorb at the same frequency producing an excited state at 2ω. Doppler-free two-photon absorption is illustrated in Figure 25.5.

Doppler-free two-photon excitation has been studied for excited d and s states for rubidium ($5s\,^2S_{1/2} \rightarrow 5d\,^2D_{5/2}$)[13] and ($5s\,^2S_{1/2} \rightarrow 7s\,^2S_{1/2}$).[14] The experiments below describe the work presented by Olsen et al.[13] and Ko et al.[14]. In both cases two-photon excitation occurs from the ground s state to the upper d and s states, respectively (please refer to Chapter 16 and 18 for a brief explanation of the selection rules for two-photon excitation). Rubidium is often

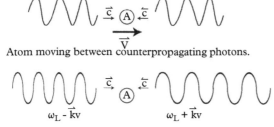

Atom moving between counterpropagating photons.

$\omega_L - \vec{k}v$ $\omega_L + \vec{k}v$

Frequencies as seen in the rest frame of the atom.

Figure 25.5 *Pictorial illustration of Doppler-free two-photon absorption. The frequency ω_L refers to the laser angular frequency and kv is the wavevector of the moving atom.*

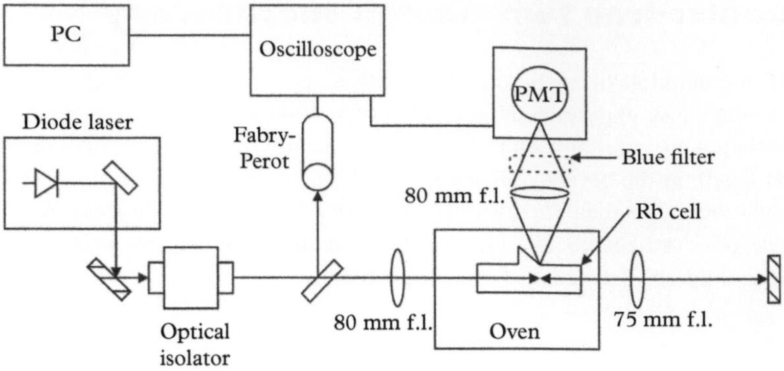

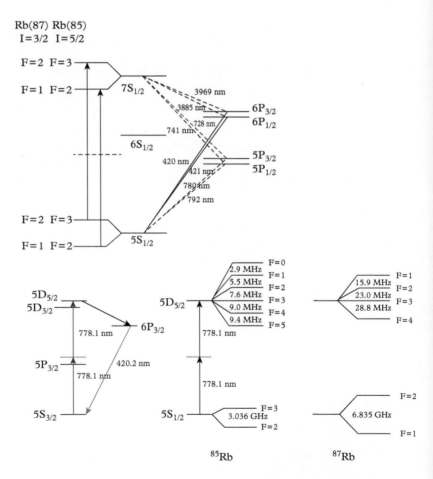

Figure 25.6 *Diagram of experiment for studying Doppler-free two-photon spectroscopy in rubidium (redrawn from [13]).*

Figure 25.7 *Diagram of the hyperfine energy levels of the 7s (top) and 5s (bottom) states of rubidium.*

used as the test atom since diode lasers operating in the wavelength region corresponding to the excited states involved are readily available. In an optical heat pipe giving sufficient vapor density, these states can be seen easily in absorption. However, in a simple cell at low densities the excited states were detected by fluorescence. A generic experimental arrangement for studying Doppler-free two-photon absorption is shown in Figure 25.6.

The two-photon energy levels for the $5d$ hyperfine levels of both isotopes of rubidium, along with the fluorescence detection scheme, is shown in Figure 25.7. The fluorescence from the excited states for these two levels is shown in Figure 25.8. In both of these experiments,[13,14] the use of Doppler-free two-photon spectroscopy tuning over a range of 4 GHz allows for determination of the

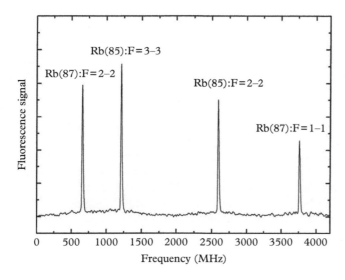

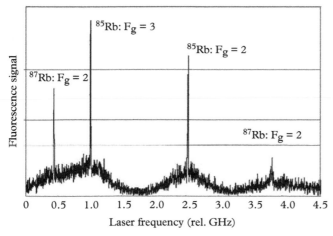

Figure 25.8 *Two-photon Doppler-free spectroscopy of rubidium vapor. The top spectrum corresponds to two-photon excitation of the 7s states [14] (reprinted with permission from Opt. Lett. 29, 1799 (2004), Copyright 2004 Optical Society of America) whereas the bottom spectrum corresponds to two-photon excitation of the 5D states [13] (reprinted with permission from Am. J. Phys. 74, 218 (2006), Copyright 2006 American Association of Physics Teachers).*

hyperfine splitting of the ground state of rubidium. In the experiments of Olsen et al.,[14] the use of linearly and circularly polarized light also allows the student to investigate the selection rules for light absorption.

....

REFERENCES

1. R. S. Conroy, A. Carleton, A. Carruthers, B. D. Sinclair, C. F. Rae, and K. Dholakia, "A visible extended cavity diode laser for the undergraduate laboratory," *Am. J. Phys.* **68**, 925 (2000).
2. A. S. Arnold, J. S. Wilson, and M. G. Boshier, "A simple extended-cavity diode laser," *Rev. Sci. Instrum.* **69**, 1236 (1998).
3. C. E. Wieman and L. Hollberg, "Using diode lasers for atomic physics," *Rev. Sci. Instrum.* **62**, 1 (1991).
4. K. B. MacAdam, A. Steinbach, and C. E. Wieman, "A narrow-band tunable diode laser system with grating feedback, and saturated absorption spectrometer for Cs and Rb," *Am. J. Phys.* **60**, 1098 (1992).
5. G. N. Rao, M. N. Reddy, and E. Hecht, "Atomic hyperfine structure studies using temperature/current tuning of diode lasers: An undergraduate experiment," *Am. J. Phys.* **66**, 702 (1998).
6. K. G. Libbrecht, R. A. Boyd, P. A. Willems, T. L. Gustavson, and D. K. Kim, "Teaching physics with 670 nm diode lasers - construction of stabilized lasers and lithium cells," *Am. J. Phys.* **63**, 1 (1995).
7. C. Leahy, J. T. Hastings, and P. M. Wilt, "Temperature dependence of Doppler-broadening in rubidium: An undergraduate experiment," *Am. J. Phys.* **65**, 367 (1996).
8. C. Shang-Yi, "Pressure effects of homogeneous rubidium vapor on its resonance lines," *Phys. Rev.* **58**, 884 (1940).
9. O. S. Heavens, "Radiative transition probabilities of the lower excited states of alkali metals," *J. Opt. Soc. Am.* **51**, 1058 (1961).
10. J. R. Brandenberger, *Lasers and Modern Optics in Undergraduate Physics*, Lawrence University Press, Appleton, WI, 1989, pp. 43–58.
11. For a thorough discussion of saturation spectroscopy, see W. Demtröder, *Laser Spectroscopy*, Springer Series in Chemical Physics, Vol. 5, Springer, New York, 1982, p.484.
12. D. W. Preston, "Doppler-free saturated absorption: Laser spectroscopy," *Am. J. Phys.* **64**, 1432 (1996).
13. A. J. Olson, E. J. Carlson, and S. K. Mayer, "Two-photon spectroscopy of rubidium using a grating-feedback diode laser," *Am. J. Phys.* **74**, 218 (2006).
14. M.-S. Ko and Y.-W. Liu, "Observation of rubidium $5S_{1/2} \rightarrow 7S_{1/2}$ two-photon transitions with a diode laser," *Opt. Lett.* **29**, 1799 (2004).

Vacuum Ultraviolet Spectroscopy using THG in Rare Gases

Introduction

We have seen in Chapters 16 and 18 how multiphoton excitation and multiphoton ionization (MPI) using focused lasers in the visible and infrared part of the E&M spectrum can be employed to access states of atoms and molecules at higher energy and even into the ionization continuum. Direct excitation (i.e., with one photon) of these high-lying states using laser light requires special techniques. One-photon spectroscopy in the vacuum ultraviolet region of the E&M spectrum is essential to the study of electronic states of atoms and molecules near and above their ionization limit. The term "vacuum ultraviolet" (VUV) is used since this radiation would be attenuated by passing through air and the experiments must be performed in a vacuum or using a cell that will pass such radiation. The windows holding the sample must also allow transmission of the radiation. The VUV region of the spectrum begins at 200 nm (50,000 cm^{-1}, ~6.2 eV) and extends to about 30 nm, where the soft X-ray region begins. This region is sometimes subdivided so that 200 – 100 nm (6.20 – 12.40 eV) is called the VUV region and the range below 100 nm is called the extreme ultraviolet (XUV).

In this experiment we describe the generation and use of VUV "laser-like" light for studies of excitation and ionization of molecules in the VUV part of the spectrum. We use the term "laser-like" since the light generated has spatial and temporal coherence identical to that of a laser. In Chapter 27 we see how tunable visible lasers can be down-shifted to the infrared region of the spectrum using Raman shifting or Stimulated Electronic Raman Scattering (SERS). Tunable lasers can also be up-shifted into the UV region. The range of tunable laser sources has been extended through the near UV by using the nonlinear response of certain crystals (BBO, etc.) to frequency double visible dye laser radiation. The use of frequency doubling and mixing of the harmonics of the Nd:YAG fundamental at 1064 nm produces 532, 355, and 266 nm light with 20–50% efficiency at each step. Doubling dye laser light also allows spectroscopic studies from 217–360 nm. Frequency doubling was discussed at length in Chapter 5 on nonlinear optics. Further extension of tunable laser light into the VUV, however, poses special problems since frequency conversion crystals become opaque in this region.

Third harmonic generation (THG) can be employed to frequency triple laser light into the VUV. The rare gases have been typically used to generate such coherent tunable light (see [1–9]). A number of books and articles have also treated this subject (see [10–12]). These studies have demonstrated tunable THG in the ranges of 140.3–146.9 nm for Xe, 120.3–123.6 nm for Kr, and 110.0–106.5 nm for Ar, with conversion efficiencies of 10^{-5} to 10^{-6} and fluxes in the range of 10^7 to 10^{10} photons/pulse. The laser is tuned such that three photons are near resonance with an allowed *s–s* transition in the rare gas. Due to phase-matching conditions the THG is generated to the blue (shorter wavelength) side of the resonant intermediate. Figure 26.1 illustrates the scheme for generating THG in the rare gases Xe, Kr, and Ar and Table 26.1 shows the wavelength ranges that can be produced with different rare gases.

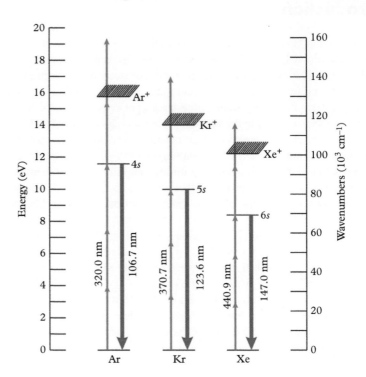

Figure 26.1 *Energy level diagram for argon, krypton, and xenon showing the lowest s-state, and intermediate and lowest ionization limit for each rare gas. The heavy purple line illustrates the wavelength for the beginning of the THG.[9]*

Table 26.1 *The wavelength ranges available for VUV generation using a nitrogen-pumped dye laser in krypton and xenon gas.*

Rare gas	Laser tuning range (nm)	Wavelength (nm)	VUV tuning range (cm⁻¹)	VUV tuning range (eV)
Xe 6s	420.8–440.6	140.3–146.9	68,063–71,275	8.44–8.84
Xe 6s'	375.0–388.8	125.0–129.6	77,160–88,000	9.57–9.92
Kr 5s	360.9–370.5	120.3–123.5	80,972–83,126	10.04–10.31

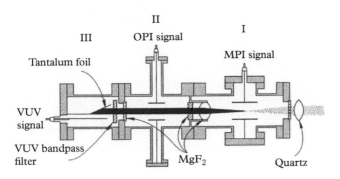

III — Tantalum foil

II — OPI signal

I — MPI signal

VUV signal

VUV bandpass filter

MgF$_2$

Quartz

Figure 26.2 *Schematic diagram of the VUV THG apparatus. Reading right to left: the laser is focused into chamber I to generate the THG. The electrodes in chamber I are also used to detect the (3 + 2) REMPI from Kr or Xe gas at low pressure. The THG light is defocused into chamber II which contains the sample gas to be studied along with the ionization collecting electrodes. Chamber III also contains a tantalum foil which is employed to detect the THG light. Each chamber also acts as a proportional counter providing large signal pulses to process with a boxcar integrator.*

In this chapter we describe a variety of spectroscopic studies using VUV light produced by THG in krypton or xenon gas. This kind of experiment has been thoroughly described in a paper by Miller et al.[12] We select two experiments involving VUV absorption for carbon monoxide and VUV photoionization of iodobenzene.

The apparatus consists of three chambers, each of which functions as a proportional counter. Figure 26.2 shows the basic apparatus.

Electrons produced in such a counter are accelerated toward an electrode maintained at a positive bias voltage (10–500 V) and these electrons undergo ionizing collisions with other gas molecules causing an electron avalanche. The amplification produced is proportional to the ratio of electric field to pressure (E/P). The gain is finally limited by dielectric breakdown. The first chamber (I) consists of a stainless steel Varian six-way cross-connected to a mechanical vacuum pump. The beam of a nitrogen-pumped dye laser (Molectron UV-24 with Molectron DL400) is focused into chamber I by a 3.8 cm quartz lens to a spot ≤ 20 μm, giving a power density of the order of 5×10^9 W/cm^2. Krypton or xenon gas (Matheson) was used as the nonlinear medium and any electrons produced by MPI in the tripling gas were detected with the biased flat-plate electrode. Multiphoton ionization created when the laser frequency is in three-photon resonance with various atomic levels of the rare gases, gives signals which are useful for coarse wavelength calibration.* The VUV light produced by THG in the focal spot is collimated with a MgF$_2$ lens (focal length $= 4$ cm) and passed through a MgF$_2$ window into the second chamber (II). This section is a stainless steel tube (3.8 cm I.D.; 7.6 cm long) with three side arms for two electrodes and a pump-out port. Again, any electrons resulting from one-photon ionization (OPI) or MPI are amplified in the sample gas and detected at the electrode. The VUV beam exits chamber II through a MgF$_2$ window and passes through a dielectric VUV bandpass filter (Acton Research) to remove the blue pump light. This filter could also be placed between chambers I and II when necessary. Although various filters were available, most experiments used a filter with 46% transmission at 141.5 nm and a bandwidth (FWHM) of 55.0 nm (Acton 145-B), which passed $\leq 1\%$ of the blue light. Chamber III contains a VUV photon detector consisting of a 3.8 cm Varian nipple (8.5 cm long) containing a tantalum foil and a single flat-plate electrode. Photons whose energy exceeds the work function of tantalum (~4.1 eV) eject electrons which are detected in the proportional counter. Xenon gas at pressures between 300–500 mTorr is used as the counter gas. The whole system uses standard flanges with copper gaskets and is thus bakeable. Each chamber has a thermocouple vacuum gauge and in addition can be connected to a capacitance manometer (MKS Baratron) with a 1 or 1000 Torr head for more precise pressure measurement. Signals from charge-sensitive preamplifiers connected to any two of the three chambers were averaged in a dual channel boxcar integrator (Princeton Applied Research, No. 162/165) and displayed on a dual-channel x-y recorder.

Greater third harmonic light intensity can be obtained by adding phase-matching gases to the Xe or Ar. This was not done in the experiments described here but the interested student could explore this aspect (see [1–11]). The bandwidth of the third harmonic light is expected to be at most one-third that of the input laser beam (~0.01 nm). If a Gaussian pump beam is assumed, an additional factor of $\sqrt{3}$ reduction in width results from the I^3 dependence of the THG. The expected bandwidth is thus 0.002 nm (0.9 cm^{-1}) at 147.0 nm.

We demonstrate the VUV absorption method described above by first showing a portion of the absorption spectrum corresponding to the $X\ ^1\Sigma^+(v=0) \rightarrow A\ ^1\Pi$ $(v=3)$ band of CO (for a description of this band see [13]).

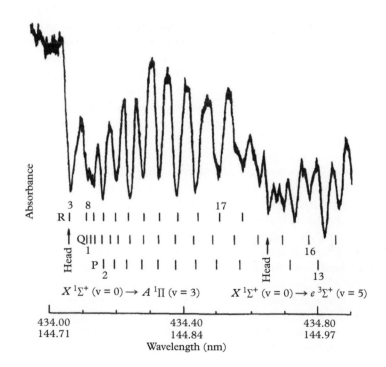

Figure 26.3 *VUV absorption spectrum of the $X\ ^1\Sigma^+(v=0) \rightarrow A\ ^1\Pi$ $(v=3)$ band of CO.[12] The upper wavelength scale corresponds to the laser wavelength in air. The lower scale is the VUV wavelength corrected for the refractive index of air (reprinted with permission from J. Chem. Phys.* **76**, *3967 (1982), Copyright 1982, AIP Publishing LLC).*

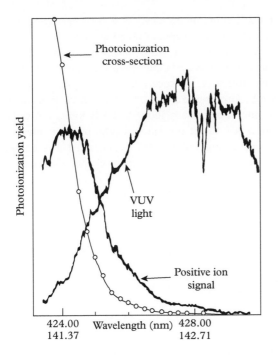

Wavelength (nm)

424.00   428.00
141.37   142.71

Figure 26.4 *Determination of the one-photon, photoionization threshold of iodobenzene using THG in xenon.[12] The curve labeled VUV light shows the THG tuning curve. The directly recorded ionization signal is also shown. The rising ionization cross-section (circles with line) represents the positive ion signal divided by the VUV light transmission (relative cross–section). The upper wavelength scale is the laser wavelength in air and the lower scale is the VUV wavelength corrected for the refractive index of air (reprinted with permission from J. Chem. Phys. 76, 3967 (1982), Copyright 1982, American Institute of Physics).*

In addition to VUV spectroscopy, the apparatus described above can be employed to study photoionization thresholds for many molecules. This is illustrated for the case of iodobenzene in Figure 26.4 which shows the one-photon ion yield at the threshold region for iodobenzene at 250 mTorr pressure.

The VUV light curve shows the tuning curve of the THG produced in 300 Torr of Xe which extends from 423 to 432 nm (141–144 nm in the VUV). The wavelength scale again corresponds to the laser wavelength in air given above and the vacuum-corrected VUV wavelength below. The dip in the THG near 428 nm is due to absorption of the VUV generated light in a two-photon absorption (1 VUV + 1 blue) to several *5f* levels of Xe (see [9]). The ion yield curve for the iodobenzene positive ion signal divided by the VUV light intensity gives the relative ionization cross-section. The first appearance of ions occurs at a visible laser wavelength of 429.0 nm, or 143.0 nm in the VUV, corresponding to 8.670 eV. This is a little lower than the expected ionization potential, as ionization in this region is dominated by hot bands and collisional ionization of highly excited Rydberg states. A better estimate of the IP can be obtained by extrapolating the steep part of the curve to the axis at 425.1 nm (141.7 nm in the UV). This corresponds to a value of 8.75 ± 0.01 eV and compares well with several previous IP values of 8.685 [14] and 8.73 [15] measured in mass spectrometers. This method could be used to obtain very accurate ionization potentials for many other molecules. As a final example, similar analysis for *o*-xylene gave an ionization potential of 8.54 ± 0.03 eV.

Other experiments involving VUV spectroscopy can be found in the paper by Miller et al.[12] The experiments described here were performed using a nitrogen-pumped dye laser. However, improved results could be obtained with a higher powered YAG-pumped dye laser or Optical Parametric Oscillator (OPO).

*To observe the three-photon allowed (3 + 1) ionization resonances it is necessary to perform REMPI at pressures slightly below 1 Torr. This is necessary because at higher pressures, when THG is generated at resonance, the third harmonic wave cancels the three-photon excitation and no excitation or ionization is observed.[9,16]

....

REFERENCES

1. A. H. Kung, "Generation of tunable picosecond VUV radiation," *Appl. Phys. Lett.* **25**, 653 (1974).
2. A. H. Kung, J. F. Young, and S. E. Harris, "Generation of 1182-Å radiation in phase-matched mixtures of inert gases," *Appl. Phys. Lett.* **22**, 301 (1973).
3. J. F. Ward and G. H. C. New, "Optical third harmonic generation in gases by a focused laser beam," *Phys. Rev.* **185**, 57 (1969).
4. R. Mahon, T. J. McIllrath, and D. W. Koopman, "Nonlinear generation of Lyman-alpha radiation," *Appl. Phys. Lett.* **33**, 305 (1978).
5. D. Cotter, "Conversion from 3371 to 1124 Å by nonresonant optical frequency tripling in compressed krypton gas," *Opt. Lett.* **4**, 134 (1979).
6. D. Cotter, "Tunable narrow-band coherent VUV source for the Lyman-alpha region," *Opt. Commun.* **31**, 397 (1980).
7. R. Wallenstein, "Generation of narrowband tunable VUV radiation at the Lyman-α wavelength," *Opt. Commun.* **33**, 119 (1980).
8. J. Reintjes, "Third-harmonic conversion of XeCl-laser radiation," *Opt. Lett.* **4**, 242 (1980); "Generation of coherent tunable VUV radiation near the Ly-β transition of atomic hydrogen," *Opt. Lett.* **5**, 342 (1980).
9. J. C. Miller and R. N. Compton, "Third-harmonic generation and multiphoton ionization in rare gases," *Phys. Rev. A* **25**, 2056 (1982).
10. R. W. Boyd, *Nonlinear Optics*, Academic Press, New York, 1993.
11. Y. R. Shen, *The Principles of Non-linear Optics*, John Wiley and Sons, New York, 1984.
12. J. C. Miller, R. N. Compton, and C. D. Cooper, "Vacuum ultraviolet spectroscopy of molecules using third-harmonic generation in rare gases," *J. Chem. Phys.* **76**, 3967 (1982).

13. S. G. Tilford and J. D. Simmons, "Atlas of the observed absorption spectrum of carbon monoxide between 1060 and 1900 Å," *J. Phys. Chem. Ref. Data* **1**, 147 (1972).

14. J. Momigny, G. Goffant, and L. D'or, "Photoionization studies by total ionization measurements. I. Benzene and its monohalogen derivatives," *Int. J. Mass Spectrom. Ion Phys.* **1**, 53 (1968).

15. Y. L. Sergeev, M. E. Akopyan, F. I. Vilesov, and V. I. Kleimenov, *Opt. Spektrosk.* **29**, 63 (1970).

16. J. C. Miller, R. N. Compton, M. G. Payne, and W. R. Garrett, "Resonantly enhanced multiphoton ionization and third-harmonic generation in xenon gas," *Phys. Rev. Lett.* **45**, 114 (1980).

27

Raman Shifting and Stimulated Electronic Raman Scattering (SERS)

Introduction

In the previous chapter it was demonstrated that fixed and tunable frequency laser light can be shifted to shorter wavelength (higher energy) using third harmonic generation in atomic gases. By focusing a pulsed laser into an atomic gas (rare gas, alkali vapor, etc.) the emerging light from the focal volume contains both the original laser beam plus a third harmonic beam, which is three times the frequency (energy) of the incident light. Tunable laser light in the vacuum ultraviolet (VUV) can be generated over a number of ranges of wavelength, although the efficiency for this process is low. Second harmonic generation in crystals is a very important method employed to generate shorter wavelength light from pulsed solid-state lasers (see Chapters 4 and 5). For example, second harmonic light from a Nd:YAG laser can be generated with high efficiency (up to 50%). In this instance, the frequency of the fundamental 1064 nm light can be doubled to produce 532 nm light, and 532 nm can be doubled to produce 266 nm light. In addition, sum-frequency mixing of the 1064 nm fundamental and 532 nm second harmonic can lead to 355 nm light. It is also possible to combine harmonic generation in crystals with third harmonic generation in gases to produce pulsed VUV light. For example, numerous researchers have coupled harmonic generation in crystals (355 nm) with third harmonic generation in gases to produce reasonably intense light at 118 nm. The Raman effect discussed in Chapter 19 can also be employed to "shift" laser light into longer and shorter wavelength regions, thereby extending the range of usefulness of laser light. In this chapter we show how Raman scattering in atomic and molecular gases can be employed to both up- and down-shift fixed and tunable laser light to produce useful Raman scattered light for many spectroscopic applications. The discussion is divided into two parts: (i) Raman shifting in hydrogen, deuterium, and methane gas; and (ii) Stimulated Electronic Raman Scattering (SERS) in alkali vapors.

Laser Experiments for Chemistry and Physics. First Edition. Robert N. Compton and Michael A. Duncan.
© Robert N. Compton and Michael A. Duncan 2016. Published in 2016 by Oxford University Press.

Raman shifting in hydrogen and methane gases

The Raman scattering effect has been treated in some detail in Chapters 19 and 22. Normal Raman scattering involving molecular gases can be discussed with reference to Figure 27.1, in which photons of frequency ν_L are scattered from a molecule, resulting in no change in frequency (Raleigh scattering) or relaxation into a different vibrational state of higher or lower energy.

It is also possible that the final state involves electronic or rotational excitation, but we do not consider such scattering here. The Raman scattering process in which molecules have gained energy through vibrational excitation and the Raman scattered light is lower in frequency is called Stokes scattering. If the molecule exists initially in an excited state and scattering occurs to a lower state of excitation, the Raman process is termed anti-Stokes Raman scattering. The Stokes and anti-Stokes radiation is scattered in various directions according to the symmetries of the molecular states involved and the polarization of the incident laser light. If the molecular species are subject to intense laser light, stimulated Raman scattering (SRS) can occur and the radiation propagates along (and against) the laser direction. If we designate the vibrational frequency of the molecule as ν_{vib}, and assume to first approximation that the vibrational levels are harmonic, we can write the Stokes scattering frequencies as $\nu_S = (\nu_L - k\nu_{vib})$ and the anti-Stokes frequencies as $\nu_{aS} = (\nu_L + k\nu_{vib})$, where $k = 1, 2, 3, 4$, etc. indicates the order of the scattering. Anti-Stokes scattering results from stimulating vibrationally excited

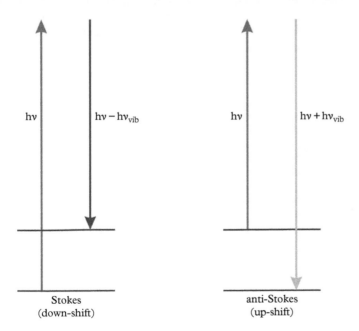

Figure 27.1 *Illustration of the Raman shift of incident radiation resulting in the down-conversion or up-conversion of laser light. The process of Rayleigh scattering (elastic scattering of light) is not shown. Up-conversion requires the presence of vibrationally excited molecules.*

(i.e., heated) molecules and thus the intensity of the anti-Stokes lines is proportional to a Boltzmann population of the levels involved. The important aspect of SRS is that the properties of the scattered Stokes and anti-Stokes beams follow that of the incident laser in both temporal and spacial coherence. Because the Raman-scattered light has all of the properties of a laser beam, SRS can be used to up- and down-convert a laser beam frequency, provided that the scattered intensities are large enough. There are many papers demonstrating the SRS for molecular spectroscopy applications. Since hydrogen gas has the highest known vibrational frequency (4159 cm^{-1}, see Chapter 14), it follows that the largest shifts are observed for H_2. For this reason, Raman shifting in hydrogen is commonly employed in commercial devices.

Bloembergen [1] has presented a rather complete description of stimulated Raman scattering. A brief summary of this treatment follows. The intensity experienced in a stimulated Raman process can be expressed through a Raman gain coefficient, g_R (see [1] and [2])

$$g_R = [2\lambda_S^2 \Delta N / (\pi h c^2 \nu_S \Delta \nu_R] d\sigma/d\Omega \tag{27.1}$$

where λ_S is the Stokes wavelength (cm), h is Planck's constant (J·s), c is the speed of light (ms^{-1}), ΔN is the population difference between the initial and the final energy levels (cm^{-3}), $\Delta \nu_R$ is the Raman linewidth (FWHM in cm^{-1}), and $d\sigma/d\Omega$ is the differential Raman cross-section (cm^2/sr). Papayannis et al., [2] among others, have examined the efficiencies for Raman scattering for the case of Nd:YAG laser wavelengths. The SRS intensity generated along the path through the hydrogen gas is a function of the pressure, P, laser pump intensity, I, and the Raman gain coefficient, g_R. Assuming that the initial pump laser is not depleted during its passage through the gas, the intensity, I(z), of the Raman beam as a function of the distance, z, along the path can be written as

$$I(z) = I(0) \, e^{g_R I z} \tag{27.2}$$

I(0) is the initial intensity of the spontaneous Raman scattering. The hydrogen pressure appears in both ΔN and $\Delta \nu_R$. As an example, for the third harmonic of the Nd:YAG laser, $\Delta \nu_R = 11.2/P + 1.58P$, where P is given in atmospheres and $\Delta \nu_R$ is in units of cm^{-1}. For values not too much above atmospheric, the SRS intensity increases with pressure. However, at higher pressures the SRS begins to fall off. Papayannis et al. examined SRS for Stokes scattering in hydrogen at 355 nm with laser pump energies of 5–105 mJ and pressures of 1–13 atm.[2] If we designate kS and kAS for the Stokes and anti-Stokes signal for order k, they reported efficiencies for 1S and 2S as 50 and 30%, respectively, at a pressure of ~7 atm using 82 mJ pump energies. Thus, SRS can be employed to produce usable frequencies outside the normal range of the laser employed. By using heated gases, anti-Stokes Raman scattering (kAS, up-shifting) is possible. In Figure 27.2 we show the ranges of frequencies available for a given pump frequency for both Stokes and anti-Stokes scattering.

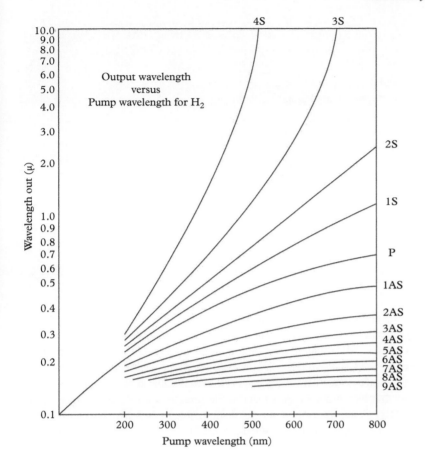

Figure 27.2 *Range of SRS wavelengths available for various pump wavelengths. S and AS refer to Stokes and anti-Stokes beams.*

Commercial Raman shifters are available which represent a nice addition to extend the wavelengths of tunable lasers. Figure 27.3 shows one of the Raman shifters from Quanta Ray/Spectra Physics employed in the laboratories at the University of Tennessee.

Gases other than H_2, such as deuterium and methane, have also been used to generate useful SRS. The SRS technique has also been employed to obtain Raman spectra at ultrashort wavelengths. Fodor et al. used SRS in H_2 gas to construct a VUV Raman spectrometer that operates down to 184 nm.[3] Observing normal Raman spectra in the UV has advantages because of the fourth power dependence on frequency and the possibility of exciting resonance Raman states in some molecules. Additionally, performing Raman spectroscopy using VUV laser light can reduce flourescence in certain cases. A major drawback of up-shifting using Raman anti-Stokes scattering is the necessity of heating the sample.

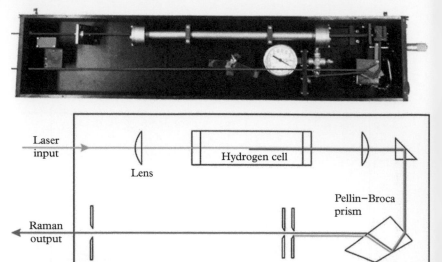

Figure 27.3 *Diagram showing the hydrogen Raman shifter and beam collimating system. In this case a blue laser beam enters the Raman cell filled with hydrogen. The left-over blue beam is dispersed using a Pellin–Broca prism and dumped at the beam slits shown after the prism. The "red" Stokes beam is directed through the output appertures as indicated.*

The construction of a Raman cell is relatively simple. One simply requires a pipe to hold the active gas (H_2, D_2, or CH_4) along with the proper windows at each end of the chamber plus the appropriate dispersion optics. Heating tapes can be used to warm the cell. However, hydrogen and methane are combustable gases and extreme care should be used to make sure that the cell is vacuum tight and that its windows can withstand the pressures involved (roughly 10 atm or 1 MPa). A well-tested commercial device is recommended. There are many vendors of Raman shifters; some using capillaries to hold the compressed gases at low volumes ($\sim$100 cm^3) for safer operation (see Appendix I, p. 397).

Stimulated electronic Raman scattering in alkali vapors

Historically, nonlinear laser interactions with solids have received the most attention for optical applications, e.g. second harmonic generation (SHG) in asymmetric crystals. Nonlinear optical studies in atomic vapors were first initiated by New and Ward [4,5] who reported third harmonic generation (THG) in alkali vapors pumped with a high-power ruby laser. In Chapter 16 we discussed THG as well as ionization in alkali metal heat pipes. Below we describe a method for efficient generation of tunable infrared light using stimulated electronic Raman scattering (SERS) in alkali metal vapor heat pipes. SERS has been treated in a number of studies.[6–9] Stimulated electronic Raman scattering in alkali atoms involves tuning a laser from the nS state of the alkali atom near an allowed $(n + 1)P$

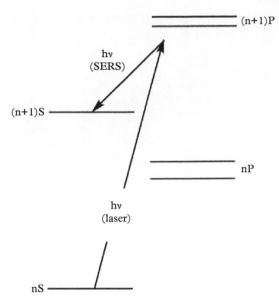

Figure 27.4 *Illustration of the generation of SERS in a typical alkali vapor. The laser is tuned near the (n+1)P state and the SERS results from emission to the lower (n + 1)S state. The $^2P_{3/2}$ and $^2P_{1/2}$ fine-structure levels are shown but not labeled.*

state followed by stimulated scattering to the lower (n + 1)S state as illustrated in Figure 27.4. SERS can be generated via tuning near all (n+1)P states, resulting in lower and lower SERS frequencies with increasing n, albeit with smaller efficiencies. The SERS experimental setup employed in the Advanced Laboratory at the University of Tennessee is shown in Figure 27.5. Light from a nitrogen laser-pumped dye laser is directed into the alkali heat pipe. A few percent of the IR light is diverted to an IR detector using a ZnSe plate. The germanium plates transmit IR but block the visible laser beam. The signal generated at the detector is used to normalize the transmitted IR laser intensity passing through the absorption cell. The ZnSe plate also serves to block the primary dye laser beam. After passing through the absorption cell, the IR beam is reflected and focused into an IR detector. With no absorbing gas in the cell, the intensity of the IR light can be recorded. Figures 27.6 and 27.7 show the IR intensity profile versus laser wavelength upon tuning above and below the $6p$ and $7p$ fine-structure levels for Rb and Cs, respectively.

The energy conversion efficiency for the SERS process is large (20–40%). However, there is a structured intensity dependence versus wavelength, as seen in Figures 27.6 and 27.7. This light can be used to measure IR absorption spectra for many gases. As in any spectrometer, it is best to record the ratio of the transmitted beam divided by the incident beam to remove the intensity variations in the IR light.

As a demonstration of the use of SERS for infrared spectroscopy, we chose the methane and ammonia molecules. Methane is of growing importance as a major

Figure 27.5 *A picture of the SERS infrared spectrometer setup is shown at the top with the various elements diagrammed below. The germanium filters block the primary laser beam, allowing the IR to be transmitted. Signal from the top IR detector is divided by that from the bottom one to give a relative absorption cross-section.*

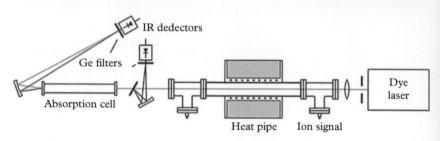

Figure 27.6 *The intensity profile of infrared laser light generated by SERS near the two $6P_{1/2}$ and $6P_{3/2}$ levels of rubidium at 4202.9 and 4216.7 Å, respectively. The range from 4170 to 4260 Å of the laser wavelength corresponds to the energy range of the SERS light from 3847 cm^{-1} to 3340 cm^{-1}.*

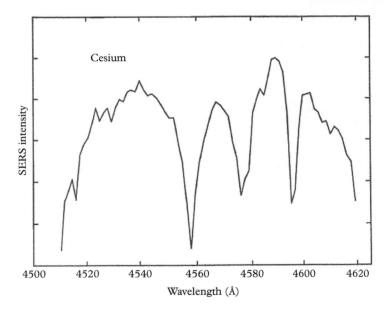

Figure 27.7 *The intensity profile of infrared laser light generated by SERS near the two $7P_{1/2}$ and $7P_{3/2}$ levels of cesium at 4202.9 and 4216.7 Å, respectively. The range from 4500 to 4620 Å of the laser wavelength corresponds to the energy range of SERS light from 3202 to 3110 cm^{-1}.*

natural resource for combustion and is also an important greenhouse gas.[10] The ammonia molecule was employed for the first stimulated-emission process, which led to the MASER (microwave amplification by stimulated emission of radiation) and eventually the LASER. Ammonia also demonstrates the effect of inversion symmetry in molecular spectroscopy, which was an important concept in the development of the MASER.

The methane molecule is a tetrahedral spherical-top molecule belonging to the T_d point group. Its infrared spectroscopy is discussed in Chapter 23. Methane has five atoms resulting in $3 \times 5 - 6$ or nine vibrational degrees of freedom. There are four normal modes, as illustrated in Figure 27.8. ν_1 is singly degenerate, ν_2 is doubly degenerate, and both ν_3 and ν_4 are triply degenerate, making a total of nine vibrational degrees of freedom. The ν_1 vibration is totally symmetric and is Raman active. The ν_2 vibration is also only Raman active. The ν_3 and ν_4 vibrations are both infrared and Raman active. Note that for both ν_3 and ν_4 the hydrogen atoms and the carbon atom move during the vibration giving rise to a fluctuating dipole moment. (The student should use the Gaussian program described in Chapter 15 to calculate the IR and Raman vibrations for methane to verify these statements.) Figure 27.9 shows a region of the P-branch of ν_2 covering eleven rotational lines of the rovibrational spectrum of methane when the visible laser scans from 4600 to 4620 Å (3202 to 3110 cm^{-1} in the infrared) in Figure 27.7.

The rotational energy levels of CH_4 are triply degenerate. However, the Coriolis interaction splits this degeneracy resulting in three lines separated by $2B_v\zeta_r(J+1)$, where ζ_r is the Coriolis coupling constant. For methane the Coriolis

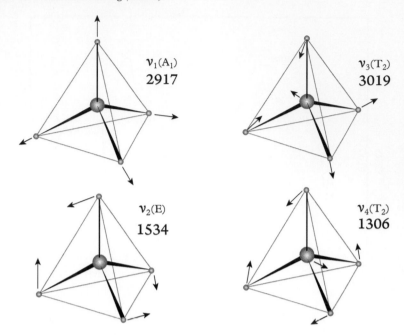

Figure 27.8 *Diagram showing the four normal modes for the methane molecule with the frequencies given in units of cm⁻¹. The frequencies quoted are from Nakamoto.[11]*

coupling constant is very small (~0.054) and as a result this splitting is not evident in Figure 27.9. This same spectrum was measured in a different way in Chapter 23.

The ammonia molecule (NH_3) has a pyramidal shape with the nitrogen atom at the top of the pyramid and the three hydrogen atoms at the base, as illustrated in Figure 27.10. The distance from the nitrogen atom to the base of the hydrogen triangle is only 0.38 Å, so the molecule is actually "flatter" than indicated in the figure. In fact, there is a small barrier for the "inversion" of the molecule leaving the nitrogen atom below the plane containing the three hydrogen atoms. Thus the nitrogen atom experiences a double well potential as indicated in Figure 27.11.[13]

Classically, if an ammonia molecule finds itself on the right (+) side of this potential it will stay there forever. However, quantum mechanically the molecule can invert or tunnel through the barrier and find itself on the left (−) side. Thus the wavefunction describing a molecule in either side of the well will contain a small probability for being on the other side. Letting ψ^+ and ψ^- represent the wavefunctions for the molecule on the right (+) and left (−) side, respectively, we can write a set of even and odd wavefunctions that satisfy the requirement of the normalized total wavefunction $\Psi(r) = \pm\, \Psi(-r)$:

$$\Psi_{\text{even}} = (1/\sqrt{2})\,[\psi^+ + \psi^-] \tag{27.3}$$

$$\Psi_{\text{odd}} = (1/\sqrt{2})\,[\psi^+ - \psi^-] \tag{27.4}$$

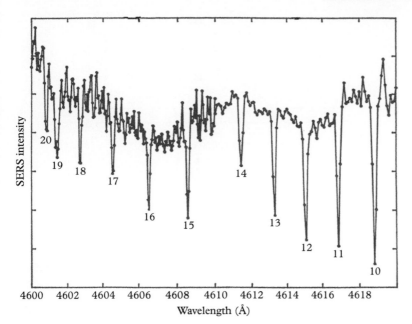

Figure 27.9 *Infrared absorption spectrum of methane in the P-branch of the ν_3 vibration.[12] The laser wavelength spans from 4600 to 4620 Å corresponding to 3202–3110 cm^{-1} of the SERS radiation.*

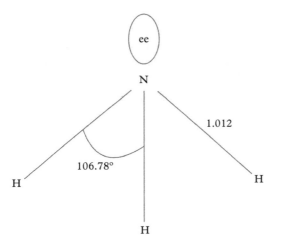

Figure 27.10 *Diagram of the ammonia molecule in which all three of the HNH angles are 106.78° and the three NH bond lengths are 1.012 Å. The lone-pair electrons accompanying the nitrogen atom are indicated.*

Solutions of the Schrödinger equation using these two wavefunctions result in the even functions having a lower energy than that of the odd function. This splitting (~0.8 cm^{-1}) is indicated in Figure 27.11 and leads to an inversion frequency of ~2.4 × 10^{10} per second. Interestingly, the barrier for the PH$_3$ molecule is ~6085 cm^{-1} and leads to an inversion frequency of 1.4 × 10^5 per second. Due to

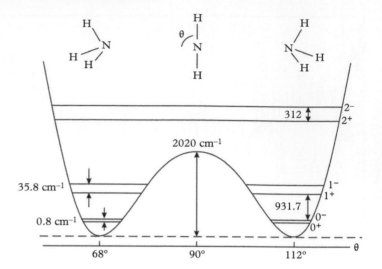

Figure 27.11 *Double well potential showing a barrier to the inversion of the two forms of NH₃ (energies not to scale).*

Table 27.1 *Fundamental vibrations of the ammonia molecule.[17]*

Vibration	Frequency (cm⁻¹)
ν_1 (a₁)	3336.2
	3337.2 strong
ν_2 (a₁)	932.5
	968.3 strong
ν_3 (e)	3443.6, 3443.9
ν_4 (e)	1626.1, 1627.4 very strong

the larger arsenic atom, the barrier for AsH_3 is 11,220 cm⁻¹ and the inversion frequency becomes ~½ inversion per day!

The energy difference between the even and odd energy levels in Figure 27.11 is in the microwave region (0.8 cm⁻¹ or 23.8 GHz).[14] This was the stimulated emission transition studied by the Charles Townes group at Columbia University leading to the development of the MASER.[15]

The inversion splitting of the upper vibrational levels becomes increasingly large. This is seen for the v = 1 level in Figure 27.11. Also transitions from the ground to the upper vibrational levels are governed by the parity selection rules (+ ↔ − or − ↔ + for IR transitions and + ↔ + or − ↔ − for Raman transitions). This is nicely illustrated in Figure 78 on page 257 of the classic book by Herzberg.[16] Table 27.1 presents the fundamental vibrations of the ammonia molecule.

The ν_2 bending vibration results from the + ↔ − and − ↔ + combinations shown in Figure 27.11. A portion of the ν_3 N–H stretch vibrational spectrum recorded using SERS radiation is presented in Figure 27.12. The inversion doublet lines are clearly evident in the recorded spectra. Many other regions of the ammonia spectrum can be examined, as well as many other molecules. Since the SERS radiation is essentially a laser beam it can be directed over long distances (and back) to record low concentrations of atmospheric gases. Although these experiments look complicated they are very straightforward and can be performed in a single afternoon class period.

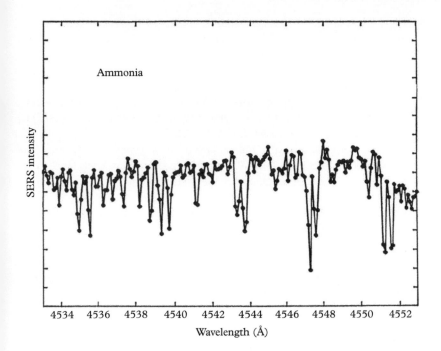

Figure 27.12 *A portion of the ν_3 vibrational spectrum of ammonia using SERS infrared light. The laser wavelength spans from 4534 to 4552 Å corresponding to 3520 to 3433 cm^{-1} of the SERS radiation.*

REFERENCES

1. N. Bloembergen, "The stimulated Raman effect," *Am. J. Phys.* **35**, 989 (1967).
2. A. D. Papayannis, G. N. Tsikrikas, and A. A. Serafetinides, "Generation of UV and VIS laser light by stimulated Raman scattering in H_2, D_2, and H_2/He using a pulsed Nd:YAG laser at 355 nm," *Appl. Phys. B* **67**, 563 (1998).
3. S. P. A. Fodor, R. P. Rava, R. A. Copeland, and T. G. Spiro, "H_2 Raman-shifted YAG laser ultraviolet Raman spectrometer operating at wavelengths down to 184 nm," *J. Raman Spectros.* **17**, 471 (1986).
4. C. H. C. New and J. F. Ward, "Optical third-harmonic generation in gases," *Phys. Rev. Lett.* **19**, 556 (1967).
5. J. F. Ward and C. H. C. New, "Optical third harmonic generation in gases by a focused laser beam," *Phys. Rev.* **185**, 57 (1969).
6. J. J. Wynne and P. P. Sorokin, *Nonlinear Infrared Generation*, Y.-R. Shen, ed., Series on Topics in Applied Physics, Vol. 16, Springer-Verlag, 1977, p. 159.
7. J. F. Reintjes, in *Nonlinear Optical Parametric Processes in Liquids and Gases*, Academic Press, New York, 1984.

8. D. C. Hanna, M. A. Yuratich, and D. Cotter, in *Nonlinear Optics of Free Atoms and Molecules*, Vol. 17 Chapter 5 of Springer Series in Optical Sciences, Springer-Verlag, Berlin, 1979.

9. Y.-R. Shen, *The Principles of Nonlinear Optics*, John Wiley and Sons, New York, 1984, p. 164.

10. See e.g. V. Boudon, J. P. Champion, T. Gabard, G. Pierre, M. Loete, and C. Wenger, "Spectroscopic tools for remote sensing of greenhouse gases CH_4, CF_4 and SF_6," *Environ. Chem. Lett.* **1**, 86 (2003).

11. K. Nakamoto, *Infrared and Raman Spectra of Inorganic and Coordination Compounds*, fifth edition, John Wiley and Sons, Inc., New York, 1997.

12. K. T. Hecht, "Vibration-rotation energies of tetrahedral XY_4 molecules: Part II. The fundamental ν_3 of CH_4," *J. Mol. Spec.* **5**, 390 (1950).

13. J. M. Hollas, *High Resolution Spectroscopy*, second edition, John Wiley & Sons, Chichester, UK, 1998, p. 253.

14. C. H. Townes and A. L. Schowlow, *Microwave Spectroscopy*, Dover Publications, Inc., New York, 1975.

15. J. P. Gordon, H. J. Zeigler, and C. H. Townes, "The MASER: New type of microwave amplifier, frequency standard, and spectrometer," *Phys. Rev.* **99**, 1264 (1955).

16. G. Herzberg, *Infrared and Raman Spectra*, D. Van Nostrand and Co., New York, 1945.

17. T. Shimanouchi, "Molecular Vibrational Frequencies" in NIST Chemistry WebBook, NIST Standard Reference Database Number 69, P. J. Linstrom and W. G. Mallard eds., National Institute of Standards and Technology, Gaithersburg MD, 20899, http://webbook.nist.gov.

Part V

Laser Experiments for Kinetics

Fluorescence Lifetime of Iodine Vapor

Introduction

When molecules absorb light of high enough energy (visible or ultraviolet wavelengths) their electrons may be rearranged to a distribution of populated molecular orbitals different from the most stable (ground state) configuration. Configurations of electrons other than the most stable one for a particular molecule are called electronically excited states, and every molecule has many such states characterized by their respective energies. Electronic spectroscopy is the study of light absorption in the visible or UV wavelength regions to measure the energies of electronic excited states. In this way, fundamental information about structure and bonding may be obtained.

Molecules that are produced in electronically excited states eventually relax to their more stable ground electronic state configuration. As discussed in Chapters 3 and 14, relaxation can occur by the emission of a photon or by any one of several non-radiative processes (internal conversion, intersystem crossing, etc.). However, the primary mechanism for relaxation of isolated small molecules, where intramolecular non-radiative processes are not efficient, is fluorescence, or re-emission of radiation, leaving the molecule in its ground electronic state along with accompanying rovibrational energy (see Figure 14.22). In addition to fluorescence, the energy in an excited state of small molecules may also be lost by collisional transfer to another species or, in some cases, by breaking a bond (dissociation) or ionization of the excited species. In all cases, however, the average time required for relaxation of the excited state is referred to as the lifetime, τ, of that state. Excited-state lifetimes are also an important way to investigate fundamental properties of molecular systems.

If fluorescence is the only relaxation mechanism for an excited species, the fluorescence lifetime is related to the absorption strength for creating that excited state. As discussed in Chapter 3, Einstein expressed this in a rate expression for spontaneous emission,

$$A_{10} = \left(64\pi^4/3hc^3\right) v^3 \, |<R_{10}>|^2 \tag{28.1}$$

A_{10} is the "Einstein coefficient for spontaneous emission" (i.e., it is a rate constant), and $R_{10}{}^2$ is the square of the dipole moment transition integral (equivalent to

Laser Experiments for Chemistry and Physics. First Edition. Robert N. Compton and Michael A. Duncan.
© Robert N. Compton and Michael A. Duncan 2016. Published in 2016 by Oxford University Press.

absorption probability). The inverse of A_{10} is the "radiative" (or "natural") lifetime of the excited state,

$$\tau_{rad} = 1/A_{10} \tag{28.2}$$

Therefore, if the absorption strength is large, the radiative lifetime will be short, and vice versa. To continue this kinetic treatment, it can be shown that an ensemble of excited molecules decaying by fluorescence follows a first-order rate law. In the usual situation, molecules are excited by a pulsed laser whose pulse duration is as short as possible (typically a few nanoseconds from a Nd:YAG, nitrogen, or excimer laser, or a dye laser pumped by one of these). The excited state population produced is allowed to relax before any subsequent laser pulses. If the number of excited molecules at any time after excitation is given by [N], then the decay rate may be written as,

$$d[N]/dt = -k_{rad}[N] \quad \text{(where } k_{rad} = A_{10}) \tag{28.3}$$

Grouping like terms and integration gives,

$$\ln [N]_0 - \ln [N]_t = -k_{rad}\, t \tag{28.4}$$

and taking the exponential of both sides gives the variation of fluorescence intensity with time:

$$[N]_t = [N]_0\, e^{-k_{rad}t} \tag{28.5}$$

Where $[N]_0$ is the number of molecules initially excited. As indicated by this expression, and verified by experiment, the fluorescence (which is proportional to excited state population) decays exponentially. The lifetime of the exponential decay is τ_{rad}.

As noted in Chapter 3, if there are other decay channels for excited molecules in addition to fluorescence, then the overall decay rate will be greater than that due to fluorescence alone, i.e.

$$k_{total} = k_{rad} + k_D + k_C + \cdots \tag{28.6}$$

where k_D and k_C are rates of dissociation and collisional energy transfer, respectively. Collectively, these and any other processes are sometimes referred to as "non-radiative" decay and combined in a single rate constant, k_{NR}. When non-radiative decay increases the overall decay rate of excited molecules, the excited-state lifetime, as observed by the decay of fluorescence, (τ_F), is shorter than the radiative lifetime. If the radiative lifetime is known from the absorption strength, this lifetime shortening can be used to measure the rate of non-radiative decay.

Collisional energy transfer, or "fluorescence quenching," is one particular form of non-radiative decay for which rates can be measured even if the natural lifetime is not known. This is because of its dependence on the collision frequency with the quencher (i.e., pressure). If τ_0 is the fluorescence lifetime at zero added gas pressure and Q is the rate constant for quenching by an added gas at concentration, [M], then the kinetic expression for excited-state decay is,[1]

$$d[N]/dt = -[N]/\tau_0 - Q[M][N] \tag{28.7}$$

After integration, this becomes,

$$[N]_t = [N]_0\, e^{-t/\tau_F} \tag{28.8}$$

The fluorescence lifetime, however, is now given by,

$$1/\tau_F = 1/\tau_0 + Q[M] \tag{28.9}$$

Note that the units of Q[M] must be $\sec^{-1}$; this may be thought of as a collision frequency. Using the ideal gas law, this expression may be written as,

$$1/\tau_F = 1/\tau_0 + QP_m/RT \tag{28.10}$$

where P_m is the pressure of added gas. The form of this equation suggests that a plot of $1/\tau_F$ versus P_m should give a straight line with slope of Q/RT and intercept of $1/\tau_0$. This is called a Stern–Volmer plot and may be used to determine the quenching rate constant Q. Fluorescence quenching rates are often studied for a variety of collision partners to determine how mass and the electronic, vibrational, and rotational states of the quencher (which must be considered in energy conservation) affect quenching efficiency.

Procedure

In this experiment, the fluorescence lifetimes and quenching behavior of excited-state iodine molecules in the gas phase are studied. For this experiment, a fast-pulsed (duration 5 nsec) Nd:YAG laser (e.g., Continuum "MiniLite" model or New Wave Research, Polaris II) operating at the green 532 nm wavelength is passed through a fluorescence cell containing the sample. Fluorescence is detected perpendicular to the laser path by a photomultiplier tube. The output of the phototube may be viewed directly on an oscilloscope to observe the exponential fluorescence decay. At the same time, the fluorescence signal is captured by a digital oscilloscope (e.g., LeCroy). The signal is transferred to the computer for plotting and curve fitting with an IEEE-488 interface cable. A schematic diagram of this setup appears in Figure 28.1. An example of experimental data is shown in Figure 28.2.

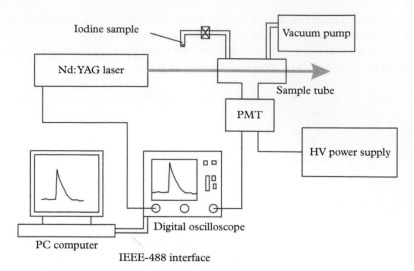

Figure 28.1 *The experimental setup for iodine fluorescence lifetime measurements.*

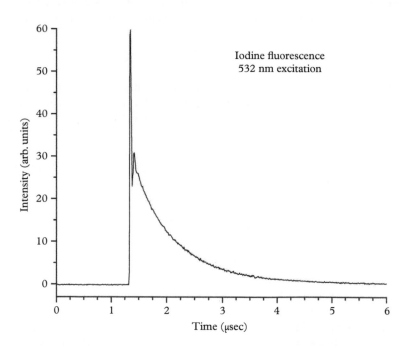

Figure 28.2 *An experimental trace of the fluorescence lifetime of iodine vapor, using excitation with the second harmonic of a Nd:YAG laser at 532 nm. The spike of signal near early time is from scattered light from the excitation laser. The lifetime of the fluorescence is about 1.0 μsec.*

The intensity versus time data can be fit with a linear function using a semi-log form as in equation 28.4, where the slope of the line is k_{rad}, or it can be fit directly to an exponential using any standard database software. The fluorescence lifetime is then $1/k_{rad} = \tau_F$. Figure 28.2 shows that real experimental data may include scattered light from the excitation laser, which appears as a spike of signal occurring sharp in time near the beginning of the decay curve. The pulsewidth of the Nd:YAG laser is about 5 nsec, explaining the short time width of this signal. Such a signal can be reduced by using a filter that blocks the excitation wavelength but transmits toward longer wavelengths. However, a scattered light signal like that here does not interfere with the experiment significantly. Because of the unique behavior of first-order exponential decays, the lifetime can be fit correctly including only that portion of the data which occurs after this spike.

The visible $X\ ^1\Sigma_g^+ \rightarrow B\ ^3\Pi_{0u}^+$ spectrum of iodine vapor has been studied extensively,[2] and is discussed in Chapters 17 and 18 of this book. The 532 nm line from the YAG laser falls in the middle of the absorption spectrum, which occurs throughout the visible wavelength region. The fluorescence lifetime with this excitation wavelength should be about 1.0 μsec, as shown in Figure 28.2. Initially, the lifetime should be measured in the absence of any added gas. However, if the pressure of iodine vapor is too high, self-quenching may be observed, producing a lifetime shorter than expected. Reduction of the pressure with a vacuum pump can eliminate this effect. Subsequently, various gases provided in the lab (Ar, N_2, etc.) may be used for fluorescence quenching by adding these one at a time via a gas inlet with a pressure gauge (e.g., baratron gauge) that measures pressure in the torr range. The pressure dependence of these phenomena should be used to determine fluorescence quenching rate constants.

Note: Iodine vapor is corrosive to stainless steel and electronic parts. All surfaces coming in contact with it should be cleaned thoroughly after use.

In addition to the iodine vapor experiment described here, another popular fluorescence lifetime experiment uses a solution containing $Ru(bipy)_3$, a well-known inorganic transition metal complex.[3] $Ru(bipy)_3$ solution absorbs strongly at the 532 nm wavelength of the YAG laser, and also has a lifetime in the μsec range. If the solution is open to air, oxygen dissolves in it, which quenches the fluorescence. A slow purge of argon, bubbled through the sample with a gas line, degases the solution, and the lifetime returns to its normal value. This process can be watched in real time on the oscilloscope to demonstrate the effect of quenching.

Fluorescence lifetimes are used widely to investigate the excited states and photochemistry of larger organic molecules.[4,5] In stable molecules such as these, the ground electronic state is closed shell, and the electron spins are paired to form a so-called "singlet" electronic configuration. When an electron is excited to a higher state, the allowed transition leaves its spin unchanged, opposite to that of its partner electron left behind in the HOMO; this produces an overall singlet excited state of the system. However, if intersystem crossing occurs, the electron flips its spin so that it has the *same* orientation as that of the unpaired electron in the HOMO, and this produces an overall triplet excited state of the system. When intersystem crossing occurs, which is quite common in larger organic molecules, it reduces the fluorescence quantum yield and lifetime. However, the triplet states produced can also emit light to relax, although it is a much less efficient process than fluorescence. The triplet → singlet emission is forbidden by spin selection rules, and is therefore very inefficient and slow; its corresponding lifetime is thus long (msec to sec). This long-lived emission from triplet states is known as phosphorescence, which is familiar from many "glow-in-the-dark" materials. Experiments measuring phosphorescence in the undergraduate lab were described many years ago.[6] Because the emission is slow, the pulse light source used need not be a fast-pulsed laser, but could be a flashlamp. The phosphorescence of naphthalene derivatives discussed in [6] could be studied with the same apparatus employed here by switching the Nd:YAG laser excitation to the fourth harmonic wavelength at 266 nm where those molecules absorb. Excimer laser excitation at the KrF wavelength of 248 nm would also work.

Fluorescence of larger organic molecules is the basis of dye laser emission, as discussed in Chapter 4. For this application, molecules must have high quantum yields and fast lifetimes, with minimal intersystem crossing. An interesting aspect of the emission of such large molecules in solution is that their fluorescence emission is the same regardless of the excitation wavelength. In the case of laser dyes, such as the common rhodamine 6G molecule (Figure 4.7), the emission is basically the same when it is pumped with the green output from a Nd:YAG laser or from ultraviolet sources such as an excimer. This phenomenon was observed and explained many years ago by Professor Michael Kasha, and is known as "Kasha's Rule."[7] Even if more highly excited states are produced than that corresponding to populating the LUMO, rapid relaxation in solution happens before emission, and it is usually the lowest excited state that fluoresces.

Safety notes

1. Be sure to read the laser safety section of this book and receive appropriate laser safety training before beginning this experiment! The green Nd:YAG laser beam is quite bright and there is much reflection as it passes through windows and general glare from the optics. It is strongly recommended that the entire laser area be enclosed under a black box to block the glare and reflections.

2. This experiment uses pressurized gases. If the fluorescence tube is over-pressurized, windows, gas fittings, or other pieces can be launched at high velocity from the experiment. Because of this, laboratory safety glasses must be worn at all times.

Questions for consideration

1. How does the observed fluorescence lifetime compare with the natural (radiative) lifetime for the *B* state of iodine? If it is different, discuss why.

2. What other processes besides quenching of added gases should be considered to understand the observed fluorescence lifetimes?

3. Compare the relative efficiencies of different added gases for fluorescence quenching. What aspects of the quencher gas structures are responsible for efficient or inefficient quenching?

....

REFERENCES

1. G. A. Capelle and H. P. Broida, "Lifetimes and quenching cross sections of I_2 ($B\,^3\Pi_{0u}{}^+$)," *J. Chem. Phys.* **58**, 4212 (1973).

2. K. P. Huber and G. Herzberg, *Constants of Diatomic Molecules*, Van Nostrand Reinhold Co., New York, 1979, p. 332.

3. J. N. Demas, "Luminescence decay times and bimolecular quenching," *J. Chem. Educ.* **53**, 657 (1976).

4. M. Klessinger and J. Michl, *Excited States and Photochemistry of Organic Molecules*, VCH Publishers, New York, 1995.

5. N. J. Turro, V. Ramamurthy, and J. C. Scaiano, *Modern Molecular Photochemistry of Organic Molecules*, University Science Books, Sausalito, CA, 2010.

6. T. R. Dyke and J. S. Muenter, "An undergraduate experiment for the measurement of phosphorescence lifetimes," *J. Chem. Educ.* **52**, 251 (1975).

7. M. Kasha, "Characterization of electronic transitions in complex molecules," *Disc. Faraday Soc.* **9**, 50 (1950).

29

Raman Spectroscopy Applied to Molecular Conformational Analysis

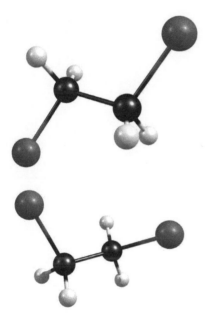

Figure 29.1 *Anti (top) and gauche (bottom) conformers of 1,2-dichloroethane. Sometimes these enantiomers are referred to as trans and cis conformers, respectively.*

Introduction

In this chapter we describe an experiment that employs Raman spectroscopy to examine the structure of molecular conformers. The measurements are compared to the results of calculations performed with the Gaussian program. Conformational isomers represent one form of stereoisomer in which the two structures can be interconverted by simple rotations about a single bond, usually a σ bond. Such conformational isomers or conformers are also called rotamers. Conformational isomers are different from other stereoisomers in which one or more bonds need to be broken in order to obtain another isomer.

The 1,2-dichloroethane molecule can exist in either of two rotamer forms, as shown in Figure 29.1. The conformation in which the two chlorine atoms occupy distances as far apart as possible is called the *anti* conformation while that in which the chlorine atoms are as close as possible is called the *gauche* conformation, as shown in the figure. The geometrical difference between the two conformers can best be seen by viewing along the axis of the carbon–carbon single bond. This is called a Newman projection and is depicted in Figure 29.2.

One might correctly expect that the anti conformer is lower in energy (more stable) than the gauche conformer since there will be fewer chlorine–chlorine repulsive interactions (see any modern organic chemistry textbook [1,2]). The student should perform calculations of the total energy for the anti and gauche 1,2-dichloroethane molecule using Gaussian and the B3LYP/6-311+G* method/basis set described in Chapter 15 to test this. A rotational energy barrier of ~3.5 kcal/mole exists for the conversion from the anti to gauche configuration. The student can estimate the barrier height by calculating the energy of the eclipsed form of the molecule. This low barrier means that at room temperature the gauche and anti conformers are in rapid equilibrium, rotating back and forth over the barrier, spending time in both configurations. At low temperatures the molecule resides primarily in the lower energy anti form and at high temperatures the two conformers are approximately equal in abundance except for the fact that the gauche form has two distinct positions, g^+ and g^-, as shown in

Laser Experiments for Chemistry and Physics. First Edition. Robert N. Compton and Michael A. Duncan.
© Robert N. Compton and Michael A. Duncan 2016. Published in 2016 by Oxford University Press.

Figure 29.3. The two conformers, g⁺ and g⁻, are mirror image molecules and are called enantiomers (see Chapter 20).

Experimental details

Using Raman spectroscopy together with calculations, it is easy to show that at low temperatures only the anti form exists to any appreciable extent. Referring to Chapter 19 on Raman spectroscopy under liquid nitrogen (RUN), one can see that the RUN spectrum should contain only the anti form and the room temperature Raman spectrum should exhibit vibrational peaks due to both conformers. Figure 29.4 illustrates this nicely in the room temperature and RUN spectrum of 1,2-dichloroethane.

The Raman spectrum for the anti and gauche conformers can be calculated separately using the Gaussian program with the results shown in Figure 29.5. Comparing the spectra in Figure 29.5, the assignments for the two conformers can be readily verified. From a measurement of the Raman spectrum and an assumption that the Raman intensities are representative of the population at a given temperature, one can use the well-known van't Hoff relationship

$$\ln\left[\frac{P_{gauche}}{P_{anti}}\right] = -\left(\frac{\Delta H_f}{R}\right)\left(\frac{1}{T_2} - \frac{1}{T_1}\right) \tag{29.1}$$

to determine the enthalpy of formation ΔH_f. $\left[\frac{P_{gauche}}{P_{anti}}\right]$ are the ratios of the populations (intensities) of the gauche to anti conformers and R is the gas constant. By plotting $\ln\left[\frac{P_{gauche}}{P_{anti}}\right]$ versus $1/T$ a line is obtained with a slope of $-\Delta H_f/R$. There are many ways to carry out this experiment, some more precise and accurate than others. A suggested way is to obtain a block of aluminum and bore a small indentation (~1/4 in. diameter) in which to place the 1,2-dichloroethane sample. Another hole is drilled into the block at a right angle to the sample hole to a depth just below the position of the indentation for the sample. A thermocouple (Radio Shack) is inserted into the second hole to monitor the temperature. The block and sample can be cooled to approximately dry ice temperature and allowed to warm to room temperature while recording Raman spectra. Cooling the sample to a temperature sufficiently low for the purposes of the analysis requires dry ice. The sample was cooled to ~0 °C for each measurement, but no lower. Instead, the 2 × 2 × 2 cm aluminum block described above was cooled to dry ice temperature inside a sealed container before being loaded with the "semi-cooled" liquid sample. 1,2-dichloroethane has a freezing point of −35 °C (significantly higher than the dry ice temperature of −78.5 °C) and a liquid sample was desired for analysis in this experiment. A digital multimeter/thermocouple was inserted into the aluminum block (with loaded sample) via a long wire. Finally, the block was placed under the objective lens of the Raman spectrometer. The experimental setup (without the thermocouple) is shown in Figure 29.6.

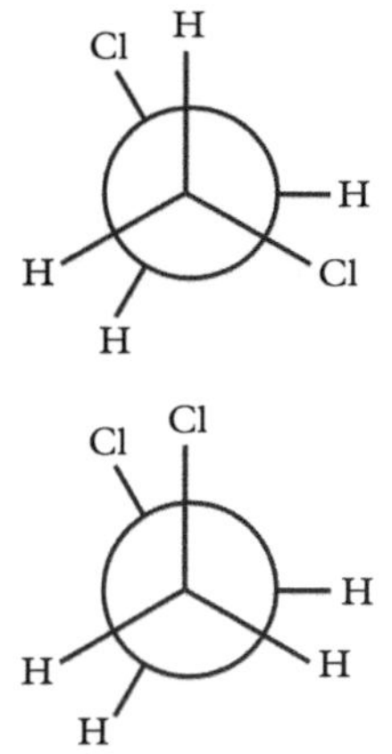

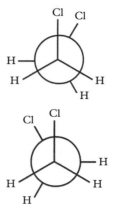

Figure 29.2 *Newman projection of the anti and gauche forms of 1,2-dichloroethane.*

Figure 29.3 *Newman projections showing the two (g⁺, left and g⁻, left) gauche forms of 1,2-dichloroethane.*

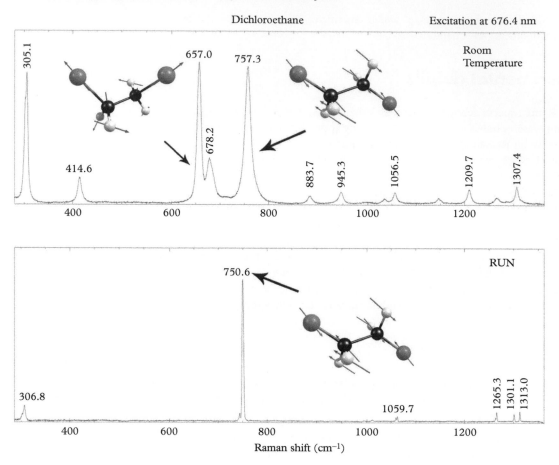

Figure 29.4 *Raman spectrum of 1,2-dichloroethane at room temperature (top) and at liquid nitrogen tempera-ture (bottom). Notice the disapearance of the second (reading from the left), third, and fourth peaks under liquid nitrogen which are attributed to the higher energy gauche conformer. The first peak is common to both the anti and gauche conformer.*

Figure 29.7 shows the Raman spectra of 1,2-dichloroethane at room temper-ature and dry ice temperature. The peak at ~650 cm⁻¹, attributed to the gauche conformer, is barely visible at dry ice temperature but is roughly equal to the anti peak at room temperature. As the sample warms up, the ratio of peaks changes accordingly.

It should be noted that P_{gauche} and P_{anti} are the relative populations of gauche and anti conformers, respectively, and they may be used to determine the equilibrium constant for the *cis/trans* isomerization. Finally, Figure 29.8 shows representative data obtained in a van't Hoff plot.

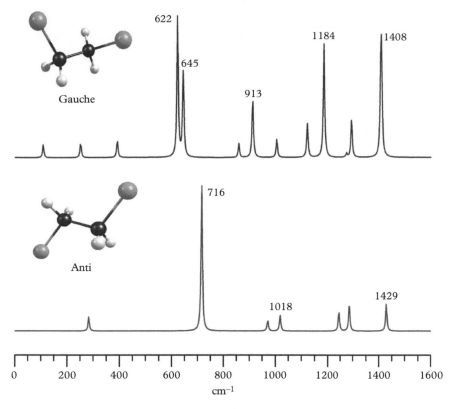

622
645
Gauche
913
1184
1408

716
Anti
1018
1429

0   200   400   600   800   1000   1200   1400   1600
cm^{-1}

Figure 29.5 *Raman spectra calculated for the gauche and anti conformers of 1,2-dichloroethane using the Gaussian program and the DFT/B3LYP method with the 6–311+G** basis set. The vibrations were scaled by a factor of 0.96.*

Figure 29.6 *Experimental arrangement for the measurement of the Raman spectrum of 1,2-dichloroethane. The thermocouple is placed into the hole at the front of the aluminum block. The tip of the thermocouple lies just below the indentation holding the liquid/solid sample.*

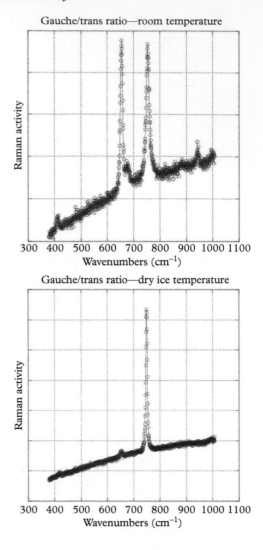

Figure 29.7 *Raman spectra of 1,2-dichloroethane at room temperature (top) and at dry ice (bottom) temperature. Note that the gauche peak is barely visible in the spectrum obtained for dry ice temperature (from undergraduate student N. Torrico, 2010).*

The average enthalpy difference between the gauche and anti forms, ΔH_f, was found to be 1.4 kcal/mol, with an R-squared value of 0.98. This value is compared to other values in Table 29.1.

A number of the values presented in Table 29.1 are energy differences obtained in various solutions. The energy of each conformer in a given solvent depends on the energy gained upon solvation. Since the gauche conformer of 1,2-dichloroethane has an appreciable dipole moment (calculated as 2.88 D with B3LYP/6-311+G*) one would expect that the gauche would be lower in energy than the anti conformer in a polar solvent because of solute–solvent dipole–dipole attractive interactions. Inspecting Table 29.1 reveals that the difference in the

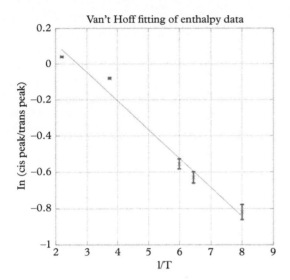

Figure 29.8 *Van't Hoff plot (equation 29.1) used to determine the ΔH_f for the energy difference of anti-gauche 1,2-dichloroethane.*

anti and gauche conformers for 1,2-dichloroethane in acetonitrile (dipole moment = 3.92 D) is only 0.15 kcal/mole, which agrees with these expectations. The interested student could make ΔH_f measurements for 1,2-dichloroethane in other polar solvents. Of course, the gauche conformer could be self-stabilized in the neat solution so that the value for ΔH_f derived in this experiment could be less than the actual isolated molecule value or that from theory.

Table 29.1 *Compilation of calculated and measured difference in the energy levels ΔH_f, between anti and gauche 1,2-dichloroethane.*

Scenario	Value (kcal/mole)
energy diff. for pure form Wiberg et al.[3]	1.91 (6–31 G* basis set)
energy diff. in acetonitrile Kobayashi et al.[4]	0.15 (observed)
energy diff. in n-hexane Kolling [5]	1.81 (observed)
calculated energy diff. of gas phase sample	0.98 (PCModel molecular mechanics simulation)
energy diff., Jorgensen and Tirado-Rives [6]	1.11 (OPLS-AA calculation)
energy diff., Murphy et al. [7]	1.31 (local MP2 basis set) 1.20 (observed)

It is interesting to note that the measured intensities of the two peaks assigned to the anti and gauche have a degeneracy factor of two since there are twice as many ways to get the gauche as the anti conformer. If this factor of two is included in the calculations, the slope is still the same but a ln2 additive factor displaces the curve upward. The factor of ln2 is the entropy change in increasing the size of a box by two (twice as much volume in which to put things).

The above experiment has also been performed by another student as a senior project at UTK for 1,2-dibromoethane with similar success to that found for 1,2-dichloroethane. The intensity of the anti rotational isomer (I_a) at 660 cm^{-1} was measured relative to that of the gauche rotational isomer (I_g) at 550 cm^{-1}. Again, the energy difference between the rotational rotomers, anti to gauche, was determined from the slope of the $\ln(I_a/I_g)$ versus $1/T$ and the result was similar to that reported in the literature (0.910 ± 0.080 kcal/mol).[8] There are many other gauche and anti conformers which could also be studied employing this method.

· ·

REFERENCES

1. F. Carey and R. Sundberg, *Advanced Organic Chemistry: Structure and Mechanisms*, third edition, Plenum Press, New York, 1990.
2. S. Ege, *Organic Chemistry*, second edition, D. C. Heath and Company, Lexington, MA, 1989. Pages 148–52.
3. K. E. Laidig, P. J. MacDougall, M. Murcko, and K. B. Wiberg, "Origin of the gauche effect in substituted ethanes and ethenes," *J. Phys. Chem.* **94**, 6956 (1990).
4. M. Kobayashi and H. Sato, "Conformational analysis of ethylene oxide and ethylene imine oligomers by quantum chemical calculations: Solvent effects," *Polymer Bulletin* **61**, 529 (2008).
5. O. W. Kolling, "Solvent reaction field effects upon the trans/gauche conformational equilibrium of 1,2-dichloroethane in aprotic media," *Trans. Kansas Acad. Sci.* **101**, 89 (1998).
6. W. L. Jorgensen and J. Tirado-Rives, "Potential energy functions for atomic-level simulations of water and organic and biomolecular systems," *Proc. Natl. Acad. Sci.* **102**, 6665 (2005).
7. R. Murphy, W. T. Pollard, and R. Friesner, "Pseudospectral localized generalized Møller-Plesset methods with a generalized valence bond reference wave function: Theory and calculation of conformational energies," *J. Chem. Phys.* **106**, 5073 (1997).
8. J. Hiraishi and T. Shinoda, "Energy difference between rotational isomers of 1,2-dibromoethane," *Bull. Chem. Soc. Japan* **48**, 2385 (1975).

Diffraction of Light from Blood Cells

30

Introduction

A rather complete description of physical and geometrical optics relating to diffraction can be found in the textbooks by Hecht [1] or Guenther.[2] The diffraction of light occupies a central position in the description of electromagnetic radiation, as discussed in Chapter 2. In what follows, we focus on those aspects relevant to the experiments described in the text. We include an experiment which describes the determination of the mean diameter of blood cells (erythrocytes) using diffraction together with Babinet's principle.

Diffraction can be described loosely as the bending of light around an obstacle. In normal geometrical optics, the light observed passing through a slit would appear as a bright line on a distant object having the same dimensions as the slit. However, if one views light from a distant source (e.g., fluorescent light) through the crack in one's fingers, the light through this "slit" is seen as a series of alternating bright and dark lines. The effects of diffraction are also seen in many everyday occurrences, such as the colored rings of light reflected from an oil slick or the appearance of "floaters" in the eye, as explained in this chapter. Pressure applied to two thin pieces of glass in contact will also produce colored "rings," which were named Newton's Rings after their discoverer, Sir Isaac Newton. Credit for the original discovery of diffraction is given to Grimaldi (1618–63). However, prehistoric man would certainly have observed and marveled at the phenomenon. A physical description of light was of primary importance in the 1600s. Central among the figures in this quest were Huygens (1629–95) and Newton (1642–1727). Huygens based his theory of light on a geometrical construction that involves the concept of "wavelets." For example, the propagation of a plane wavefront is described by requiring that all points on such a wavefront be considered as secondary sources of wavelets. After some time t the wavelets progress to a point d ($d = ct$, where c = speed of light) and the new wave will be the surface that is tangent to these wavelets (see Figure 30.1).

Augustin Jean Fresnel (1788–1827), a French mathematician and physicist, applied Huygens ideas to explain diffraction as an interference pattern between waves. The patterns generated by waves on a lake were understood at that time to be the constructive and destructive interference between matter waves. In those days, light was also believed to comprise mechanical waves traveling on the

Laser Experiments for Chemistry and Physics. First Edition. Robert N. Compton and Michael A. Duncan.
© Robert N. Compton and Michael A. Duncan 2016. Published in 2016 by Oxford University Press.

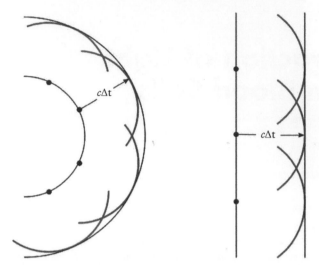

Figure 30.1 *Illustration of the Huygens "wavelet" construction of electromagnetic radiation propagation.*

all-pervading "ether." Of course Einstein later showed that such an ether need not be postulated (or rather that it could not be detected if it does exist). Maxwell demonstrated that light waves are electromagnetic waves. As discussed in Chapter 1, Maxwell's equations unite electricity and magnetism into one electromagnetic theory of light. In principle, diffraction can be explained through the solution of Maxwell's equations for a given geometry. There are two important general types of diffraction: (1) Fresnel and (2) Fraunhofer. Fresnel diffraction describes the diffraction of spherical light waves, e.g. those that emanate from a point source, whereas Fraunhofer diffraction treats the case of diffraction of coherent plane waves.

Light emanating from a star (approximately a point source of light) becomes very nearly a plane wave upon reaching the Earth. Thus diffraction of light by an object on Earth (a slit, for example) could be treated by either Fresnel (exact) or Fraunhofer (approximate) diffraction. The mathematics of the former are so intractable that Fresnel diffraction patterns have few analytical solutions. The experiments we perform employ parallel laser beam rays, and thus we consider only the more tractable Fraunhofer diffraction. In Chapter 2 we discussed the diffraction of light through a slit; below we consider diffraction through a circular aperture as an introduction to the diffraction of light by blood cells.

Diffraction through a circular aperture

Fraunhofer diffraction through a circular aperture is probably the most important case for diffraction in optics and the student should become familiar with this topic. In this case, the intensity of scattered light on the back side of the aperture depends upon θ and φ as shown in Figure 30.2. The intensity of the radiation is

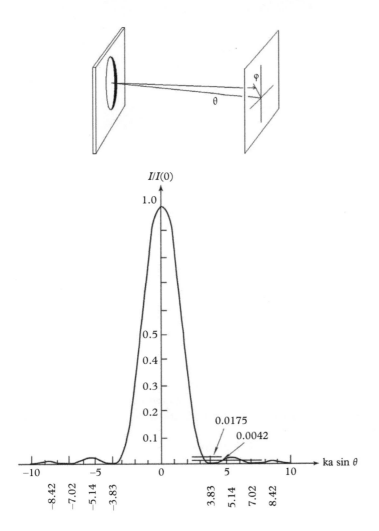

Figure 30.2 *Theoretical solution for light transmission through an aperture of radius a shown above as a function of the angle θ. The intensity is symmetric about the angle φ.*

symmetric about the angle φ, and the intensity as a function of the angle θ is given by a Bessel function (J_1), i.e. $I(\theta) = I(0)\,[(2J_1\,ka\sin\theta)/ka\sin\theta]^2$.

It can be shown that the first minimum in the Fraunhofer diffraction pattern from a circular aperture of diameter d is given by

$$d\sin\theta = 1.22\,\lambda \qquad\qquad (30.1)$$

One will notice the similarity to the case of Fraunhofer diffraction from a slit of width a, i.e., $a\sin\theta = \lambda$ (see Chapter 2). The factor of 1.22 comes from the first Bessel function.

The bright central disc is termed the "Airy disc," after Sir George Airy, who first solved the problem in 1835. Measurements of the minima in the diffraction of

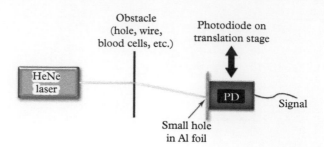

Figure 30.3 *Experimental arrangement used to obtain a quantitative description of light diffraction through various objects.*

a pin hole using a He-Ne laser can be used to obtain the diameter of the hole. This method is especially useful in determining the diameter of the holes in nozzles used in producing molecular jets.

Diffraction by an opaque disc (a BB for example) shows a pattern with a bright "dot" at the center. This pattern was first predicted by Simeon Poisson and later verified experimentally by Dominique Arago; it is sometimes called "Poisson's bright spot."

The image of a distant star formed by a converging lens also shows the diffraction pattern of an aperture. If we consider two stars so close that the maximum (Airy disc) of one star falls on the minimum of the other star, one has what is called the Rayleigh criterion for the resolving power of a microscope. That is, two objects separated by a distance d that are barely resolvable by Rayleigh's criterion must have an angular separation θ_R given by $\sin \theta_R = 1.22 \, \lambda/d$. Since the angles involved are rather small, $\sin \theta_R$ can be replaced by θ_R and therefore $\theta_R = 1.22 \, \lambda/d$.

Experiments demonstrating a quantitative description of diffraction can be carried out using the simple geometry shown in Figure 30.3. The experiment presented in Figure 30.3 was employed to measure scattered light intensities when He-Ne laser light is transmitted through various obstacles, as shown in Figure 30.4. The measured diffraction patterns agree very well with the measurements. Also note that the pattern for a slit and a wire of equal width and diameter, respectively, are the same. This is a vivid demonstration of Babinet's principle, as discussed below.

Determination of the mean diameter of blood cells (erythrocytes)

Babinet's principle states that the diffraction from an array of objects is the same as that from a corresponding array of opaque objects. For example, "Poisson's bright dot" seen in the diffraction of a small sphere is the analogy of the Airy disc seen from a small aperture. Babinet's principle is best demonstrated by comparing the diffraction pattern from a slit as described above with the diffraction from a

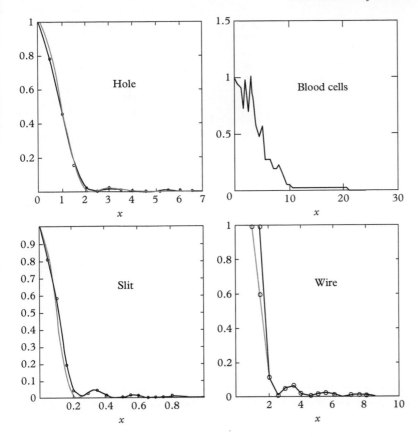

Figure 30.4 *Light intensities transmitted through a hole, slit, a wire, and dried blood cells. Theoretical predictions using the equations above are also plotted as dots, except for the blood cells (discussed later).*

wire of the same diameter. The diffraction pattern is the same. The diameter of a human hair can be easily determined from such a diffraction pattern. Simply tape a human hair on the end of a laser pointer and observe the Airy disc on a distant wall. Knowing the distance to the wall and using equation 30.1 with m = 1, one can calculate the width of the hair. The comparison of diffraction through a slit and around a wire is shown at the bottom of Figure 30.4.

Babinet's principle is illustrated in Figure 30.5 for the case of diffraction through a hole and through an arbitrary array of small discs with the same diameter as the holes. Babinet's principle has many novel applications, as shown below.

An important application of Babinet's principle can be demonstrated in the determination of the diameter of blood cells. A small sample of blood can be smeared on the surface of a microscope slide. A diffraction pattern results from shining a He-Ne or other laser through the film. A measurement of the Airy disc and minimum of this pattern provides an estimate of the diameter of the blood cell (see Figure 30.4 above). A nice summary of the data obtained from such an

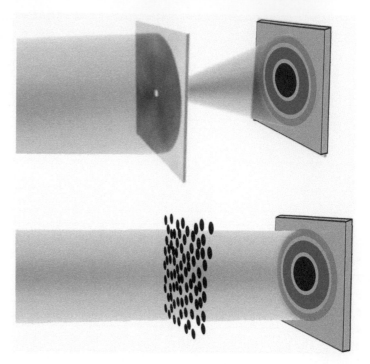

Figure 30.5 *Diffraction of light from a hole (or surface of holes) in the top panel compared to the diffraction from a multitude of opaque discs of the same diameter as the hole in the bottom image. This is an illustration of Babinet's principle.*

experiment is described in the paper by Bowlt in 1971.[3] For safety purposes, students are required to extract their own blood sample and should not involve another student in handling the sample. A beautiful interference fringe pattern of a dried blood sample from Bowlt's experiments is reproduced in Figure 30.6.

The conditions for diffraction from a hole or discs, using Babinet's principle, are very similar to that for a slit, $\sin \theta = m\lambda/d$, except that the order m is noninteger (see Figure 30.6), i.e., m = 1.22, 2.23, 3.24, 4.24, 5.24, etc. In this experiment

Figure 30.6 *Interference fringes produced by diffraction of He-Ne laser light by red corpuscles (left). The inset shows a photomicrograph of a smear of dried blood. (Right) Plot of the order* m *as a function of the fringe diameter (reproduced with permission from G. Bowlt, Phys. Educ. 6, 13 (1971), copyright 1971, Institute of Physics Publishing).*

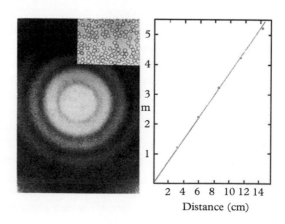

the distance, d, from the sample to the screen was 20 cm. Undergraduate student experiments at the University of Tennessee have been successful in producing such diffraction patterns from both wet and dried blood to estimate cell size, but the quality of the diffraction patterns, while clear and discernible, has not equaled that presented by Bowlt,[3] particularly for the higher order diffraction peaks. The upper right example in Figure 30.6 was taken from one laboratory experiment in which an average size of ~7 µm was obtained. Bowlt also points out that the dried blood cells are ~10% smaller than in blood itself. Also, it is interesting to note from the photomicrograph in the inset of Figure 30.6 that the blood cells are all uniform, implying that they are lying flat on the microscope slide.

As pointed out by Bowlt,[3] Isaac Newton attributed haloes around both the Sun and Moon to the effects of diffraction from small water globules. Such scattering was studied by Young many years later. Young also considered using this diffraction to obtain the diameter of red blood cells.[4] In 1928 Alan and Ponder [5] also used monochromatic light scattering from blood samples as a method of cell size determination but considered the method unreliable. The invention of the laser has improved the quality of these measurements and they have become more routine. Krakau [6] has theoretically considered the differences in the diffraction pattern from dense and sparse smears as well as deviations from spherical cells. It would be interesting to examine the diffraction pattern for sickle cells. In this connection, recently Rao et al.[7] have applied this diffraction method to determine the size and shape of various cancer cells. The results of these interesting studies are shown in Table 30.1.

Normal red blood cells may vary in size from 6 to 8 µm. Blood deficiencies are known to result in a range of cell sizes as seen in Table 30.1.

Diffraction of light from red blood cells in the eye: "floaters"

Many people experience what are termed "floaters" appearing in the eye, especially when looking at a bright light source. They can appear as black/gray dots, squiggly lines, or threadlike strands which slowly drift across the field of view as the eye moves. In some cases these objects can appear to stop in the direct line of vision. The frequency of these floaters increases with age, and in the event of retinal damage larger clumps of these dots can appear. These objects exist in the vitreous humor of the eye and are called *Muscea volitantes*, which is Latin for "flying flics." Floaters can be as small as a single red blood cell or a collection of these. Sometimes floaters are a result of a protein (collagen) in the vitreous humor.

The well-known physicist, Harvey White, and his surgeon, Paul Levatin, published an article in the *Scientific American* describing the surgical correction of White's detached retina.[8] In that article they proposed that "floaters are usually

Table 30.1 *Diameter of normal and cancerous erythrocytes determined from diffraction.[7]*

Cell type	Diameter (µm)
normal, mean	7.12
cervical cancer	10.5
breast cancer	9.53
rectal cancer	8.77
cheek cancer	9.38
esophageal cancer	8.59
ovarian cancer	8.97
lung cancer	9.68
thyroid cancer	8.97
tongue cancer	9.30

diffraction patterns cast on the retina by red blood cells." They describe a simple experiment that allowed them to estimate the size of the tiny dots and show that their size is approximately 8 μm (0.008 mm), essentially that found in the more controlled experiment described above.

White and Levatin [8] describe a simple procedure to observe diffraction from these individual red blood cells. To do this, tape a piece of aluminum foil over the open hole in a spool of thread. Next, make a small hole in the aluminum foil using a needle. One should make the hole as small and round as possible. Now look

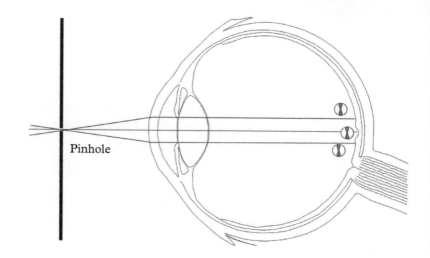

Figure 30.7 *Light being projected through a small pinhole into the eye. The nearly parallel rays are diffracted onto the retina after being diffracted by the red blood cells "floating" in the vitreous humor. The blood platelets are shown as spherical (see the text). (Adapted from [7])*

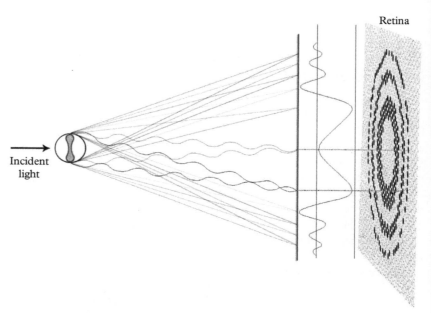

Figure 30.8 *Illustration of diffraction of light in the eye due to red blood cell floaters. On the right the diffraction pattern from a single red blood cell is illustrated, and on the far right is the pattern seen on the retina (the dots represent the rods and cones on the retina). (Adapted from [7]).*

through the open end of the spool at the light exiting the small hole from a diffuse ("frosted") light bulb, or better the bright light from the Sun (do not look directly at the Sun, of course, but rather the light from the bright blue sky), as shown in Figure 30.7.

Figure 30.8 demonstrates the diffraction of light around a single red blood cell and the image appearing on the retina.

The red blood cells are depicted as round in both figures, whereas they are actually shaped like a dented disc in blood. When the red cell is immersed in a solution (e.g., the vitreous humor) that has a lower osmotic pressure than its own protein-rich contents, the cell becomes spherical. The diffraction from a single red cell looks very similar to that shown in Figure 30.6 from a collection of cells.

....

REFERENCES

1. E. Hecht, *Optics*, fourth edition, Addison-Wesley Publishing Co., San Francisco, CA, 2002.
2. R. Guenther, *Modern Optics*, John Wiley & Sons, Hoboken, NJ, 1990.
3. G. Bowlt, "Measurement of red blood cell diameters using a laser," *Phys. Educ.* **6**, 13 (1971).
4. T. Young, "The Bakerian Lecture: Experiments and calculations relative to physical optics," *Philos. Trans. Roy. Soc. (London)* **94**, 1 (1803).
5. A. Allen and E. Ponder, "The determination of the diameter of erythrocytes by the diffraction method," *J. Physiol.* **6**, 37 (1928).
6. C. E. T. Krakau, "The diffusion pattern of dried blood smears," *Biophys. J.* **6**, 801 (1966).
7. C. R. Rao, K. A. Jaeeli, B. S. Bellubbi, and A. Ahmad, "A study of the size and shape of erythrocytes of cancer patients using a laser diffraction technique," *J. Pure & Applied Phys.* **21**, 141 (2009).
8. H. E. White and P. Levatin, "Floaters in the eye," *Sci. Am.* **206**, 119 (1962).

31

Inversion of Sucrose by Acid-Catalyzed Hydrolysis

Introduction

This experiment involves the measurement of optical rotation for the determination of the rate at which the optically active disaccharide sucrose molecule is converted into the monosaccharides, fructose and glucose, by acid-catalyzed hydrolysis. Before we consider this classic experiment in physical organic chemistry, let us first examine the concept of chirality and racemization.

As discussed in Chapter 20, the atomic positions of a chiral molecule are non-superimposable on that of its mirror image. The energies of these two isomers are the same, and as a result all chiral molecules will interconvert between their mirror images by tunneling through the potential barrier separating the two. This is shown pictorially for the amino acid alanine in Figure 31.1.

Since the two enantiomers can be interconverted, they cannot be stable. In fact Friedrich Hund regarded the existence of a stable chiral molecule as paradoxical. This paradox is often referred to as the *Hund Dilemma*. When the two are equally populated, one has a racemic mixture of the two enantiomers. Of course the rate at which the two enantiomers interconvert is dependent upon the height of the barrier to racemization and the conditions under which the molecule finds itself (pH of the solution, temperature, presence of metal ions, etc.). Racemization times vary greatly from one molecule to another. For example, the gauche form of 1,2-dichloroethane (Chapter 29) interconverts on a very short timescale (on the order of a rotational period) because of the small barrier height separating the two enantiomers, whereas most chiral molecules are stable over the age of the Earth. The long lifetimes for tunneling between enantiomers for most chiral molecules provide a simple answer to the Hund Dilemma. The amino acids that make up all of life are exclusively so-called *L* conformers and have a rather long lifetime against racemization. The *L* amino acids in living organisms are replenished by the organism eating material that also contains *L* amino acids. However, upon the death of the organism, the *L* amino acid begins to slowly approach equilibrium with the *D* enantiomer. The determination of the ratio of *L* to *D* can be used to estimate the time of death of the organism, provided enough is known about the conditions (temperature, pH, water, and ion content) under which the decaying material existed.

Following this brief introduction to chirality, we suggest an experiment in which a chiral liquid sample is heated or placed in an acidic solution and the

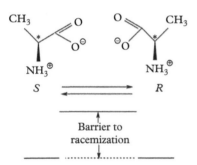

Figure 31.1 *Illustration of the racemization of the S- and R- zwitterionic forms of the amino acid alanine under conditions of neutral pH. The H atom is not shown at the position of the chiral carbon (C*). It is the motion of this hydrogen that leads to the lowest barrier to racemization.*

Laser Experiments for Chemistry and Physics. First Edition. Robert N. Compton and Michael A. Duncan.
© Robert N. Compton and Michael A. Duncan 2016. Published in 2016 by Oxford University Press.

rate of racemization is measured by recording the change in optical rotation as a function of time. The rotation can be measured with a commercial polarimeter or the simple polarimeter described in Chapter 20. The interested student may also repeat an experiment that has recently been reported in which a microwave oven was used to completely racemize an enantiomerically pure amino acid in solution.[1] Quoting from the abstract of this paper:

> Rapid racemization of optically active amino acids has been developed in a Teflon vial reactor by using the commercial microwave oven as heating source. Compared with the conventional methods of racemization, there are two main advantages: (I) The whole process of racemization by this method can be completed within 2 min. (II) It can prevent the decomposition of some labile amino acids at high temperature.

This is an interesting experiment in light (microwave)-induced racemization which can easily be carried out in any laboratory, or kitchen for that matter. The rate of the racemization reaction is followed by recording the optical rotation as a function of the time of irradiation. Using circularly polarized microwaves it should be possible to convert one enantiomer into its mirror image, providing an enantiomeric excess without destruction of either enantiomer. More commonly, circularly polarized laser light has also been used to asymmetrically photodissociate or convert one enantiomer into its mirror image. This produces an enantiomeric excess, which is easily monitored with polarimetry. For a review of this subject and its relation to the origins of life, see Compton and Pagni [2,3] and others cited therein.

We now turn to an experiment involving the classic *invert* sugar reaction. The sucrose molecule is a disaccharide, i.e. it consists of two simple sugars or monosaccharides, fructose and glucose, as shown below. In an aqueous solution the sucrose molecule can be split into fructose and glucose by heat, or more effectively by the addition of protons (acid addition) as shown in Figure 31.2.

In either case sucrose is "split" in two by adding a hydrogen to the linking oxygen, a reaction described as hydrolysis. The hydrolysis reaction is commonly

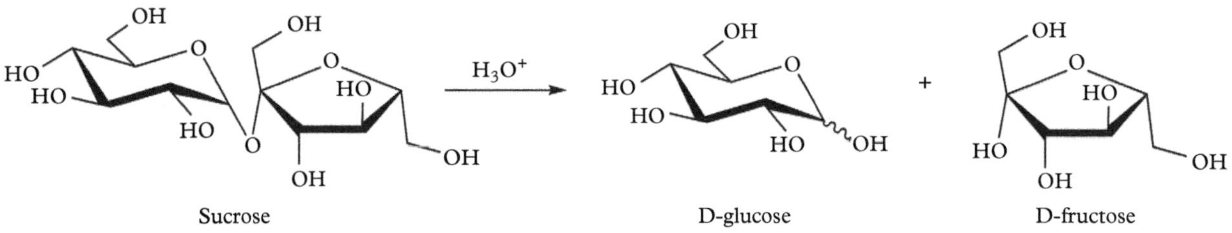

Figure 31.2 *Hydrolysis reaction of sucrose into glucose and fructose.*

accelerated by the addition of biological catalysts, which are called sucrases from animals or inverases from plants. The term *invert* is used to indicate that upon addition of the catalyst over time the optical rotation of the solution will change sign (+66.3° to – 36.5°). This is because the optical rotation of the fructose molecule is opposite to and greater than that of glucose, as illustrated in the reaction below:

$$C_{12}H_{22}O_{11} + H_2O \quad \xrightarrow{H^+} \quad C_6H_{12}O_6 \quad + \quad C_6H_{12}O_6$$

$$(\text{sucrose}, \alpha_D = +66.3°) \qquad (\text{glucose}, \alpha_D = +52.5°) \qquad (\text{fructose}, \alpha_D = -89°)$$

Wienen and Shallenberger [4] carefully studied the inversion kinetics and developed a useful relationship between the time required for complete inversion as a function of the pH and temperature of the solution:

$$\log(\text{hours to 99.99\% completion}) = (\text{pH} - 0.5) + \frac{39}{\text{temperature}} \qquad (31.1)$$

After a pH and temperature are selected, the progress of the reaction can easily be followed by optical rotation. Many studies of the kinetics of sucrose inversion have been made, and it has been determined that the reaction proceeds according to a first-order rate law when performed in aqueous solution:

$$[\text{sucrose}]_t = [\text{sucrose}]_0 e^{-kt} \qquad (31.2)$$

Experimental

Optical rotations observed in this experiment are obtained using a Perkin–Elmer auto polarimeter operating with a sodium vapor lamp and producing incident light at 589 nm. In preparation for observing the inversion reaction by polarimetry, a 1-dm polarimeter cell containing deionized water should be placed into the polarimeter, and the zero for the instrument obtained. The cell can then be filled with samples from the sucrose solutions and placed into the instrument to obtain initial rotations for both solutions. The concentrations of sucrose and acid solutions used in studying the reaction kinetics are shown in Table 31.1.

For each trial solution, combine 5 ml of sucrose solution with 1 ml of diluted hydrochloric acid to yield a total volume of 6 ml. Fill a polarimeter cell with a sample from the reaction mixture and place this into the instrument. Record rotations at fixed intervals across 90 minutes. After the last measurement, allow the samples to stand for 24 hours before recording a final equilibrium rotation. Note the temperature for the measurements. The optical rotation of invert sugar is ∼ −36.5° at 20°C.

Table 31.1 *Reactant concentrations.*

Trial	Sucrose (M)	Hydrochloric acid (M)
1	0.584	2.0
2	0.292	2.0
3	0.584	1.0

Analysis and results

Equation 31.3 describes the rotation of a sample solution in terms of the concentration of sucrose, s, and invert sugar, i:

$$\alpha_t = [\alpha]_s \times d \times c_s + [\alpha]_i \times d \times c_i \qquad (31.3)$$

The rotation of the sample at time t, α_t, is equal to the sum of the specific rotations of sucrose and invert sugar, $[\alpha]_s$ and $[\alpha]_i$, each multiplied by their respective concentrations in grams per milliliter, c_s and c_i, and the sample cell path length in decimeters, d.

If the terms for sucrose and invert sugar rotation are taken as equivalent to the measured initial and equilibrium rotations, equation 31.3 can be arranged in terms of the fractional concentration of sucrose remaining at a time t as shown in equation (31.4).

$$\frac{C_{s,t}}{C_{s,0}} = \frac{\alpha_t - \alpha_\infty}{\alpha_0 - \alpha_\infty} \qquad (31.4)$$

The fraction of the initial sucrose concentration remaining at time t is equal to the ratio of rotation at time t minus the equilibrium rotation to initial rotation minus the equilibrium rotation.

As the reaction is first order, the result of equation 31.4 can be used to give the relation shown in equation 31.5.

$$kt = \ln\left[\frac{\alpha_0 - \alpha_\infty}{\alpha_t - \alpha_\infty}\right] \qquad (31.5)$$

Therefore, a plot of the natural logarithm of the inverse of fractional sucrose concentration versus time yields a line with a slope equal to the reaction rate constant, k. Figure 31.3 shows such plots and lines of best fit for three trials performed in such an experiment. Rate constants derived from this measurement using polarimetry vary from experiment to experiment. Literature values for initial sugar concentrations of 0.12 g/cm^3 at 30 °C are ~0.025/min.

The inversion of sucrose by acid-catalyzed hydrolysis can also be performed using the experiment described in Chapter 20 (Optical Rotary Dispersion of a Chiral Liquid) in which a sugar solution is placed in a vertical tube with an open end. The light from a polarized laser is observed to "spiral around inside the tube" along its direction of path (see Figure 20.1). The bands appear to move upward as one walks counterclockwise around the tube. Upon adding the acid to the tube, the distance between the bands will increase until it reaches infinity (mixture of glucose and fructose is such that the rotations for the two are equal and opposite, giving no rotation). As time progresses the bands will appear again but with the opposite sense of helicity. Recording the temporal change in rotation with time gives the rate of inversion.

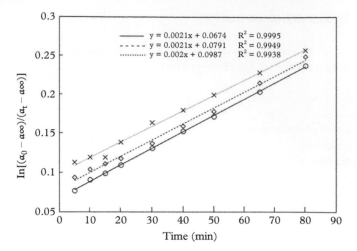

Figure 31.3 *Natural logarithm of inverse fractional sucrose concentration versus time. The slopes of the lines of best fit are equivalent to the observed rate constant for the respective experimental trial.*

The invert sugar reaction described above has been a classic experiment in physical and organic chemistry over the years [5,6] and combines concepts in physics, physical chemistry, reaction rates, and organic chemistry.

REFERENCES

1. S.-T. Chen, S.-H. Wu, and K.-T. Wang, "Rapid racemization of optically active amino acids by microwave oven-based heating treatment," *Int. J. of Pep. Prot. Res.* **33**, 73 (2009).
2. R. N. Compton and R. M. Pagni, "The chirality of biomolecules," *Adv. Atom. Mol. Opt. Phys.* **48**, 219 (2002).
3. R. M. Pagni and R. N. Compton, "Is circularly polarized light an effective reagent for asymmetric synthesis?" *Mini-Rev. Org. Chem.* **2**, 203 (2005).
4. W. J. Wienen and R. S. Shallenberger, "Influence of acid and temperature on the rate of inversion of sucrose," *Food Chem.* **29**, 51 (1988).
5. F. Daniels, *Experimental Physical Chemistry*, seventh edition, McGraw-Hill, New York, 1970.
6. C. W. Garland, J. W. Nibler, and D. P. Shoemaker, *Experiments in Physical Chemistry*, eighth edition, McGraw-Hill, New York, 2003, p. 271.

Appendix I
Recommended Components and Equipment

1. Optics

Windows, prisms, lenses, etc.:

Newport Corporation
http://www.newport.com/

Thorlabs
https://www.thorlabs.com/

ESCO Optics
https://www.escooptics.com/

Edmund Optics
http://www.edmundoptics.com/

2. Optical mounting hardware

Newport Corporation
http://www.newport.com/

Thorlabs
https://www.thorlabs.com/

ESCO Optics
https://www.escooptics.com/

3. Detectors

Photodiodes:

Hamamatsu
http://www.hamamatsu.com/

Thorlabs
https://www.thorlabs.com/

Photomultiplier tubes:

Hamamatsu
http://www.hamamatsu.com/

Newport Corporation
http://www.newport.com/

ET Enterprises
http://www.et-enterprises.com/

Electron multiplier tubes:

Hamamatsu
http://www.hamamatsu.com/

ETP Electron Multipliers
http://www.etp-ms.com/

Microchannel plates:

Hamamatsu
http://www.hamamatsu.com/

Photonis
http://www.photonis.com/

Del Mar photonics
http://www.dmphotonics.com/

4.   Electronics

Digital oscilloscopes:

LeCroy
http://teledynelecroy.com/oscilloscope/

Tektronix
http://www.tek.com/

Boxcar averagers:

Stanford Research Instruments
http://www.thinksrs.com/

Fast preamplifiers:

Stanford Research Instruments
http://www.thinksrs.com/

Lock-in amplifiers:

Stanford Research Instruments
http://www.thinksrs.com

5. Lasers

He-Ne lasers:

Newport Corporation
http://www.newport.com/

Melles Griot
http://mellesgriot.com/products/

Edmund Optics
http://www.edmundoptics.com/

Laser pointers (with power levels high enough for some experiments):

Dragon Lasers
http://www.dragonlasers.com/

Nd:YAG lasers:

Spectra Physics Lasers (a division of Newport Corporation)
http://www.newport.com/

Continuum
http://www.continuumlasers.com/

Quantel
http://www.quantel-laser.com/

New Wave Research (a division of ESI)
http://www.esi.com/

Excimer lasers:

Lambda Physik (a division of Coherent)
http://lasers.coherent.com/lasers/lambda-physik/

GAM
http://www.gamlaser.com/

Diode lasers:

Toptica
http://www.toptica.com/

Moglabs
http://www.moglabs.com/

Lumina Power
http://luminapower.com/

Thorlabs
http://www.thorlabs.com/

Coherent
http://www.coherent.com/

CO$_2$ lasers:

GAM
http://www.gamlaser.com/

Dye lasers:

Lambda Physik (a division of Coherent)
http://lasers.coherent.com/lasers/lambda-physik/

Spectra Physics Lasers (a division of Newport Corporation)
http://www.newport.com/

Continuum
http://www.continuumlasers.com/

Sirah
http://www.sirah.com/

6. Time-of-flight mass spectrometers (modular units)

Comstock
http://www.comstockinc.com/

R. M. Jordan Co.
http://www.rmjordan.com/

7. Vacuum fittings

A & N Corporation
https://www.ancorp.com/

MDC Vacuum Products
http://www.mdcvacuum.com/

Kurt J. Lesker
http://www.lesker.com/

MKS
http://www.mksinst.com/

8.   Monochromators

Ocean Optics
http://www.oceanoptics.com/

Thorlabs
http://www.thorlabs.com/

9.   Raman shifters

Avesta
http://www.avesta.ru/

Sirah (Lasertechnik)
http://www.sirah.com/laser/frequency-conversion/raman-shift-cell/

Appendix II
Fast Signal Measurements

Chemical and physical events occur over a wide dynamic time range. Many atomic and molecular processes occur at "fast" rates in the microsecond (10^{-6} sec), nanosecond (10^{-9} sec), or even picosecond (10^{-12} sec) time domains. The experimental observation of these processes, therefore, requires fast observation methods, usually involving fast time-dependent electronic-signal processors. Time-dependent signal detection is employed throughout this book in every experiment detecting light with photodiodes or PMT detectors, or detecting ions or electrons with electron multiplier tubes or microchannel plates. Digital oscilloscopes, photon counters, and boxcar integrators are all examples of fast electronic-signal processors. It is therefore important to consider the appropriate equipment operation and techniques for these measurements. [1]

Detectors used in electronics experiments produce an electrical current, I, when a signal is detected. For example, when a photon strikes a photodiode or a photomultiplier tube, a photo-current is produced. However, electronic devices such as oscilloscopes usually display or measure the corresponding voltage, V, resulting from driving this current into a load resistor, R. The current and voltage are related according to Ohm's Law, V = IR. This relation applies exactly when the units of the quantities are volts, amps, and ohms, respectively. Thus, if a signal current of 3 microamps is detected with an oscilloscope into a 1 megohm resistor (a typical value), the voltage intensity displayed on the oscilloscope will be

$$V = \left(3 \times 10^{-6}\right)\left(1 \times 10^{6}\right) = 3 \, \text{volts}$$

Smaller load resistors produce a corresponding smaller voltage for the same input current. To make the signal appear larger on an oscilloscope, therefore, a larger load resistor should be used.

Additional considerations about the signal are required when this current or voltage changes rapidly with time. This is because all electronic equipment, including the processors themselves and even the wires and connectors used to carry the signal, have characteristic time-response limitations. For example, a 100 MHz oscilloscope cannot detect any signals varying with a frequency greater than this value. Its response is then limited to events occurring on a timescale τ_c, such that

$$\tau_c = 1/\nu = 1/(100 \times 10^6 \text{ sec}^{-1}) = 1 \times 10^{-8} \text{ sec}$$
$$= 10 \text{ nsec}$$

Voltage signals changing faster than 10 nsec will therefore not be measured accurately by this instrument.

The time limitation in electronic equipment occurs because of the capacitance, C, in its components. Capacitance occurs whenever conductors and non-conducting materials (insulators) are used together, as in all electronic components. Because of capacitance, the movement of electrical charge (current) is limited to a response time given by,

$$\tau_c = RC$$

where R is the load resistance in ohms and C is the capacitance in Farads. This is often referred to as the RC time constant of the circuit. The flow of charge through wires or other components is always limited by this charging time, which affects both the rise and fall of electrical signals. The response is exponential in nature (following first-order kinetics), so that the fall of a current can be expressed as

$$I(t) = I(0)e^{-t/\tau} = I(0)\, e^{-t/RC}$$

As usual, for exponential decays, the value τ_c occurs when the signal intensity (current) has dropped to $1/e$ of its initial value $I(0)$.

The coaxial cables used in typical laser labs have a capacitance of approximately 10 picoFarads per foot. A typical cable of length 10 ft. therefore has about 100 pFd capacitance. If this is used with a typical load resistor on an oscilloscope of 1 megohm, the time constant for the scope/cable combination is,

$$\tau_c = RC = (1 \times 10^6)\,(100 \times 10^{-12})$$
$$= 1 \times 10^{-4} \text{ sec} = 100\,\mu\text{sec}$$

Signals faster than 100 μsec will therefore not be measured accurately. If a faster time-dependent signal is to be measured, a *larger* load (*smaller* resistor) must be used to *decrease* the response time. Usually, a 50 ohm load resistor is used. Then,

$$\tau_c = (50)\,(100 \times 10^{-12}) = 5 \times 10^{-9} \text{ sec}$$

Notice, however, that reducing the load resistor also *reduces* the measured voltage (via V = IR). Therefore, there is always a compromise between increasing the speed of a measurement and decreasing the voltage amplitude observed. In other words, conditions that allow fast signals to be measured result in their appearing as small voltages on an oscilloscope.

Standard oscilloscopes usually have a setting for 1 megohm input impedance for the detection of signals whose time response is not fast, and a setting for 50 ohm input impedance for detecting fast signals. Unless this setting is chosen correctly, detection of fast signals from photon, ion, or electron detectors cannot be accomplished properly.

....

REFERENCE

1. D. A. Skoog, F. J. Holler, and S. R. Couch, *Principles of Instrumental Analysis*, sixth edition, Thomson Brooks/Cole, Belmont, CA, 2007, chapter 2.

Index